CHARLES DARWIN

Also by A. N. Wilson

NON-FICTION

The Laird of Abbotsford: A View of Sir Walter Scott

A Life of John Milton

Hilaire Belloc: A Biography

How Can We Know?

Tolstoy

Eminent Victorians

C. S. Lewis: A Biography

Paul: The Mind of the Apostle

God's Funeral

The Victorians

Iris Murdoch as I Knew Her

London: A Short History

After the Victorians

Betjeman: A Life

Our Times

Dante in Love

The Elizabethans

Hitler: A Short Biography

Victoria: A Life

The Book of the People

The Queen

etc.

FICTION

The Sweets of Pimlico

Unguarded Hours

Kindly Light

The Healing Art

Who Was Oswald Fish?

Wise Virgin

Scandal: Or, Priscilla's Kindness

Gentlemen in England

Love Unknown

Stray

The Vicar of Sorrows

Dream Children

My Name is Legion

A Jealous Ghost

Winnie and Wolf

The Potter's Hand

Resolution

THE LAMPITT CHRONICLES

Incline Our Hearts

A Bottle in the Smoke

Daughters of Albion

Hearing Voices

A Watch in the Night

CHARLES DARWIN

Victorian Mythmaker

A. N. WILSON

NEW YORK • LONDON • TORONTO • SYDNEY • NEW DELHI • AUCKLAND

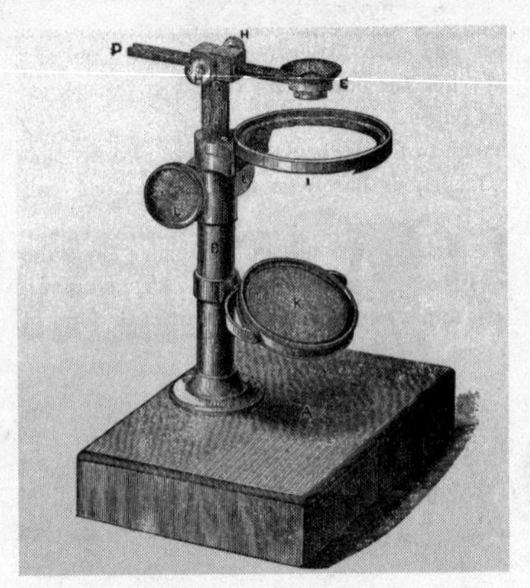

HARPER PERENNIAL

Originally published in Great Britain in 2017 by John Murray (Publishers), an Hachette UK Company.

A hardcover edition of this book was published in 2017 by HarperCollins Publishers.

HarperCollins books may be purchased for educational, business, or sales promotional use. For information, please email the Special Markets Department at SPsales@harpercollins.com.

FIRST HARPER PERENNIAL EDITION PUBLISHED 2018.

Library of Congress Cataloging-in-Publication Data has been applied for.

ISBN 978-0-06-243350-3 (pbk.)

18 19 20 21 22 LSC 10 9 8 7 6 5 4 3 2 1

Contents

Death to the weak! That is the watchword of what we might call the equestrian order established in every nation of the earth, for there is a wealthy class in every country, and that death-sentence is deeply engraved on the heart of every nobleman or millionaire . . . Take a few steps farther down the ladder of creation: if a barnyard fowl falls sick, the other hens hunt it around, attack it, scratch out its feathers and peck it to death.

Honoré de Balzac, *The Wild Ass's Skin*,

Prelude

DARWIN WAS WRONG. That was the unlooked-for conclusion to which I was inexorably led while writing this book.

To write the life of this Victorian Titan was an ambition which had been at the back of my mind for a quarter of a century. In 1999, I published a book about the Victorian crisis of faith, whose title was borrowed from a poem by Thomas Hardy – *God's Funeral*. Subsequently, I wrote a general survey of the period – *The Victorians* – and later a biography of their monarch and figurehead, Victoria. It was irresistible to return to their most famous intellectual revolutionary, the more so since the last major biographies appeared a good while back, and had been published before the monumental Cambridge Darwin project, the publication of his complete correspondence, had got far under way.

It was certainly not my intention when I began detailed reading for this book to part company from the mainstream of scientific opinion which still claims to believe, and in some senses does believe, the central contentions of Darwin's most famous book, *On the Origin of Species*.

There were a number of reasons why it did not even cross my mind that I would come to disbelieve in Darwin's theories. The first reason is that I am not a scientist, and I am inevitably dependent on scientists for what I know about the subject. So are we all – including scientists. In fact, if you are a professional scientist, you will have even less time than a layperson to read at all deeply in areas of science other than your own, since there is so much, in every branch of science, being explored and discovered every year.

In the last half-century, we have lived through a neo-Darwinian Golden Age: another reason why it would be a bold person – or so

I supposed before I started reading up the subject – who came to question Darwin's two central claims. These are, first, that by a gradual process of evolution one species evolves into another (the process is *always* gradual); the core of the theory is that nature does not make leaps – *Natura non facit saltum*. Secondly, Darwin believed that nature is in a state of perpetual warfare and struggle; that progress in evolution, and the perfecting of a species, takes place as a result of everlasting fight. 'The Preservation of Favoured Races in the Struggle for Life' is the second half of Darwin's title for *The Origin of Species*. Even if you realize that by 'Favoured Races' he was not, at this stage, referring to human beings, the phraseology has a strange ring to it.

The Golden Age of Darwinism saw the publication of Julian Huxley's *Evolution in Action* (1963) which I remember reading when I was a schoolboy, and which seemed to me then entirely convincing. It was a shorter version of his 1942 groundbreaking book *Evolution: The Modern Synthesis*. This was a synthesis between basic Victorian Darwinism and the discoveries of modern genetics which, when Huxley wrote, were still in their infancy relative to our level of knowledge today. The most coherent expression of this belief in the synthesis is *The Theory of Evolution* by John Maynard Smith, which was revised in 1993. Since those classics, there has been an abundance of books, articles, television programmes expounding the ideas of Darwin. Although Darwin's original ideas are in fact, when you come to read much of this material, very heavily revised and indeed changed, in the writings of the Darwinians the central contention remains the same: Darwin was right; species evolve by a series of micro-changes, and this explanation is sufficient for everything. The Darwinian Process explains all. Perhaps the most readable and pugnacious proponent of this viewpoint is Richard Dawkins, who since publishing such books as *The Selfish Gene* and *The Blind Watchmaker* in the 1970s has seemed to put the truth of the Darwinian position beyond question.

Nothing, however, is beyond question. As I quarried the history of Darwin himself, it was inevitable that I should wish to see how his ideas stood up in the light of contemporary scientific knowledge. This book, I quickly came to see, was very different from a biography of a painter or a politician. If our tastes have changed, today, and we no longer admire G. F. Watts, for example, as much as his contemporaries did, it does not mean he was a bad painter. Lord Palmerston's way of being Prime Minister might not work today, but we can still esteem him in his own time and place.

Science is not, however, a matter of passing taste. It is a matter of verifiable fact. With any scientist in the past, we are bound to ask whether their insights and theories still seem plausible. And in the case of Darwin this is doubly true, since his name is invoked so frequently by current evolutionary theorists.

I soon came to realize, when I started my reading, that in fact there is no consensus among scientists about the theory of evolution. Most would recognize that Darwin was a great pioneer of evolutionary biology. And everyone must recognize that he was a prodigiously wide-ranging and observant naturalist, whose voyage in HMS *Beagle*, when he was a very young man, brought back a wealth of specimens and changed many branches of natural science. Everyone must recognize that he was among the foremost experts on the earthworm. And his book on *The Expression of Emotions in Man and Animals* is a masterpiece.

When it comes to the two books which give the world the adjective 'Darwinian', however, *The Origin of Species* (1859) and *The Descent of Man* (1871), opinions vary enormously in the scientific world. Until I got down to doing my reading, I had assumed that, broadly speaking, scientific opinion accepted the truth of Darwin's central theories, and that objections to it were motivated not by scientific doubts but by some other set of ideas – most likely religious ones. In so far as Darwin's contemporaries rejected him on solely religious grounds, that is part of the story. (Very few did.) In so far as people today question Darwin, on religious grounds, that – it seemed to me – belonged to a different book. What interests me is whether he got it right scientifically. And there it is obvious that we are entitled to judge him not merely by the assessments made of his work by scientific and professional contemporaries – nearly all of them rejected it. We are entitled to ask how much our contemporary state of knowledge would lead us to question *The Origin of Species*.

And this is where I was so astonished. One of the modern scientists whose work I had followed, lightly, over the years was my namesake (no relation) E. O. Wilson, one of the foremost entomologists in the history of science. Not only was this Harvard professor a great scientist, but he also wrote sociobiology. I am not sure how convincing I have found the Social Darwinism. Certainly, Wilson has received many brickbats, accusing him of racism, misogyny and so forth.

Sociobiology and Social Darwinism, in particular, are contentious areas, sure enough. Darwin was beyond question a racist in modern-day terms, but I would be cautious about judging men and women of the nineteenth century by the standards of the twenty-first. Even before I came to write the Victorian story of Darwin himself, however, I started to become aware of the violent dissent within the ranks of the Darwinians of our time.

Perhaps the most arresting example happened in the early stages of my work, when I realized that E. O. Wilson, who had, forty years previously, been broadly supportive of Richard Dawkins's belief in a 'selfish gene', had broken ranks and begun to think that, far from evolutionary progress being dependent on selfishness and struggle, it sometimes owed quite a bit to co-operation. Ants don't build anthills by fighting one another; nor bees hives. In *The Social Conquest of Earth*, E. O. Wilson dared to put forward a 'multi-level selection theory', as opposed to the idea of genes themselves being 'selfish'. 'There is no such thing as a good or bad gene,' he opined. Dawkins cited 137 other scientists who agreed with him, and denounced Wilson for 'an act of wanton arrogance'[1] for daring to depart from orthodoxy. Wilson responded in a BBC TV interview by claiming that Dawkins was no longer a scientist, but a . . . 'journalist'.[2] Ouch.

Scientists were not, then, at one over the status of Darwin's ideas. If they were reduced to trading insults with one another, it surely suggested a theory which had collapsed: one camp accused the other of making 'unsubstantiated assertions resting on the surface of a quaking marsh of unsupported claims'. The other retorted that the enemy had 'ideas so confused as to be hardly worth bothering with'.[3]

When I read more of what the evolutionists had to say, both about their subject and about one another, I realized that Darwin's position as the great man of life-sciences looked uncertain. Geology has moved on since his friend and mentor Sir Charles Lyell pioneered the subject in the early nineteenth century. But we do not dismiss Lyell, since he discovered so much. What, exactly, did Darwin discover? Or is his theory just that – simply a theory?

Wilson's slow but inexorable journey to disbelief in Darwin's 'struggle for life' theory gave me pause. Then I came to read the work of Stephen Jay Gould. His book *The Structure of Evolutionary Theory* (2002) was based on a lifetime's palaeontological research. Gould and his colleague Niles Eldredge developed a theory of what they called 'punctuated equilibrium'. The little two-word phrase is a deadly one for the orthodox believer in Darwin. In *The Origin*, Darwin admitted that the fossil evidence to support his theory was sparse. Gould, one of the foremost palaeontologists of modern times, revealed that it was not sparse: it was non-existent. What the fossil evidence demonstrated, beyond doubt and contrary to what Darwin had claimed, was that nature *did* make leaps. It hopped. Species did not, apparently, evolve little by little, though micro-mutation did indeed take place within species. The famous finches and their beaks, gradually evolving to adapt themselves to different Galápagos Islands, are a case in point.

Gould wrote in 1980, 'The absence of fossil evidence for intermediary stages between major transitions in organic design, indeed our inability, even in our imagination, to construct functional intermediates in many cases, has been a persistent and nagging problem for gradualistic accounts of evolution.'[4] Eldredge, in 1995, would write,

> No wonder paleontologists shied away from evolution for so long. It never seemed to happen. Assiduous collecting up cliff faces yields zigzags, minor oscillations, and the very occasional slight accumulation of change over millions of years, at a rate too slow to really account for all the prodigious change that has occurred in evolutionary history. When we do see the introduction of evolutionary novelty, it usually shows up with a bang, and often with no firm evidence that the organisms did not evolve elsewhere![5]

Then I read Michael Denton's now classic work *Evolution: A Theory in Crisis* (1985). Where necessary, in the pages that follow, I have summarized some of Denton's arguments. Denton does not deny that evolution occurs. He points out that we observe it taking place *within species*. Darwin's theory that all species have emerged by a series of gradual, infinitely slow, infinitely small mutations is simply not borne out by the evidence. Moreover it is not merely scientists who have provided us with reasons to doubt Darwin. Thomas Nagel is only one philosopher, possibly the most distinguished, to question the plausibility of Darwinism (*Mind and Cosmos: Why the Materialist neo-Darwinian Conception of Nature is Almost Certainly False*, 2012). Do we need catch-all theories, such as Darwin's, which seem to explain everything at once?

It is probably worth saying that Nagel is not a religious believer. His fields have been political philosophy and the philosophy of mind. What he has come to question is the extremely simplistic, reductionist 'explanations' of complex phenomena – above all, *consciousness* – which Darwinism offers. 'The world is an astonishing place, and the idea that we have in our possession the basic tools needed to understand it is no more credible now than it was in Aristotle's day.'[6]

I did not, and do not, want to write a destructive book, but I now found it was not possible to tell the story from the position of a simple belief in *The Origin of Species*. After a few months of agonizing over the matter, I decided that this makes my task much more difficult than that of my predecessors in the Darwinian field, but also more interesting. The more I thought about it, the more I saw that Darwin's errors sprang from the mindset of his particular age and milieu. They were programmed to have a particular view of the world by the economic and social world in which they lived. Darwin was frank enough to say that his theory came to him, not as a result of his biological researches, but from reading an economist – Thomas Malthus.

In time, I came to see Darwin as two men. One was the observant naturalist, who spent nearly a decade writing a book about barnacles. The other was the theorist. The theorist is the one whose name is invoked in our day to justify the theories of those who espouse them. Often these theories have nothing to do with science. I therefore find myself writing, the biography not merely of a man, but of his idea; and not merely of his idea, but of his age.

I
A Symbol

ANY TOURIST IN London is likely, at some point, to visit the Natural History Museum in South Kensington. Standing in its vast hall, you could almost be in a Romanesque cathedral. You are greeted, however, by the statue of a man who was credited by his closest associates with having undermined the very grounds for religious belief: Charles Darwin. The huge museum, with its skeletons of creatures long since extinct from the earth's surfaces and its prodigious collection of specimens, continues to grow, and to be adapted, as a reflection of the state of modern scientific knowledge. Yet it remains an essentially Victorian museum – Victorian not merely in its architectural style, but in its aims and purpose.

The collections of specimens – of insects, birds, skeletons, fossils, many of them made and donated by Darwin himself – were originally housed in the British Museum in Bloomsbury. The construction of a separate museum constituted to celebrate the advances made in biology and botany and geology and zoology in the nineteenth century was the inspiration of Richard Owen (1804–92). Not only was Owen one of the great naturalists, and great anatomists, of his age, he was also an administrator of genius, a wheeler-dealer, a getter-of-things-done. It was the energy and resolve of this former poor apothecary's assistant, later Hunterian Professor of Anatomy, which led to the accumulation of enough funds to build the museum, and to cajole the government into recognizing what this building was. It was a monument to the fact that the Victorian Age had seen advances in scientific knowledge which were without historical parallel.

The statue of Darwin which sits in the hall at the bottom of the stairs in Owen's great museum was made by the sculptor Sir Edgar

Boehm. It was unveiled a year after the museum opened, and the speech was made by the President of the Royal Society, Thomas Huxley. 'Whatever be the ultimate verdict of posterity upon this or that opinion which Mr Darwin has propounded,' Huxley said; 'whatever adumbrations or anticipations of his doctrines may be found in the writings of his predecessors; the broad fact remains that, since the publication and by reason of the publication, of "The Origin of Species", the fundamental conceptions and the aims of the students of living Nature have been completely changed. From that work has sprung a great renewal, a true "instauratio magna" of the zoological and botanical sciences.'[1]

Huxley's words were both true and provocatively paradoxical. They were erecting the statue in the museum which was the creation of Owen, a man who had helped Darwin hugely in his early career but whom Darwin was to describe in his *Autobiography* as a bitter enemy.[2] The placing of Darwin's statue in this particular place was an act of deliberate renunciation not merely of Owen personally, but of his whole attitude to science. Owen, dubbed the 'British Cuvier' by Huxley (who meant it as an insult), was committed to the museum as the primary institution through which science moved forward.[3]

To each nation, their own pioneer. Many French people, sixty years before Darwin published *The Origin of Species*, would have echoed Honoré de Balzac's hymn to the great Parisian museologist, and seen in Georges Cuvier's assemblage of dinosaur-skeletons and palaeontological relics the true beginnings of modern evolutionary science, and with it their changed perception of what it meant to be a human being on the planet earth.

> Is not Cuvier the greatest poet of our century? Certainly Lord Byron has expressed in words some aspects of spiritual turmoil; but our immortal natural historian has reconstructed worlds from bleached bones, has, like Cadmus, rebuilt cities by means of teeth, peopled anew a thousand forests with all the wonders of zoology thanks to a few chips of coal and rediscovered races of giants in a mammoth's foot. These figures rise from the soil, tower up and people whole regions whose dimensions are in harmony with their colossal stature. He writes poems in numbers, he is sublime in the way he places cyphers after a seven [that is, he greatly increases the biblical number of seven days

> of creation] . . . And suddenly . . . lost worlds are unfolded before us! After countless dynasties of gigantic creatures, after generations of fishes, innumerable clans of molluscs, comes at last the human race, the degenerate product of a grandiose type whose mould was perhaps broken by the Creator Himself . . . We wonder, crushed as we are by so many worlds in ruin, what can our glories avail, our hatreds and our loves, and if it is worth living at all if we are to become, for future generations, an imperceptible speck in the past.[4]

For Huxley, however, and for his generation of English scientists, it made sense to claim that evolution had been the discovery of an Englishman. Darwin was their magic genie, whom they had conjured, not out of dead bones in a glass case, but out of their dissecting rooms and laboratories. Darwin's theories had been concocted, by his own confession, from reading the standard textbook of selfish capitalist economics, Malthus's *Essay on the Principle of Population*, but he had tested his idea by observing living beings – pigeons, worms, dogs, apes incarcerated in the zoological gardens. Logically speaking, the French novelist should have been right: if you believe in 'evolution', you must believe that all species, including our own, are heading for a transformative process which amounts to extinction – unless you concede, which Darwin doggedly never would, that there is something 'special' about human beings which makes them different from other species. Strangely enough, however, Huxley and the other Victorian Darwinians, unlike Balzac and the followers of Cuvier, did not see their genus and class as heading for extinction. Rather, they saw science as confirming their position as lords of the universe. For Huxley, by contrast with Cuvier and Owen, it was the laboratory, and not the museum, where science had made its strides.

It is not only in the churches at the time of the Reformation, nor yet alone in the former Soviet Union, that statues find themselves on the move, often with deep symbolic effect. In the Natural History Museum, for example, Darwin's statue on the staircase was moved less than a decade after it had been placed there. Owen died in 1892, ten years after Darwin, and his statue replaced that of the celebrated naturalist of Down House. There seemed justice and logic in the decision at the time; after all, the museum was largely Owen's creation. Moreover, by then, Darwin's reputation was on the wane, so much so that by the early twentieth century it had almost the nature of a family cult, kept going by Darwins and Huxleys and their Cambridge friends. Darwin in his lifetime had changed his mind so often about the details of his theories that the scientific world had moved on. Evolution was accepted as a given by most scientists. How it operated, however, remained mysterious until the rediscovery of Mendel's genetics in the early part of the twentieth century.[5]

Then came neo-Darwinism, which is still the orthodoxy in most academic scientific faculties in the world. This is the melding of Darwinism with Mendelian genetics. The weirdly Victorian battle between creationists and neo-Darwinians needed a patron, and the neo-Darwinians could scarcely, without absurdity, have enlisted for this purpose the genial Bohemian friar who had pioneered the science of genetics. Since Darwin and his champion Huxley were among the relatively few scientists in history who entered the sphere of religious controversy, Darwin was the obvious such figurehead. The bicentenary of Darwin's birth, 2009, seemed a timely moment to heave the 2.2 ton statue back to its original place on the landing overlooking the gigantic skeleton of a diplodocus. It was rather as if the statues of Lenin had been re-erected in the central squares of the now democratic capitals of the former Warsaw Pact countries. Darwin's hero Charles Lyell liked to quote Constant Prévost, 'Comme nous allons rire de nos vieilles idées! Comme nous allons nous moquer de nous mêmes.'

Classification was Owen's métier. For Huxley, by contrast, it was not the museum but the laboratory which was the place where good science could progress. Owen, in Huxley's view, had committed the ultimate sin for a man of science: he had compromised the truth, in order to appease his conservative-minded patrons. He had, in Darwin's and Huxley's view, shilly-shallied over the truth of evolution, appearing, when he spoke to conservatives, to disbelieve that species could mutate, while hinting, when he spoke to Darwin, that they could. Darwin punished Owen, in *The Origin of Species*, by numbering him among those diehards who believed species were fixed.

As this shows, Darwin was no saint, for he knew perfectly well what Owen thought about evolution, and knew he was not a simple-minded anti-evolutionist. Huxley, however, egged Darwin on in his quarrel with Owen. And long before Darwin was an icon sculpted by Boehm, and gazing benignly at visitors to the Natural History Museum, Huxley had made his friend into a Type of the Perfect Man of Science, the secular-materialist equivalent of a saint.

Whereas Owen was a man of the world, who had to convince politicians and churchmen that his museums would not upset the apple-cart, Darwin was single-hearted and single-minded in his pursuit of the only thing which counted: the Truth. Moreover, Darwin, toiling for years in spite of bad health, to establish the truth, was rooted neither in the cigar-scented committee rooms of clubs, nor in the corridors of Whitehall, nor in the common rooms of academe. Rather, he did his work in that sacred Victorian place, the home, where his faithful wife, mother to nine children, dabbed her eyes with grief as her husband dismantled the grounds for religious belief. Here, again, was a story which bore some relationship to reality, but which Huxley had moulded creatively.

Huxley coined the word 'agnostic' for his own religious position. By the time he unveiled the Darwin statue, it was pleasing to consider oneself daring or anti-establishment by not being Christian, but in fact, in the sort of circles in which Huxley moved – metropolitan and intellectual – agnosticism was commonplace. Though there were intellectuals in public life – the poet Browning, the politicians Gladstone or Salisbury – who were articulate Christian believers, the Sea of Faith, as it ebbed in Matthew Arnold's great poem 'Dover Beach', had made its melancholy, long, withdrawing roar. G. K. Chesterton was probably right to say that by the end of the nineteenth century atheism was the religion of the suburbs. Materialism, as expressed by Herbert Spencer, the most famous and popular philosopher of the Victorian Age, defined that age. Thomas Hardy, wistful and lyric, had described God's Funeral in a poem. And Darwin, who with his copious white beard resembled one of the biblical prophets, was the prophet of the irreligious position. This was partly of his own doing, but it very much suited Huxley's vision of things to have the figure of Darwin to display to the public. Behold the Man! Here was the man who, in the words of one of his most fervent disciples in our own times, had legitimized unbelief. 'Although atheism might have been logically tenable before Darwin, Darwin made it possible to be an intellectually fulfilled atheist.' So wrote Richard Dawkins, in his classic distillation of Darwinism *The Blind Watchmaker*.[6]

Darwin, therefore, from the publication of his most famous book, was always something more than a scientist. Huxley concluded his speech in the Natural History Museum with a plea that Darwin should be seen not merely as a man. 'We beg you to cherish this Memorial as a symbol by which, as generation after generation of students of Nature enter yonder door, they shall be reminded of the ideal according to which they must shape their lives . . .'[7]

In a survey conducted by the *New Statesman* magazine in 2011, various public or intellectual figures were asked their religious views. Richard Swinburne, Emeritus Professor of Philosophy at the University of Oxford, delivered one of the more succinct answers.

> To suppose that there is a God explains why there is a physical universe at all; why there are the scientific laws there are; why animals and then humans have evolved; why human beings have the opportunity to mould their character and those of their fellow humans for good or ill and to change the environment in which we live; why we have the well-authenticated account of Christ's life, death and resurrection; why throughout the centuries millions of people (other than ourselves) have had the apparent experience of being in touch with and guided by God; and so much else. In fact, the hypothesis of the existence of God makes sense of the whole of our experience and it does so better than any other explanation that can be put forward, and that is the grounds for believing it to be true.[8]

I choose to quote this statement by a twenty-first-century Oxford professor, at the beginning of a book about a nineteenth-century scientist, for two reasons. One is that it is the best and simplest account of the Christian, theistic position which I know. The second is that you would be hard pressed to find any philosopher, any highly intelligent person, in the entire intellectual history of Europe, from the fall of Rome in the fifth century AD to the time of the French Revolution, who would not have agreed with Professor Swinburne's words. The atheist or non-believing or materialist tradition existed, of course it did, and can be found among the Epicureans of Dante's time, the followers of Spinoza in the seventeenth century and the atheist forerunners of modern thought in eighteenth-century France.

And there was David Hume. It remains true that the broad mainstream of intellectual opinion would have supported the Swinburne line, not for fear of popes or patriarchs, not for dread of the Inquisition, but because it would appear to be a sensible and reasonable statement. One aspect of his credo, however, would have puzzled some – though not all – of the thinkers and philosophers for that long period of one and a half millennia. And that is his belief that God explains 'why animals and then humans have evolved'.

Plato, Aristotle, St Augustine, all believed in evolution in some form or another. It was not a view entirely peculiar to the nineteenth century. But in the decades before Darwin, and during his lifetime, evolution of the species had been questioned – more, really, on theological than on scientific grounds. Various philosophical and scientific mistakes account for this phenomenon. So it was that by Darwin's day there were those who supposed that you had to choose between believing in the fixity of species and losing your religious faith. Darwin's faith evaporated slowly, and for a number of reasons which a biography such as this might be able to unpick. The battle over his *Origin of Species* theory, a battle which Huxley so relished and encouraged, was a useful weapon against religion only if you supposed – as many people appeared at that date to suppose – that God could have created the universe only by placing unaltered and unalterable species *in situ*, rather like those gardeners in large municipal parks who prick out the flower beds with plants ready-grown in potting sheds, rather than allowing the exuberant chaos of the herbaceous border to form a life of its own. There are still those who hold that to believe in evolution is incompatible with religious faith. They include those on both sides of the divide – both materialist atheists and the creationists, so called. That their strange battles still occupy so much of the antagonists' energy in the twenty-first century is testimony to Huxley's instincts: this is certainly a show that will run and run. For such as these, Charles Darwin is either the hero who 'made it possible to be an intellectually fulfilled atheist' or he is the demon who made so many people lose faith in the Bible. Both the creationists and the Darwinists here seem to be saying the same thing. In any event, it gives Darwin a strange position in the history of science. True, Galileo got into trouble with the Roman Catholic Church for asserting what Copernicus had worked out mathematically seventy years earlier, that we live in a heliocentric, not an earth-centred, universe. But the Church recovered its equilibrium, and nowadays the Copernican–Galilean–Newtonian universe is a fact which it would be insane to deny. No other major scientific discovery or breakthrough has 'made it possible to be an intellectually fulfilled atheist' – only Darwin's. This will prompt some people to wonder whether Darwin's distinctive twin doctrines – that evolution occurs gradually by means of natural selection, and that this process necessitates an everlasting struggle for existence – are not in fact scientific statements at all, but expressions of opinion. Metaphysical opinion at that. To conceal this fact, Darwin's ardent disciples in our own day are quite happy for it to be supposed – as it is by very many of the people who look at his image on a British ten-pound note – that Darwin was the person who discovered the phenomenon of evolution. We shall come to see that the story is not so simple. The story of evolution, and how some scientists resisted it and others came to accept it, is, obviously enough, intimately connected with the story of Darwin. Long before Darwin wrote *The Origin of Species*, however, scientists were aware that there was a process at work within species which enabled them to adapt themselves to their environment. French biologists had been divided between two opposing camps usually known as functionalists and structuralists. Cuvier was the great champion of the functionalists. He believed that flora and fauna had developed their characteristics as a way of surviving. That is why – let us say – a bat has wings, in order that it can live in trees and high places to escape predators. Cuvier's great opponent was Etienne Geoffroy Saint-Hilaire. He believed that the best way of studying species was to examine their basic structures rather than their functions.

Taxonomy, the subdivision of living plants, animals, insects and so on, was basically a study of structure. Cuvier noted, for example, the *furcula*, a bone which enabled birds to fly. Saint-Hilaire found a corresponding bone in fish. This showed, for the structuralists, that different species often had cognate characteristics which lay dormant or unused, but which revealed relationships between the species.

Neither Cuvier nor Saint-Hilaire propounded a theory of how the species evolved. Both accepted that the essential building-blocks of species were a sort of given. Owen believed he had found the way to reconcile functionalist and structuralist biology. In his groundbreaking discourse *On the Nature of Limbs*, given as a paper at an evening meeting of the Royal Institution of Great Britain on 9 February 1849, Owen accepted that taxonomy's task was to classify the building-blocks, what he called *homologues*. A mammal, for instance, possesses certain characteristics – it is amniote, it has hair, it has tetrapod limbs, at the end of which we find a pentadactyl pattern. These are the given homologues of the taxa. Owen noted however that every species of mammal has a different 'adaptive mask', as he called it. Depending on its needs, the basic building-blocks for each species of mammal, will *adapt*, and this is most obvious in the limbs, where in a bat you see wings, in a man useful fingers and thumbs, and legs which enable him to stand upright, in the horse a series of adaptations which allows the pentadactyl 'hands' and 'feet' to morph into hoofs. Owen did not offer an explanation for how the species assumed their various 'adaptive masks'. Nor did he offer an explanation of how the building-blocks of nature arose in the first instance. He did conclude his discourse with words which make it quite plain that science had, by 1849, come to accept changes within the taxa.

> To what natural laws or secondary causes the orderly succession and progression of such organic phaenomena may have been committed we as yet are ignorant. But if, without derogation of the Divine power, we may conceive the existence of such ministers, and personify them by the term 'Nature', we learn from the past history of our globe that she has advanced with slow and stately steps, guided by the archetypal light, amidst the wreck of worlds, from the first embodiment of the Vertebrate idea under its old Ichthyic vestment, until it became arrayed in the glorious garb of the human form.[9]

The scientific study of how species adapt themselves 'with slow and stately steps' would have advanced – as this glorious paragraph shows – had Charles Darwin never been born. The distinctive and Darwinian idea, of course, was that one species changed into another, that the building-blocks themselves – the carpels of angiosperms, for example, enabling thousands of different plant forms to burst into flower, or the feathers of birds – had somehow themselves evolved by a mysterious, impersonal process. Adaptation of the kind demonstrated by Cuvier and Owen in their museums is a phenomenon which we can watch at work, through the discoveries of palaeontology. These would eventually demonstrate, for instance, almost every adaptive stage by which a small foxlike animal that walked on four toes, and whose fifth dactyl was redundant thirty to fifty million years ago in the Eocene Epoch, adapted itself in a progressive line producing *Orohippus*, *Epihippus*, *Mesohippus* and so on, until it came to stand on one toe enclosed in a protective hoof. The redundant fifth dactyl was evolved out of existence, and eventually *Equus caballus*, our horse, stands before us.

Darwin's theory was that the basic building-blocks – the amniotic sack, the feather, the carpel – had themselves evolved in the way that a horse's limbs had adapted. For this, palaeontology has never provided any evidence whatsoever. Stephen Jay Gould described the total absence of any transitional forms as 'the trade secret of palaeontology'.[10]

There is surely a reason for this, and when we identify the reason, we see why Charles Darwin occupies so unusual a role in the history of science.

The story of his life is not quite what you might expect. The first principle of hydrostatics was plausibly attributed to a particular brainwave, occurring to a particular man, Archimedes, when he stepped into the bath and realized that the mass of the overflowing water was equivalent to his own. Darwin's relation to the science of evolution was not of this order. But, once he had been made into a symbol by Huxley (and he was very happy to be a symbol), Darwin could take the credit for an idea which had, appropriately, evolved in many minds over many decades. Moreover, although, to read the writings of modern-day Darwinians, you might suppose that Darwin's version of evolution has been irrefutably proved, or that it is now contested only by religious bigots, this is decidedly untrue. Boyle's Law, on the inverse proportionality of volume and pressure in gases, remains in place until another chemist comes along and refutes it by persuasive experiment. The same is true of all real scientific discoveries. Darwin's theory of natural selection, and his claim that the process works as a result of an everlasting warfare in nature, are not laws like Boyle's Law or Newton's Law of Gravity. Although their extreme unlikelihood (especially of the 'struggle for existence' idea) can be demonstrated, they are not strictly speaking verifiable, or falsifiable, and in this sense they are not scientific statements at all. It is for this reason that there has never been a time, since *The Origin of Species* was first published in 1859, when Darwin's theory was universally accepted by the scientific academy. And even among Darwinians, down to our own day, there remain deep fissures between the varying sects – for example, those who hold to the true faith of each evolutionary change having come about by an infinitesimally slow gradualism or micromutation, and those Darwinians such as Stephen Jay Gould and Niles Eldredge whose punctuated equilibria provided a kind of Fast Forward button on the evolutionary Remote which allowed a species to cut out some of the boring waiting-time and jolt forward to the next stage. So it is in many senses surprising that Darwin is seen as a symbol, or as the Man Who Discovered Evolution. How this came to pass will be explored in the story that follows. It certainly did not happen by accident. It happened by determined and ruthless self-promotion on Darwin's part, by cultivation of his own image, and by enormous good fortune in his choice of enemies. In his own lifetime, Darwin attracted little support among his scientific colleagues for his own distinctive take on evolution. His loyal 'Bulldog',[11] Huxley, however, was usually able to minimize the intellectual and scientific objections to Darwinism by one simple device. He made it seem as if Darwin's enemies objected to his theory only for reasons of religious bigotry. If the Darwinists had not managed to represent themselves as single-minded warriors for truth against obscurantists from the Dark Ages, the unsatisfactoriness of their science might have been made clearer. It is this book's contention that Darwinism succeeded for precisely the reason that so many critics of religions think that religions succeed. Darwin offered to the emergent Victorian middle classes a consolation myth. He told them that all their getting and spending, all their neglect of their own poor huddled masses, all their greed and selfishness was in fact *natural*. It was the way things were. The whole of nature, arising from the primeval slime and evolving through its various animal forms from amoebas to the higher primates, was on a journey of improvement, moving onwards and upwards, from barnacles to shrimps, from fish to fowl, from orang-outangs to silk-hatted Members of Parliament and leaders of British industry. It was all happening without the interference or tiresome conscience-pricking of the Almighty. He, in fact, had been conveniently removed from the picture, as had the names of the many other thinkers and scientists, including Darwin's own grandfather, who had posited theories of evolution a good deal more plausible than his own. Copernicus had removed the earth – and by implication the human race – from the centre of the universe. Darwin in effect put them back. For all the brave, Darwinian talk of natural selection being non-purposive and impersonal, it breathes through the pores of everything which Darwin and Darwinists write that natural selection in fact favours white middle-class people, Western people, educated people, over 'savages'. The survival of the fittest was really the survival of the Darwin family and of their type – a relatively new class, which emerged in the years after the Napoleonic Wars in Britain and held sway until relatively recently. It remains to be seen, as this class dies out, to be replaced by quite different social groupings, whether the Darwinian idea will survive, or whether, like other cranky Victorian fads – the belief in mesmerism or in phrenology, for example – it will be visited only by those interested in the quainter byways of intellectual history.

2

The Old Hat

CHARLES DARWIN, BORN 12 February 1809 at The Mount, Shrewsbury, was a war baby. When he came into the world, as the child of a very prosperous doctor-cum-banker, Britain had been at war with France over fifteen years. For the previous fifteen months, since Napoleon's issue of the Milan Decrees, there had been an effectual ban on products coming directly from Britain into the continent. Goods which were known to have come from Britain were confiscated at the European ports. There was a food blockade. British trade was at an all-time low. Imports of raw materials and of food had dropped disastrously. In the months before Darwin's birth – the last quarter of 1808 – grain imports fell to one-twentieth of the previous year. There was widespread industrial disquiet. In Manchester, there were frequent and violent strikes.[1] Wages fell. There was real hunger, and the fear of actual starvation. The population of Britain had risen sharply and now stood at around ten and a half million – compared with seven million in 1760. No wonder in these totally abnormal circumstances that the Revd T. R. Malthus, whose writings were to have so profound an influence upon Darwin, believed that the fight for food was the key to economic history, and that when the limited food supply ran out, the population would eliminate itself by violent struggle or starvation.

Malthus was not alone in fearing that the British population might starve. Way back in 1798, when the war with France had lasted only five years, the Prime Minister, William Pitt, had established the Board of Agriculture. Its prime object was to increase food production.[2] On the whole, during the eighteenth century, artificial stock improvement in Britain had been unknown, except among horse-breeders. An exception had been Robert Bakewell

(1725–95) of Dishley Grange, near Loughborough in Leicestershire, who had used techniques followed by the breeders of racehorses to improve farmstock. He began with the Leicester breed of sheep and the Longhorn breed of cattle. He realized that by careful selection of the progeny of favoured animals, a breed could be changed dramatically. Long-woolled sheep from Leicestershire and Lincolnshire were London's chief source of mutton. Bakewell's New Leicesters were bred to build up fat deposits while bone and muscle were still developing, and they were ready for the butchers a full year younger than their predecessors. Bakewell's methods were followed by beef farmers, starting with Herefords. (The Aberdeen Angus was a later development first shown in the 1820s. The Ayrshire was a cross between local and imported cattle, and first recognized as a breed in 1814. Herefords had been pioneered since the 1720s and in the 1870s would be the major bull for export to the American market.[3]) In the middle of the eighteenth century, there were only two types of native domestic pig in Britain. The importing of Asian pigs in the middle of the century literally saved Britain's bacon during the French wars, leading to the development of modern breeds such as the Berkshire and the Saddleback. (The Tamworth was imported from Barbados. Gloucester Old Spot was an artificial hybrid.)[4]

Even as the baby Darwin lay in his cradle, then, Britain, in the highly artificial conditions of the Napoleonic blockade, was learning to feed itself by processes of highly artificial hybrid selection. As things turned out, Malthus's prediction of a struggle for survival, followed by cataclysm, could not have been less accurate. Instead of blind struggle, there was ingenuity; instead of selfish grab, there was co-operation; with an increase in population, there actually followed an increase of food.

Gloucester Old Spot pigs were not the only triumphant result of ingenious breeding. The upper-middle classes, to which Darwin and his family belonged, had recently emerged as a unique British hybrid. Quite as much as the occasional periods of hunger in France, it was the rigidity of the French system, the inflexibility of aristocratic privilege, which had fired the Revolution. In Britain, the bourgeoisie, the professional classes and the emergent merchant and manufacturing classes were united: partly by innumerable strands of interrelation and marriage, partly by shared money-interest, partly by (broadly speaking) shared values. Although, in origin, they came from a variety of classes – including the humblest – they were united by cleverness. 'Thus they gradually spread over the length and breadth of English intellectual life, criticizing the assumption of the ruling class above them and forming the opinions of the upper middle class to which they belonged. They were the leaders of the new intelligentsia.'[5] As Noel Annan, one of the most astute observers of this species noted, 'they all regarded themselves as gentlemen'; they devoted themselves, during the 1860s – that is, in the decade immediately following Darwin's *Origin of Species* – to two great aims: intellectual freedom in the universities and 'the creation of a public service open to talent'. They ruled Britain, in effect, from the 1850s until the 1950s. Writing in 1953, Osbert Lancaster, who belonged to this class himself, reflected upon the fact that the other classes – the aristocracy, the landed gentry, the lower-middle and working classes – had all survived the twentieth-century convulsions of two world wars and social revolution. The casualty had been the class who possessed the huge houses in Kensington and Notting Hill in London.[6] His analysis of the class – which was on the whole London based – repays scrutiny, but one point, for those of us who live after the demise of the species, needs underlining.

> In Victorian times writers and artists, save one or two of the most exalted, living remote and inaccessible on private Sinais in the Isle of Wight or Cheyne Row, had conformed to the pattern of the upper-middle-class to which most of them belonged. Matthew Arnold, Browning, Millais were all indistinguishable in appearance and behaviour from the great army of Victorian clubmen, and took very good care that this should be so. The *haute Bohème* did not exist and the Athenaeum rather than the Closerie des Lilas shaped the social life of the literary world.[7]

The Athenaeum is one of the London clubs. If you enter it today, you will find a large portrait of Charles Darwin hanging over the bar. It is hard to explain the importance of 'clubland' in Victorian London to a generation of the twenty-first century for whom clubs, if they impinge at all on the consciousness, are merely places to have a bit of lunch or dinner.[8] In Victorian England, they were places where political life – to left and to right – was discussed and forged; where chaps – for they were male preserves – decided who should be the next Regius Professor of Greek, the next Bishop of Bath and Wells, the next Chancellor of the Exchequer. In short, they were places which changed human destinies. Darwin grew up to become – in his own estimation, in legend and to a certain extent in reality – a recluse. He was also a clubman. His membership of the Athenaeum – and his exclusion from the Athenaeum of those who did not accept his ideas – mattered to him. The Athenaeum, as its names implies, was an intellectual place. Bishops, university professors, poets, the higher journalists could mingle here. When Osbert Lancaster said that the Athenaeum shaped the literary world, he could also have added the scientific world.

Though politically a Liberal, Darwin was profoundly a small-c conservative, and much of his life-story is incomprehensible unless the twenty-first-century reader is acclimatized to the fact. Outside this enclosed family grouping, and this relatively new class-stockade, Darwin's story could not have happened.

Charles Darwin was the fifth of six children. (There were two boys and four girls.) Their grandfather Erasmus Darwin was not merely a sought-after physician. He was also an inventor. Addicted to ever-faster travel, he invented a steering mechanism for his phaeton which, in essence, is still used in motorcars. Heaving his vast bulk into his phaeton, Dr Erasmus would whizz round from patient to patient, his lecherous hands happily exploring the bodies of female clients while his mind buzzed, not only with technological invention, but also with speculative thought of a truly revolutionary colouring.

After two marriages, many dalliances and umpteen children, this vast, stammering, prodigy of a man had retired to Derby. The Derby Philosophical Society was largely devoted to a discussion of his ideas. Oxford intellectual Thomas Mozley (1806–93), who was at Oriel College in the 1820s, remembered the daring 'Darwinians' of Derby during his adolescence. These 'Darwinians', steeped in the works of Erasmus Darwin, maintained that 'creatures were created by themselves' – that species appeared on earth not because God had made them, but by evolution.[9]

Erasmus Darwin's evolutionary ideas had been expressed in *Zoonomia* (Greek for the Law of Life), a scientific treatise published in 1794. His ideas were also popularized in such long poems as the 4,384-line *The Botanic Garden* (1789) or *The Temple of Nature* (1803), which expounded a theory of impersonal evolution.

> But REPRODUCTION with ethereal fires
> New Life rekindles ere the first expires . . .
> Organic forms with chemic changes strive,
> Live but to die, and die but to revive!
> Immortal matter braves the transient storm,
> Mounts from the wreck, unchanging but in form.[10]

The young Wordsworth and Coleridge objected to Darwin's conservative poetic technique – it was in reaction against Darwin that they published the *Lyrical Ballads* in 1798. The middle-aged Wordsworth and Coleridge objected even more to the non-conservative complexion of Darwin's ideas. Once they had become Anglican Tories, the implied atheism of Darwin's poetry was anathema to them, though many of *Zoonomia*'s central ideas were borrowed by Wordsworth for his masterpiece on 'Tintern Abbey'.[11] It is hard for us to realize that in the 1790s very few people had heard of Wordsworth and Coleridge, whereas Erasmus Darwin – as well as being a medical, technological and scientific prodigy – had been the most famous poet in England.

In *The Botanic Garden*, Darwin had apostrophized the moon which shone on their famous dining group, the Lunar Men:

> And pleased on WEDGWOOD ray your partial smile,
> A new Etruria decks Britannia's isle.
> Charmed by your touch, the kneaded clay refines,
> The biscuit hardens, the enamel shines.[12]

Both Josiah Wedgwood and Erasmus Darwin had, from the first, been supporters of the French Revolution. Government spies kept a watch on Wedgwood, fearful that he, who sent his sons to Paris on a trading mission in 1791, might have been plotting the overthrow of monarchical government. Nothing could have been further from the truth. Wedgwood was not a profoundly political person. He merely belonged to that sunny, optimistic category of human being who thought that political change was likely to be for the better. Erasmus Darwin was more political than this, and fundamentally believed in Voltaire's *écrasez l'infâme!* – the infamy being not only the old aristocratic order, but the orthodox Christianity which supported it. Dr Erasmus was quite happy for altars, as well as thrones, to topple. In 1798, the year of the *Lyrical Ballads*, the Under Secretary for Foreign Affairs in Pitt's government, George Canning (who later became Prime Minister himself), actually thought it was worth blackening Dr Erasmus's name by getting two hack writers to publish a parody of Darwin's verse. In this feeble imitation-poem, which Canning hoped would be passed off as Darwin's work, Pitt dies on the guillotine.

> Down falls the impatient axe with deafening din;
> The liberated head rolls off below,
> And simpering freedom hails the happy blow![13]

Erasmus was not arrested, but the perception remained that if you believed in revolutionary *scientific* ideas, you were also likely to be a political revolutionary. Erasmus had died in 1802, seven years before Charles Robert Darwin saw the light of day in Shrewsbury. Charles played down any influence which his grandfather's ideas might have had upon him. One reason was, he wanted to be considered the originator of the idea of evolution. Another contributory factor, however, was the simple awkwardness of the Darwin name being associated in any way with the anarchy and bloodshed of revolution. There was an absolute need, as the new, aspirant upper-middle class emerged from the eighteenth-century chrysalis, to put a distance between the blood-spattered, cobbled streets of Paris and the sloping lawns and beautifully constructed hothouses of The Mount, Shrewsbury.

Charles's mother, Sukey (Susannah), was the eldest (and favourite) child of Josiah Wedgwood – potter, inventor, entrepreneur, millionaire phenomenon of Etruria in north Staffordshire. When George Stubbs painted the Wedgwood family in 1780, Sukey took centre-stage, mounted on a bay pony and looking indistinguishable from a member of the landed gentry. Her parents, who were cousins, sit to the side of the group portrait. Her mother, in an enormous hat which might have adorned the head of a Gainsborough countess, looks downwards, an indication of her depressive temperament, while Owd Wooden Leg, as the manufacturer was dubbed by his workforce, sits beside her, the prosthetic limb disguised as a stockinged and shoe-buckled calf casually crossed over a leg of flesh and blood; the source of the family wealth – the Etruria Works – discreetly blows a small puff of smoke into the bucolic air at a safe distance behind some trees. Sukey, like her mother and father, was a clever person of broad intellectual interests. Her sharp eyes, thick hair and thin lips were all inherited by Charles's sister Caroline. Her appearance was a little ethereal, but also humorous, animated and affectionate. It has been plausibly assumed that affection for the old potter – who was ill for much of his last decade of life – was Sukey's reason for delaying matrimony until she was thirty-one years old. This was distinctly late by eighteenth-century standards, though her parents, Sally and Jos, had been made to wait until Sally was nearly thirty before marrying. The Wedgwoods of that generation were cautious and financially shrewd. Old Richard Wedgwood, Sally's father, had been a prosperous cheese-merchant in Cheshire, who forced her to hold back before marrying her potter-cousin – having no notion of the prodigious fortune which Jos Wedgwood's genius would bring him. Another reason for the delay, and perhaps the likeliest, is that Sukey was in no hurry to marry Robert Darwin, who was vastly fat and had a filthy temper. A love match it was not.

When Jos died in 1795, he left Sukey £25,000 and a fifth share of the Etruria Works. It was a colossal fortune. In Jane Austen's *Pride and Prejudice*, the immense wealth of the aristocratic Mr Darcy is indicated by the fact that his sister Georgiana is worth £30,000 – plainly intended to convey to the reader riches beyond the normal dreams of avarice (though not beyond the dreams of the rascal Mr Wickham).[14]

It was something of a triumph for Jos Wedgwood, who had suffered smallpox and a major amputation, to live sixty-five years in the world. His son Tom, all but the inventor of modern photography, made it only to thirty-four, not sharing his father's high tolerance for alcohol and opiates. Eighteenth-century medicine was a rough business and they were lucky to have as their physician and family friend Erasmus

Darwin, who stood by Wedgwood's side through many a crisis, including the amputation.

Wedgwood and Darwin were not merely the literal ancestors of many of the Victorian intellectual 'aristocracy'. They were also its spiritual godfathers. The Lunar Men, of whom Erasmus Darwin was a leading luminary, met in Birmingham, Lichfield or Derby, to exchange world-changing thoughts. Joseph Priestley would tell his friends around the dinner-table of what he called 'dephlogisticated air'. Priestley was within an inch of discovering oxygen. His experiments would enable Antoine-Laurent Lavoisier to do so. Priestley could tell his dinner-neighbour, though not quite using this language, of the properties of water – H_2O. Since his neighbour at table happened to be Scottish engineer James Watt, they were fast on their way to seeing the uses of steam. Since their host was Birmingham manufacturer Matthew Boulton, there was money to construct machinery using steam, and since one of their friends was the ever-adventurous entrepreneur Wedgwood he could pay for a small steam engine for use in his works. The cleverness of these men stimulated each to new heights of invention. Wedgwood, as a nonconformist, could not attend a university or join a profession, but his sons and daughters could marry into the landed gentry. Cleverness turned into money which turned into new social positionings. Such transformations could not happen in the France of the old regime.

It was not an 'arranged' marriage in the formal sense of that word, as a royal marriage might be, or as oriental marriages might be, in our own day. It was, however, all but arranged. Josiah and Erasmus had been the best of friends, and a marriage between Robert Waring Darwin and Susannah Wedgwood did slightly more than cement this friendship. It established a dynasty, and it confirmed that, with their great wealth, the Wedgwoods were now safely rescued from the class of mere manufacturers. Josiah's son, Josiah II, Sukey's brother, had established himself at Maer Hall, some twenty-five miles from Shrewsbury, a large Elizabethan house, with extensive park laid out by Capability Brown and woodland full of game. Jos II (Uncle Jos to Darwin) and his elder brother John had married two of the daughters of a Welsh squire – John Allen of Cresselly in Pembrokeshire. Socially speaking, they had arrived, and, while they led the life of gentlemen, they left the old Etruria Works, with its bottle-ovens belching smoke into the Burslem air, in the incompetent hands of their cousin Tom Byerley. None of them possessed any of old Josiah's business acumen – not Byerley, who had tried his hand at acting before being rescued by old Jos and brought into the firm; not Tom Wedgwood, the youngest son, who had died of a drug-and-drink overdose; not John, who had tried to establish himself as a banker and lost a fortune; not quiet, gentle, amiable Jos II. Indeed, by the time Charles Darwin was born, the fortune could very easily have been dissipated had it not been for Robert Waring Darwin, who, as well as being a competent doctor, was also an astute money-man. It was Robert Darwin who had bailed out John Wedgwood when he got into difficulties. It was Robert Darwin who had put up the money for Jos II to take Maer Hall on a thirty-year lease for £30,000, and to pay for his sons to go to Eton. It was Robert Darwin who tried to control the excesses of Byerley's children, whose residence at Etruria Hall, the fine old Palladian house built near the works by Josiah I, gave them delusions of gentrified grandeur. Indeed, when Charles Darwin was only a year old, his father had gone to Maer to a crisis meeting with Jos. Continental and American trade had more or less ceased because of the war, and the Wedgwoods were still spending money as if they were aristocrats. John's London bank was not the only bank to fail because of the economic crisis. He was insolvent.

Robert Waring Darwin, who had inherited considerable sums from Erasmus, was the man who kept the Wedgwoods from bankruptcy. Like his father, he was huge – tall and fat – with a stammer; and, unlike his father, who was on the whole good-natured, Robert was depressive and irascible. His marriage to Sukey cannot have been one which made her very happy, even if she had not inherited her own mother's depressive temperament. Hardly a week passed without furious outbursts by her husband, whose only consolation, after such eruptions, was to visit the plants in his hothouses. He was not able to conceal from her the fact that, had it not been for him, her father's business, snobbishly neglected by her brothers, who did not wish to sully their hands with trade, would probably have gone under.

Charles Darwin, as well as growing up with a father whose anger was terrifying, also grew up with a sense that the Darwins were a cut above the Wedgwoods. In grown-up life, he would absorb himself, when not contemplating the breeding of pigeons or orchids or the ancestry of baboons, with his own pedigree, but it was always with the Darwin and not the Wedgwood side of the story that he was concerned. Unlike Owd Wooden Leg, who had spent a mere couple of years at a school in Newcastle under Lyme, and who had served an apprenticeship as a humble potter, the Darwins had gentle pedigree. A William Darwin had possessed an estate at Cleatham. His son, born in 1620, had served in the royalist army and had afterwards been called to the Bar. His eldest son, William Darwin, had married an heiress, after whom Charles Darwin's father was named, Miss Waring, who had brought the manor of Elston, Nottinghamshire, into the Darwin family. Erasmus Darwin, whose brothers included a poetical botanist and the rector of the family living (Elston), had been to Cambridge before studying medicine at Edinburgh. They were miles apart from the nonconformist artisans from whom Sukey sprang. Only Josiah I's money, and the mutation we have described, by which a new species, the upper-middle class, came into being, made possible Charles Darwin's parentage.

Whatever the nature of Robert Waring Darwin's beliefs, it must have been with a mixture of emotions that he witnessed his wife taking her children each Sunday, not to the parish church, but to the Unitarian meeting-house in Shrewsbury.

Had they visited the meeting-house in 1798, they might have heard the young Coleridge preaching there, when he was briefly the minister to the Shrewsbury congregation. It was in this meeting-house that the teenager William Hazlitt heard Coleridge speaking – 'his forehead was broad and high, light as if built of ivory, with large projecting eyebrows'.[15] The poet's career as a Unitarian preacher was of short duration partly because of the generosity of Sukey's brothers, Jos and Tom, who gave Coleridge an allowance of £100 a year to pursue his philosophical journeyings. Unitarianism was new to Coleridge, who was the child of a parson in the West Country, but it ran deep in the Wedgwoods. Old Jos supported the meeting-house at Newcastle under Lyme, where among other luminaries Joseph Priestley, no less, was a preacher. Jos and Priestley did their best to support the Warrington Academy – known as 'the cradle of Unitarianism',[16] and Sukey's brothers were educated there. This was at a time when scientific education was non-existent in such establishments as Eton and Winchester. Priestley helped ensure that the students were abreast of the latest scientific learning. They had some interesting teachers, including Dr Marat, who was subsequently murdered in his bath in Paris by Charlotte Corday on account of his bloodthirsty support of Jacobinism.

Unitarianism was an attempt to marry Christianity with Reason. The doctrine of the Trinity was discarded, though the person of Christ was venerated. Sukey was imbued with many of the values of the meeting-house, not least a sense that education was as vital for women as for men, and a belief in science. A passion for botany was one of the things which she had in common with her husband. Darwin's schoolfriend William Leighton remembered Mrs Darwin teaching him 'how by looking at the inside of the blossom the name of the plant could be discovered'.[17] It is a strange fact that in his *Autobiography* Darwin makes no mention of his parents' intense interest in plants. Nor does he tell us that his later fascination with pigeon-breeding was copied from his parents, who reared a great variety – the 'Mount pigeons' were well known in Shrewsbury.[18] '*All my recollections seem to be connected most with self*'[19] (my italics). This sentence could, and indeed must, be said by anyone attempting to write an autobiography, even one as rudimentary as Darwin's, which was penned for purely private consumption by his children. Nevertheless, the self-absorption of Darwin is by any standards remarkable. One slightly questions, for example, his claim that he, of all the Darwin children, was alone in forming collections of coins, seals and minerals. It is clear that he and his siblings shared their parents' love of botanizing, for example, and one wonders whether his sisters had flower albums which he simply failed to notice. Something to which he does own up is having been a compulsive liar. The same Leighton who recalled Mrs Darwin's teaching him about plants (and Leighton went on to become a well-known lichenologist and botanist) was told by the infant Darwin that he 'could produce variously coloured polyanthuses and primroses by watering them with certain coloured fluids, which was of course a monstrous fable, and had never been tried by me. I may here also confess that as a little boy I was much given to inventing deliberate falsehoods, and this was always done for the sake of causing excitement.'[20]

The solipsism and the dishonesty would scarcely be worth mentioning in so small a child were it not that both characteristics were carried on into grown-up life. *All my recollections seem to be connected most with self.* It would be fascinating to know, say from interviewing his sisters, whether these characteristics predated Sukey Darwin's death, or whether they are in some way connected with it.

Sukey Darwin evidently suffered from a variety of symptoms, difficult at this distance to diagnose. There were frequent blinding headaches, and intestinal disorders. Like her mother before her, she often went to Bath to take the curative waters. In July 1817, she was taken very ill, with severe vomiting. Dr Robert Darwin realized that her condition was serious enough to warrant summoning her unmarried sister Kitty from their Staffordshire cottage Parkfields, Tittensor – where she lived with her mother and her sister Sarah. Kitty found Sukey in a bad way. Apart from brief visits from her eldest and youngest daughters, Marianne and Caroline, she had not seen her children for some days. She could scarcely talk. Writing back to Parkfields, she added a postscript that she did not think she would live through the night. In fact, she did so, but twenty-four hours later – probably of peritonitis – she died. Darwin had not been allowed into the sickroom. In later years, he would say that nearly all memory of his mother had totally faded from his mind. 'I can remember hardly anything about her except her death-bed, her black velvet gown, and her curiously constructed work-table,' he wrote.[21] He remembered his father's tears, and Dr Robert's inability to convey his grief; but of his mother, he said, he remembered almost nothing.

Darwin was eight and a half years old when Sukey died, aged fifty-two. I recall once asking the psychiatrist Anthony Storr what was the worst possible age at which to lose a parent. He replied without hesitation, 'Nine,' going on to say that you were old enough to be able to be fully aware of what had happened, but lacking any of the emotional equipment which enables you to grieve. John Bowlby, another eminent (child) psychiatrist, fixed upon Darwin's habit, even as a child, of taking long solitary walks in a condition known as the 'fugue' state, without any idea what he was thinking. This, said Bowlby, suggests 'a state that is known to occur in persons who have failed to recover from a bereavement'.[22] It would clearly be a mistake to say that because Darwin failed to remember his mother she was therefore of no importance to him. Equally mistaken would be to say that, because Darwin's childhood was prosperous and superficially carefree, it was happy. From the age of eight and a half onwards, a number of behaviour patterns emerge from which Darwin never escaped until his dying day. One is for a series of psychosomatic illnesses and stress-related symptoms to manifest themselves. Almost as soon as his mother had died, Darwin began to develop eczema, which usually erupted all over his hands, but also encrusted his face and lips. When this happened, it was permitted for him to forgo the society of strangers and *remain at home*. The need to withdraw into the family, and for extended periods not to see anyone except his family, was, once again, a compulsion which never left him. Another factor, observable from the start, was a need to prove himself, to win, and to be seen as the only player in the field. Thus others could remember Mrs Darwin teaching her children botany, but in his *Autobiography* Darwin wished to represent himself, even as a child, as self-taught. The desire to *get on*, to assert the will, was at the same time concealed, partly by involuntary bouts of illness and melancholy, and partly by compulsive time-wasting: by apparently mindless brooding and by sport.

So, in Shrewsbury, a new life began for the six of them, living with the vast, obese, irascible Doctor: Marianne, who was nineteen; Susan, fourteen, Erasmus (Ras), thirteen, Catherine, nine, Bobby – as Darwin was known in the family – eight, and Caroline, seven. From now on, the family bond, both with his siblings at The Mount and with his Wedgwood cousins at Maer, was absolutely central, and he had every opportunity for fishing, shooting, botanizing. Yet there was the great Absence – that of his mother. There were the telltale psychosomatic disorders. There were the long solitary walks, his mind blank. And there was the alarming, gigantic, angry father. Because Darwin was a 'gentleman' – and because family was all to him – he would never have given himself over to the heretical, subversive thoughts of his fellow Salopian (twenty-six years younger) Samuel Butler, who in a note 'On Wild Animals and One's Relations' wrote: 'If one would watch them and know what they are driving at, one must keep perfectly still.'[23] Butler, even more than Freud (certainly as far as the English-speaking world was concerned, up to the First World War), was the great subversive pointing out, in his heretical novel *The Way of All Flesh* and in his explosive posthumous notebooks, what for many of those great, self-enclosed Victorian families was the hideous truth about the Family:

> I believe that more unhappiness comes from this source than from any other – I mean from the attempt to prolong family connection unduly and to make people hang together artificially who would never naturally do so. The mischief among the lower classes is not so great, but among the middle and upper classes it is killing a large number daily. And the old people do not really like it much better than the young.[24]

The reason for it was simple: money. In the past, among the upper classes, younger sons had to leave their estates or lands and seek a fortune, either in the armed services or the professions, or in the Empire. The classes beneath, likewise, were either taken up with the need to work or went in search of it, leaving their kinfolk behind. Only the new class, the Victorian rentier class, lived in one another's pockets as the Darwins and Wedgwoods were to do. They had to do so, for fear of the glue which held them together – their money – slipping out of their grasp, and their being obliged, horror of horrors, to enter a trade. Butler again: 'Next to sexual matters there are none upon which there is such complete reserve between parent and child as those connected with money. The father keeps his affairs as closely as he can to himself and is most jealous of letting his children into a knowledge of how he manages his money.'[25] This is deeply true of the Darwins. It was not until he married that Charles Darwin had anything approaching a candid conversation about money with his prodigiously rich father. Only when the old man died did Darwin realize quite how rich they were. But it was always there in the background. Money produces the greatest security a human being can have in the capitalist world, but – witness the chaotic affairs of Darwin's uncle John Wedgwood, for example, trying his hand at banking, going bust, and suffering from mental illnesses – it also produces the greatest insecurities. This was what that other great subversive Victorian text, *Das Kapital*, played upon constantly. Money and capital, which underpinned the rich man's flowering lawns, could also land even so venerable a figure as Sir Walter Scott in Queer Street as the result of improvident investment and the unseen movement of markets. This lurks behind the playful, yet essentially nervous, little story in Darwin's *Autobiography*, in which a pal of his played a trick on him.

> A boy of the name of Garnett took me into a cake shop one day, and bought some cakes for which he did not pay, as the shopman trusted him. When we came out, I asked him why he did not pay for them, and he instantly answered, 'Why, do you not know that my uncle left a great sum of money to the town on condition that every tradesman should give whatever was wanted without payment to anyone who wore his old hat and moved [it] in a particular manner?' And then he showed me how it was moved. He then went into another shop where he was trusted, and asked for some small article, moving his hat in the proper manner, and of course obtained it without payment. When we came out, he said, 'Now if you like to go by yourself into that cake shop (how well I remember its exact position), I will lend you my hat, and you can get whatever you like if you move the hat on your head properly.' I gladly accepted the generous offer, and went in and asked for some cakes, moved the old hat, and was walking out of the shop when the shopman made a rush at me, so I dropped the cakes and ran for dear life, and was astonished by being greeted with shouts of laughter by my false friend Garnett.[26]

Proudhon ('Property is theft') or Lenin would have appreciated the story: the mysterious tilt of the propertied person's hat being as arcane as any other indicator of the rights of property. Old Josiah's money-bags had been filled by selling exquisite wares which thousands of people actually wished to own, to adorn their dinner-tables and mantelpieces. That was Capitalism, Stage One, and of course, Victorian England was dotted with similar success stories in mills, steelworks and manufactories of all kinds. Capitalism Stage Two, however, in which money itself was making money, as in the life of Dr Robert Waring Darwin, took you into the world of Mr Merdle in *Little Dorrit* – one moment the richest and most sought-after man in London, the next a ruined swindler who cut his own throat.

3
What He Owed to Edinburgh

After the death of Sukey Darwin, the household and its habits underwent a change. Darwin remembered, a month after his mother had died, staring blankly out of a classroom window, at the little Unitarian school in Shrewsbury run by the minister, Mr Case, and seeing a horse, riderless, being led to the graveyard. The funeral of a dragoon was taking place in the very churchyard where Darwin's mother lay. 'It is surprising how clearly I can still see the horse with the man's empty boots and carbine suspended to the saddle, and the firing over the grave.'[1]

This was one of the last memories he would have of the world seen through a Unitarian window. With his wife's death, Dr Robert Waring Darwin took the children to the parish church, rather than to the meeting-house. And Bobby was soon withdrawn from the little school run by Dr Case and sent as a boarder to join his elder brother Ras at Shrewsbury School. Geographically, the school was close enough to The Mount, but gone were the comforts of home, gone the good-night kiss from his affectionate elder sisters before he went to sleep. Public schools in 1818 were alarming places, and without the protection of his brother Ras it would have been even worse. The elder Darwin boy would – with gaps – be the younger boy's protector until he grew up. The children slept in crowded, unheated dormitories. There was no modern plumbing, so the place stank. There was small opportunity to wash. Cruelties and sexual depravity among the boys were the norm. Many of the buildings were all but unchanged since the school had been founded in the reign of Edward VI. The headmaster – from 1798 until 1836 – was Dr Samuel Butler (grandfather of the satirist of the same name, and later Bishop of Lichfield). Butler was one of the best headmasters in England of the date, and Shrewsbury was, by comparison with Rugby before Dr Arnold, a good school. When Butler took it over, there were just eighteen pupils,[2] so the overcrowding witnessed by Bobby and Ras Darwin was a token of Butler's success in attracting the children of the gentry and the aspirant middle classes. The place had a distinguished history – it was Sir Philip Sidney's old school – but it was highly unsuitable for a boy of Darwin's cast of mind. In those days, the curriculum at a public school consisted of a little mathematics, and the rest of the time was devoted to learning Latin and Greek – chiefly Latin. A boy at Harrow during this period believed that he *thought in Latin* while at the school.[3]

Darwin, by his own confession, never mastered languages. 'Nothing could have been worse for the development of my mind than Dr Butler's school, as it was strictly classical, nothing else being taught except a little ancient geography and history. The school as a means of education to me was simply a blank.'[4]

In its own strange way, Shrewsbury School was a preparation for life, in so far as it gave time to Darwin to develop on his own. He was in a condition which we should recognize as suppressed bereavement, but which was not seen as such either by him or by his teachers and family – even though the observant father was struck by Bobby's solitary, blank-minded walks. Once, when walking in this state around the old fortifications of Shrewsbury, the boy careered off a path and fell – as it happened only seven or eight feet. For the rest, he was not bullied. His classmates considered him 'old for his age . . . in manner and in mind'.[5] He was an avid reader. He liked Shakespeare's History Plays, and Thomson's *Seasons* – the favourite poem of his grandfather Josiah Wedgwood. Someone gave him a book called *Wonders of the World* 'which I often read, and disputed with other boys about the veracity of some of the statements; and I believe that this book first gave me a wish to travel in remote countries, which was ultimately fulfilled by the voyage of the *Beagle*'.[6] Moreover Ras introduced him, when he got older, to the delights of chemistry. They read Henry and Parkes's *Chemical Catechism* and they made 'all the gases and many compounds'. The other boys nicknamed him 'Gas'. 'I was once publicly rebuked by the head-master, Dr Butler, for thus wasting my time over such useless subjects; and he called me very unjustly a "poco curante", and as I did not understand what he meant, it seemed to me a fearful reproach.'[7] (It means an indifferent or uninterested person.)

Many of his happiest, and most intellectually engaged, moments were spent out of doors. From an early age, he was an avid beetle-collector, and looked back with pleasure on a visit to Plas Edwards in Wales, when he was only ten, when he identified a large black and scarlet hemipterous insect, many moths (*Zygaena*) and a *Cicindela* which were not found in Shropshire.

And there was Maer, their second home, with the Wedgwood cousins, where they could escape the moods and rages of Robert Darwin. In the year following the death of Darwin's mother, the chance of going to Maer was denied him. Probably this was one of the reasons why his father sent him to board at Shrewsbury – just to get him out of the house, and to give him the companionship of other boys. Sukey's death had coincided with painful money-wrangles between her two brothers, John and Jos, each blaming the other, and each expecting Robert Darwin both to give them financial assistance and to advise them. The death of Sukey and the money-strife plunged the whole extended family into gloom, and Jos decided that the best way of putting it to one side was to take six months abroad. He made no secret of the fact that his ideal life would be one of 'wise and masterly inactivity'. So in March 1818 he took his wife, their four daughters – ranging in age from nine to twenty-four – and his twenty-two-year-old niece Eliza to Paris for six months. His sister Sarah moved into Maer. While the Wedgwoods were in Paris, they had a rather exalted life, attending the salon of Madame Récamier, and being presented to the Queen of Sweden.[8] It was only after the Wedgwood tribe had returned to Maer, when Darwin was ten and upwards, that his life there could resume. It was at Maer that he could develop his passion for field sports, shooting and fishing. So passed his teens.

When Darwin was thirteen, his elder brother was admitted to Christ's College, Cambridge. Although Darwin's classmates at Shrewsbury found him 'old for his age', it is clear from Ras's letters to him from Cambridge that Bobby was still an indulged and rather juvenile boy. 'I am getting on very comfortably here,' Ras wrote from Christ's. 'Settled in fruiterers, where ye jellies & puffs & cakes & buns &c would tempt the most obdurate sinner, quite as much as ye Lumberland pigs with knives and forks stuck in their backs' (he is alluding to Peter Breughel's 1567 painting of *Das Schlaraffenland*, or Land of Cockaigne, where such creatures appeared cooked and ready for consumption). 'I know you don't like long letters & I have nothing to say so good bye be a good boy & you shall have a sugar plum I remain yours affect E. Darwin.'[9]

From Cambridge, Ras regaled his younger brother with accounts of the lectures he was attending on chemistry. It is in these letters, too, that we first hear the name of Professor Adam Sedgwick, the Woodwardian Professor of Geology, who advised Ras where to find geological specimens for his younger brother. Sedgwick would play a large role in Charles Darwin's later life. One of the advantages of college life, compared with school, was that the young men were expected to attend chapel only once a day, rather than the twice which was the norm for Shrewsbury boys. 'Some evenings preceding Sts days we go to Chapel in surplices, wh: look for all the world like sheets, & indeed one man of St Johns went, for a wager, into chapel dressed in a sheet, & sat before the master without being discovered.'[10]

Grandfather Erasmus Darwin had been at St John's College. He had sent Robert Waring Darwin to Edinburgh University, which had a much more distinguished record at this date for the teaching of medicine, followed by a spell at the University of Leiden in the Netherlands. Probably old Erasmus also felt it was absurd to send his son to a university (Cambridge or Oxford) which would admit pupils only if they submitted to the Thirty-Nine Articles of the Church of England, even though he had done so himself in his youth. (The requirement to swear to the Articles was not lifted until after Catholic Emancipation in 1829, which allowed Roman Catholics and nonconformists to attend the older English universities, and to read for the Bar at the Inns of Court.) Ras would eventually become, like his father, a fully qualified physician, but, unlike Robert Waring, he never practised – indeed, he scarcely 'did' anything, learning to engage in 'wise and masterly inactivity' with as much aplomb as his charming Uncle Jos at Maer.

For Robert Waring Darwin, the idleness of his sons was a grief to witness. He could see that Charles was getting nothing out of Shrewsbury, and he was anxious that Ras should not become a pure idler after Cambridge. He decided, when Ras had spent three years at Christ's, to send his elder son for a year's medical training at Edinburgh. As for Charles, the irate Doctor once rebuked him, 'You care for nothing but shooting, dogs, and rat-catching, and you will be a disgrace to yourself and all your family.'[11] In a moment of inspired decisiveness, Robert Darwin decided to take Bobby out of school and send him north to Edinburgh with his brother Erasmus. He would make doctors out of both his sons, and they could follow in the family tradition of medicine.

In the early nineteenth century Edinburgh was the largest and most prestigious medical school in Great Britain, with over 900 students a year preparing to be doctors.[12] A medical degree was usually conferred after three years, most students having mastered at least one modern language and attended lectures on general science, as well as having undergone three intensive periods of six months apiece, during which they attended medical lectures. Very many students, in the early decades of the nineteenth century, found the lectures unsatisfactory, and reckoned that they learnt far more, especially about surgery, from their practical work on the wards and in the theatres of the hospitals.[13]

Darwin doubted whether Ras ever really intended to practise as a physician, but it was very clear that their father intended them to follow in his profession. Even before he was sent to Edinburgh, Charles was given practical experience doing medical visits in Shrewsbury.

> I began attending some of the poor people, chiefly children and women in Shrewsbury. I wrote down as full an account as I could of the cases with all the symptoms, and read them aloud to my father, who suggested further inquiries, and advised me what medicines to give, which I made up myself. At one time, I had at least a dozen patients, and I felt a keen interest in the work. My father . . . declared that I should make a successful physician, meaning by this, one who got many patients.[14]

Whether Darwin himself ever supposed he would complete his medical training, we cannot tell. For the time being, he was happy again, with his brother as his constant companion.

The two tall youths – Darwin still only sixteen and his elder brother just twenty-one – arrived in Edinburgh in October 1825, a week before the lectures began, to give them time to find lodgings and, as Ras put it, so that they could 'both read like horses'.[15] They found lodgings at 11 Lothian Street with a Mrs Mackay, 'a nice clean old body'.[16] They promenaded about the town, they went to church, where they were agreeably surprised to find the sermon lasted only twenty minutes, as opposed to the two hours they had feared from reading Sir Walter Scott's novels – and they often wrote home. Their sisters Catherine and Caroline took it in turns to write to them, urging Bobby to be diligent about reading French – and 'something more interesting than the *Baroness & Countess* "silly letters"' (presumably a reference to the letters of Mme de Sévigné to her daughter the Comtesse de Grignan) – and passing on such micromanaged instructions as 'Papa says Erasmus may wear Flannel next his skin in cold weather by all means & that he may *sleep* in it also, tho he does not think that very advisable – but in warm weather he very much objects to it.'[17] The Doctor who had himself studied at Edinburgh knew that there was not a strong likelihood of much warm weather in Edinburgh between the months of October and March.

Darwin was never an aesthete, but he was, among other things, an aspirant geologist. He could not have failed, from the very first, to be impressed by the skyline of Arthur's Seat, the remains of a volcano, 375 million year old (as we now know), which broods over the city, and by Salisbury Crags, to the east of the city centre, formed some twenty-five million years later, a sill, or lateral intrusion of volcanic material into the older sedimentary rock. The dense grey of this rock was quarried to build the city – the grey cobbles of the streets, the dour grey houses of the old town that cluster round the Royal Mile going up to the massive stone castle, and the elegant grey stone of the Georgian New Town, with its perfectly proportioned streets, squares and crescents which follow the wold-like roll of the hills looking down into the Firth of Forth. There is no city to compare with it for sheer majesty, and in few cities is one so aware of the relationship between geology and the human inhabitants.

Charles Darwin's *Autobiography* plays down the significance of his years in Edinburgh. This is all part of his quite deliberate attempt, as the author of *The Origin of Species*, to represent himself as a complete original, and to be silent about his influences. In fact, without the years in Edinburgh, it is doubtful whether he would have become an evolutionist at all.

At first, however, he concentrated on medicine. Some of his notes, made during lectures, survive from this period, and it is clear that he was a diligent student, despite finding some of the lectures boring. 'Your very entertaining letter', he wrote to his sister Caroline,

> . . . was a great relief after hearing a long stupid lecture from Duncan on Materia Medica – But as you know nothing of the Lectures or Lecturers, I will give you a short account of them. Dr Duncan is so very learned that his wisdom has left no room for his sense, & he lectures, as I have already said, on the Materia Medica, which cannot be translated into any word expressive enough of this stupidity. These last few mornings, however, he has shown signs of improvement & I hope he will 'go on as well as can be expected'. His lectures begin at eight in the morning. – Dr Hope begins at 10 o'clock, & I like both him and his lectures *very* much. (After which Erasmus goes to Mr Lizars on Anatomy, who is a charming Lecturer.) At 12, the Hospital, after which I attend Munro on Anatomy – I dislike him & his Lectures so much that I cannot speak with decency about them. He is so dirty in person & actions . . .[18]

Darwin was fascinated by the clinical lectures, held in the hospitals, with real patients to demonstrate upon. This was preparation for becoming a physician such as his father. About anatomy, however, he was squeamish. The better lecturers, such as the hated Alexander Monro, would offer students an ancient grizzly corpse 'fished up from the bottom of a tub of spirits'. This was the Edinburgh of the Burke and Hare scandals, and corpses were not easily obtained.[19] Darwin also, like the other medical students, attended operations. He saw two, one of which was conducted on a child – 'but I rushed away before they were completed. Nor did I ever attend again, for hardly any inducement would have been strong enough to make me do so; this being long before the blessed days of chloroform. The two cases fairly haunted me for many a long year.'[20]

Some writers on Darwin have given the impression that this squeamishness about surgery made him abandon his medical course at once.

It is true that he everlastingly regretted not learning the art of dissection at Edinburgh – though in later life he taught himself, and became good at it. He did not, however, give up either medicine or Edinburgh, however boring he found Duncan, and however distressing the operating theatres. He still had much to learn there. Both he and Ras were voracious readers, and the titles they borrowed from the library belie any idea of them being idlers. They read their way through John Mason Good's *The Study of Medicine* (1822) and Christopher Robert Pemberton's *A Practical Treatise on Various Diseases of the Abdominal Viscera* (1806), while keeping up general scientific reading and studies of natural history, including John Fleming's *The Philosophy of Zoology* (1822), William Wood's *Illustrations of the Linnaean Genera of Insects* (1821), Robert Kerr's *The Animal Kingdom* (1792) and Samuel Brookes's *Introduction to Conchology* (1815). Darwin borrowed Newton's *Optics* and Boswell's *Life of Johnson* as well. They also walked for miles in the Pentland Hills or explored the seashore at Portobello or Leith where the Firth of Forth with its mudflats and marshes teemed with wildlife.[21]

When the vacation came, in March, Erasmus left Edinburgh. He would return to Cambridge to complete his studies. Darwin stayed behind in Edinburgh. He returned to Shrewsbury in late spring for a glorious, long vacation. Accompanied by two friends, he explored North Wales. With knapsacks on their backs, they covered thirty miles a day. They climbed Snowdon. There followed a riding holiday with his sister, and many happy weeks were spent at Maer. This was the summer when shooting became a real passion. 'My zeal was so great that I used to place my shooting-boots open by my bedside when I went to bed, so as not to lose half a minute in putting them on in the morning; and on one occasion I reached a distant part of the Maer estate, on the 20th of August for black-game shooting, before I could see: I then toiled on with the gamekeeper the whole day through thick heath and young Scotch firs.'[22] He kept an exact record of every bird he shot.

Apart from the sport, the joy of Maer was the company of the Wedgwoods. He finally got his eye in with his mother's brother, the diffident Uncle Jos. 'He was the very type of an upright man, with the clearest judgement. I do not believe any power on earth could have made him swerve an inch from what he considered the right course.'[23] It was at Maer that Darwin first found himself a grown man, able to converse with the older generation of his mother's family in a way he never found easy with his own father. The Wedgwoods of Maer of his own generation were as close to him as his own siblings: and now that he was grown up, Darwin need not feel deterred by an age-gap with his cousins. He could appreciate, for example, what a very clever man was Uncle Jos's fourth son, Hensleigh – six years older than Darwin, a Fellow of Christ's College and an accomplished mathematician. There were three other boys – the first-born, Josiah III, Harry and Frank – and four daughters – Charlotte (born 1797), Elizabeth, so short as to be almost a dwarf, Fanny, who though a couple of years older than Darwin was considered a possible wife for him, and Emma, who was every inch a Wedgwood in physiognomy (brow, eyes, chin). She bore a striking resemblance to Darwin's mother, her aunt. She had thick chestnut hair and was 'scatty' – wildly untidy. Her family nickname was 'Little Miss Slip-Slop'.[24] She and her tiny sister Elizabeth ran a Sunday school in the Maer Hall laundry, teaching the sixty or so village children to read and write, and instructing them in religion.

This was one of the great eras of evangelical revival in the Church of England. John Henry Newman and his brother Francis – who had been taught mathematics at Ealing Grammar School by the father of Thomas H. Huxley – had become evangelicals before going up to Oxford. Jane Austen, three years before her death (which was in 1817), wrote to her sister Fanny, 'I am by no means convinced that we ought not all to be Evangelicals, & am at least persuaded that they who are so from Reason & Feeling, must be happiest and safest.'[25] The Evangelical Revival(s) – one adds a hesitant plural, for these revivals have been a feature of national Church life in England since the Reformation – received a particular boost at the beginning of the nineteenth century from a number of sources. One was the sense among many Christians that the Dissenters – Methodists, Baptists and others – made a more serious attempt to apply the principles of the Gospels to their daily lives than did some of the more worldly representatives of the Church of England, whose titular heads on earth were until 1820 George III, who was mad, and thereafter his dissolute and hugely unpopular son George IV. Another factor – again inspired by the thought of actually applying the Gospels to the realities of life – was the influence of those London evangelicals who came to be known as the Clapham Sect. It was a nickname given to them by the witty Sydney Smith, because so many of them attended the church on Clapham Common presided over by Parson Venn from 1792 to 1813. Many of the Clapham Sect would turn out to be the parents, uncles or grandparents of those who in later generations were famously non-evangelical, or actually irreligious. They were undoubtedly a formative group. Their earnestness, and their seriousness about trying to apply Christian values to the real world, led – particularly in the case of two of them, Granville Sharp and William Wilberforce – to the most impressive achievement of the evangelical movement: the abolition of the Evil Trade, slavery. Others, such as Zachary Macaulay (father of Thomas Babington and a co-founder of London University), James Stephen (grandfather of Virginia Woolf) and Henry Thornton (great-grandfather of E. M. Forster), were part of an impressive cast-list who founded that intellectual dynasty, or intellectual aristocracy, of which Noel Annan wrote (p. 21) and of which the Wedgwoods and the Darwins were also a part.

When Jane Austen said that she was not convinced we ought not all to be evangelicals, she spoke, not – as perhaps later evangelicals, especially in America, might speak – of the evangelical appeal to the emotions, and the 'conversion experience', in which each individual believer gives heart and soul to the Lord. Rather, she said, there was *Reason* in wishing to act 'more strictly up to the precepts of the New Testament'.[26] The Duke of Wellington, at a similar date, would have agreed with her. When a friend raised an eyebrow at the Duke giving money to a missionary organization, he replied simply, 'Orders are orders.'

Another factor which perhaps contributed to the forceful attraction of evangelicalism when Darwin was growing to manhood was the aftermath of the French Revolution and its subsequent wars. Darwin's grandfather Erasmus could continue to espouse the values of 1789, to publish beautifully crafted verses which disposed of the necessity of a Creator, and to be a (somewhat self-satisfied) independent-minded intellectual. Christianity has thrived, in (relatively) modern times, not during periods of prosperity and indifference but when the alternative to Christianity is made glaringly obvious. During the Second World War in Britain, the contrast between the 'Mere Christianity' of C. S. Lewis's radio broadcasts and the anti-human, anti-humanist genocides of the Communists and the Fascists on the continental mainland drew many to faith.

Robespierre's Terror, the wholesale demolition of cathedrals and churches, the torture and beheading of nuns and priests did not merely shock many in Britain. These horrors were seen to be the direct consequences of believing the kind of things which, in the palmy times of peace, Dr Erasmus Darwin had so freely discussed at dinners of the Lunar Men or at meetings of the Derby Philosophical Society. It was no wonder that his son Robert Waring had spoken little of his father's ideas to his children; nor that Darwin's sisters, like the Wedgwood girls of his generation, should have embraced evangelical Christianity, as practised in the Church of England. Emma Wedgwood was an especially devout girl. So, too, though, was her cousin, Darwin's sister Caroline. When Bobby returned to Edinburgh to resume his medical studies in March 1826, she directly confronted him: 'dear Charles I hope you read the bible & not only because you think it wrong not to read it, but with the wish of learning there what is necessary to feel & do to go to heaven after you die. I am sure I gain more by praying over a few verses than by reading simply – many chapters – I suppose you do not feel prepared yet to take the sacrament.'[27]

This period of Darwin's life – the very first when he was cast adrift from his family entirely, and where he found himself back in Scotland but without Ras – seems to be the only one in which he appears to have been seriously reflective about religion, though later, at Cambridge, he would make an *intellectual* commitment to orthodoxy. Clearly, when he was alone in Edinburgh, it began to dawn upon him how fully it was true that 'all my recollections seem to be connected most with self'. He wrote to Caroline that her letter 'makes me feel how very ungrateful I have been to you for all the kindness and trouble you took for me when I was a child. Indeed, I cannot help wondering at my own blind Ungratefulness. I have tried to follow your advice about the Bible, what part of the Bible do you like best? I like the Gospels. Do you know which of them is generally reckoned the best?'[28] In spite of the twice-daily exposure to chapel prayers at Shrewsbury School, or perhaps because of it, Darwin writes with the airy ignorance of the Scriptures which is the norm in intellectual households.

His return to Edinburgh, however, in the spring of 1826 would not in the event be devoted to Bible study. Alone for the first time in his life, and without the comfort-blanket of Ras to hang on to, the shy, gawky Darwin was now compelled, in his second year of medical study, to mix with students and teachers. He was about to be confronted with the scientific and philosophical questions which would engage him for the rest of his life, and with which his name would be forever associated.

Without the companionship of Ras, Darwin joined the Plinian Natural History Society (named after the most famous natural historian of antiquity). It was here that there first became apparent that quality in Charles Darwin which perhaps his Uncle Jos had hitherto been alone in noticing: that he was 'a man of enlarged curiosity'.[29] From now onwards in the story, our hindsight inevitably outshines the myopia of Darwin's father about the direction of Charles's professional life and the uncertainties Darwin himself felt about his own future. The great intellectual journey had begun.

Appropriately, in that geological phenomenon of ancient volcanic necks and sills – Edinburgh – geology was the branch of science which was dominant: in part, because Edinburgh was home to two of the most eminent and innovative geologists in the history of the subject. James Hutton, who had died in 1797, has a greater claim than Darwin to have changed the way in which the human race looks at the world. 'He discovered an intangible thing against which the human mind had long armoured itself. He discovered, in other words, time – time boundless and without end, the time of the ancient Easterners – but in this case demonstrated by the very stones of the world, by the dust and clay over which the devout passed to their places of worship.'[30]

It is not in the least clear, when the various authors of Scripture used phrases about time – such as the six 'days' of creation in Genesis, or the 'forty years' in which the Hebrews wandered in the wilderness, or the 969 'years' of the patriarch of Methuselah – whether they were making any attempt to suggest mathematical measurements. Probably not. To a later age, however, when their words had been translated, from Hebrew to Greek, from Greek to Latin, from Latin to English, and read perhaps thousands of years after they were written down, the words appeared peculiar indeed. When not merely domestic clocks but nautical chronometers had been perfected, the years of Methuselah must have seemed fantastic, not least because, slowly but inexorably since the Reformation, a habit had developed, especially in the Western world, of reading the Bible not as a series of types and allegories, but as a literal reality. So it was that when James Ussher (1581–1656) added up the years in the Bible he found that our world was of very short duration. Count back through the Hebrew Scriptures, through the generations of this king or patriarch begetting another, and you will find that human history began as recently as 4004 BC – the date when Eve made the interesting mistake of eating the apple. It was to be inferred, if you insisted upon taking this approach to the texts, that the six days of creation were also literal 'days' and that the world was therefore only about as old as the human race. Ussher was by the judgements of his contemporaries a learned and good man. Though an archbishop – not the Lord Protector's favourite category of functionary – Ussher was given by Oliver Cromwell a state funeral in Westminster Abbey,[31] so highly was he regarded.

His legacy, however, would prove problematic to Christians. Measuring the poetic Hebrew time-scale with modern chronometers was going to force many an honest doubter into thinking the unthinkable: namely that the Bible, the inspired word of God, could not possibly be true.

James Hutton saw that the earth was not a finished creation, but a geological phenomenon in a state of constant change. The dynamic forces in the crust of the earth created tensions and strains which in the course of time threw up new lands from the ocean bed. 'Thus . . . from the top of the mountain to the shore of the sea . . . everything is in a state of change; the rock and solid strata slowly dissolving, breaking and decomposing for the purpose of becoming soil; the soil travelling along the surface of the earth on its way to the shore; and the shore itself wearing and wasting by the agitation of the sea, an agitation which is essential to the purposes of the living world.'[32]

In Hutton's lifetime, geology, as studied in Europe, was explained by what has been called 'catastrophism'. That is to say that observed geological phenomena have come about as a result of some departure from an harmonious 'norm' of stillness. Broadly speaking, the catastrophists fell into two camps. The first was represented by Abraham Werner, the 'father of German geology'.[33] This theory, called Neptunist after the Roman god of the sea, supposed that the stratification of the earth's crust had come about as the result of a turbulent universal sea which had once covered the entire surface of the planet. The convenience of this viewpoint was that it was compatible with the Bible's story of Noah's flood. Opposed to the Neptunists, the Plutonists or Vulcanists were those who questioned whether all geological change took place under a great sea. This view posited the possibility that there could have been a series of worldwide catastrophes, which accounted, for example, for the fossil-bearing strata in which the constituents of life were discernible. These life-forms, revealed in the fossils, showed species which had become extinct. This was compatible with a literalist view of Genesis because the Vulcanist/Plutonist school could contort their minds into believing that after each catastrophe the Creator brought new species into being to replace the ones which had been fossilized. This, the Vulcanist viewpoint, was most conspicuously associated in Europe with Baron Georges Cuvier (1769–1832) of Stuttgart, one of the great pioneers (as we have already seen in Chapter 1) of the museum as the fundamental means of scientific inquiry. He was a prodigious palaeontologist with an unrivalled knowledge of molluscs and fish – he identified 5,000 species of fish in his *Histoire naturelle des poissons* – and hugely enriched the collections in Paris, for which he was promoted by Napoleon and ennobled by King Louis Philippe.

These two catastrophist views – Vulcanist versus Neptunist – were the geological orthodoxy in Oxford and Cambridge when Darwin was young. It was so in Edinburgh too. But Hutton's *Theory of the Earth* (1795) had really made the debate redundant, and paved the way for modern geology and the work of Charles Lyell which would so change scientific thinking, not only about geology but about biology and the origin of species. Some of Hutton's mind-changing discoveries – such as that the earth, 'like the body of an animal, is wasted at the same time that it is repaired' – had been made in Edinburgh itself. It was in Salisbury Crags that he saw that the 'dykes' had once been molten, and had been intruded into the sill long after the sill itself had cooled. It was possible actually to see how the heat of the molten lava had altered the adjacent rocks. Hutton had not discovered the whole of modern geology, but he was closer to recognizing the reality of our planet, with semi-molten rocks churning ominously only a little beneath the fragile surface of the earth, than any of his contemporaries. And he had changed the human perception of time itself.[34]

Edinburgh was without doubt a good place to be for a budding young scientist. Since Hutton's death, the most outstanding geologist there was Robert Jameson, Professor of Natural History from 1805 to 1844 and whose lectures on both geology and zoology Darwin attended. 'The sole effect they produced on me was the determination never so long as I lived to read a book on Geology or in any way to study the science,' he recalled in his *Autobiography*.[35] This was because Jameson was a dry-as-dust scholar, not because the material was itself uninteresting. Asa Gray, the American scientist who was one of the earliest to espouse Darwinism, also studied under Jameson at Edinburgh and found him an 'old dry brown stick'.[36]

In fact, Darwin absorbed the geological debate avidly. The only 'professional position' – one could not exactly call it a job – which he would ever take up was the secretaryship of the Geological Society of London when he returned from the voyage of the *Beagle*, and one can tell, from the speed with which he began to study geology on that voyage, that his time at Edinburgh had not been wasted. He was attentive at the course of a hundred lectures given by Jameson in his great creation, the Natural History Museum, with its synoptic displays of insects, birds, fish, minerals and rock types. Jameson's collection at the Natural History Museum was the largest in Europe. His manner of delivery might have been dull but his range was staggering, and it is not surprising that Darwin attended so faithfully. His course consisted of some hundred lectures, five days a week, beginning with 'the natural history of man' and, having spread over the broad range of taxonomy, ending with reflections on 'The Philosophy of Zoology'. No course of this kind was available to students at English universities and it is clear, from all Darwin's subsequent development, that these lectures of Jameson's were the foundation of his life's work. It was from Jameson's lectures that Darwin would learn breadth. The great question to which he would devote the bulk of his professional life, and the theory which still bears his name, depended upon Darwin swimming in and out of scientific disciplines, mastering (whatever he claims jokingly in his *Autobiography*) geology, as well as palaeontology, and a detailed knowledge of zoology. It was observed, by Sir Alexander Grant, that Jameson 'used to finish up with lectures on the origins of species of Animals!'[37] (Grant, Principal of the University, 1868–84, was the author of the substantial *The Story of the University of Edinburgh during its First Three Hundred Years*, 1884.)

Jameson was primarily responsible for introducing to his students the new wave of scientific thinking which had come from France. Darwin would never have heard about this if he had gone straight from school to Cambridge. It was Jameson who founded and ran the natural history museum in Edinburgh; Jameson who founded and edited the *Edinburgh New Philosophical Journal* and who, in that periodical, almost certainly expounded the new continental ideas of Étienne Geoffory Saint-Hilaire, Georges Cuvier and Lamarck. It was almost certainly Jameson who wrote an obituary of Giambattista Brocchi. Niles Eldredge, doyen of modern evolutionary palaeontology, in a recent book about Darwin's early influences, comments on the appropriateness of the name John the Baptist for two of the most influential early believers in evolution: Jean-Baptiste Lamarck and Giambattista Brocchi.[38] They shared 'the distinction of being the first to develop natural causal explanations of the origins of modern species that were both empirically and phylogenetically based'.[39] Both based their ideas on research into fossils. In many ways, Brocchi was the more revolutionary, the closer to what we can now see must be something like the truth. Whereas Lamarck, studying sixty-five-million-year-old Cenozoic fossils, concluded that there was a continuity of species which slowly changed and intergraded into other species of the same genus, Brocchi was more radical. He believed that species have births, histories and then deaths. Species die out, just as individuals do. Old species do not change into descendants; they give birth to new species, rather as individuals sire new offspring. Eldredge points out the supreme irony that Darwin, in his developed theory of evolution (from 1844 onwards), accepted that Lamarck's picture of the pattern of evolution was the right one, even though he departed from Lamarck in his idea of how these changes take place. Namely, he believed, as Lamarck did, that species intergrade, almost imperceptibly, into descendants. However, as Eldredge points out, the fossil evidence indicates almost the opposite being the case. 'When the fossil evidence seems to say the opposite, Darwin decided the fossil record itself was at fault.'[40] Lamarck, and later Darwin, propounded the idea that species evolve transformationally; Brocchi thought 'taxically'.

Both Lamarck's and Brocchi's ideas were taught to Darwin at Edinburgh, and their ideas were played against one another. At that stage of history, the fossil evidence was much patchier than it is today. A palaeontologist such as Niles Eldredge, who with Stephen Jay Gould pioneered the idea of punctuated equilibrium – evolution proceeding by leaps – demonstrated that Brocchi is a much more reliable witness. Brocchi was partly able to be more accurate through sheer luck; Lamarck's Paris basin fossils were between fifty-six and thirty-four million years old (Eocene), whereas Brocchi was able to study much more recent fossils, in five-million-year-old Upper Tertiary Miocene–Pleistocene strata. He estimated that some 50 per cent of the fossils he discovered were of species which could still be found swimming, or living, in Italian waters during the early nineteenth century. He saw that while some species change, many do not. Species do not necessarily change much over time, if at all. The cause of the origin of species – that would perhaps never be explained by science. The debate about natural processes – that was already under way when Darwin began to study science in Edinburgh. Very typically, Darwin, in the autobiography he wrote for his family to read when he was sixty-seven years old, dismissed Jameson's lectures as 'incredibly dull'.[41] It is only by a patient reassembling of the evidence – evidence that Jameson knew the work both of Brocchi and Lamarck, evidence that Charles Lyell used Brocchi's work – that modern Darwin scholars have been able to trace the origins of his ideas.

As well as attending the Jameson course, Darwin, as a member of the Plinian, had a more social life than he would have done clinging to Ras's coat-tails. He went to meetings of the Wernerian Society (geology again) and he went to hear Sir Walter Scott speak at the Royal Society of Edinburgh. They were small dinners, with about thirty men attending.[42] Darwin would have been aware, as he looked at the noble face of the great novelist – for all Edinburgh knew of it – that Sir Walter was a ruined man. It had been in January 1826, when Ras and Darwin had been in Edinburgh three months, that Scott wrote in his journal the unforgettable line, 'came through cold roads to as cold news' – the news that money borrowed by his printers from his publishers, which Scott had agreed to guarantee, had all been lost, and that he must 'with his own right hand' write off the debt, a prodigy of scribbling which exhausted Scott and killed him finally in 1832.

It has been rightly said that 'from November 1826 to April 1827 Darwin . . . led a tripartite life'.[43] There were the medical studies, his prime reason for being in Edinburgh. Secondly, there were the studies of geology and zoology at Jameson's lectures, a more or less daily occurrence, which were the vital preparation for his life as a naturalist. There was also a third life, his membership of the Plinian Society, which enabled his friendship with the first great intellectual influence upon his life and mind: Robert Edmond Grant (1793–1874).

Robert Grant, sixteen years Darwin's senior, and a bachelor, became a role model. Like Darwin, he had studied medicine at Edinburgh, but substantial family money had liberated him from the need to practise as a physician. Instead, he had devoted himself to science. He studied in Paris, where he not only absorbed much scientific skill, but also quickened his taste for radical politics. He became a distinguished zoologist, and made himself an expert on the marine life of the Scottish coast. Keith Thomson's book *The Young Charles Darwin* (2009) has brought into sharper relief than any previous biographies the debt Darwin owed to Grant, and, rather less creditably, the extent to which Darwin attempted to play this down when he had become a famous scientist. We shall see that this is an absolutely habitual trait of Darwin's who was a self-mythologizer, a man who wanted to represent himself, when the moment was ripe, as the pioneer evolutionist.

Grant, a shy, diffident bachelor, clearly took a shine to Darwin, seeing his potential as a man of science. It scarcely required second sight to perceive these qualities in the young medical student, however, given his surname, and given that Grant was an avid reader of Darwin's grandfather, above all of *Zoonomia*. One of the most misleading paragraphs in the entire *Autobiography* describes a conversation with Grant about evolution.

> He one day, when we were walking together, burst forth in high admiration of Lamarck and his views on evolution. I listened in silent astonishment and as far as I can judge, without any effect on my mind. I had previously read the *Zoönomia* of my grandfather, in which similar views are maintained, but without producing any effect on me. Nevertheless it is probable that the hearing rather early in life such views maintained and praised may have favoured my upholding them under a different form in my *Origin of Species*.[44]

Keith Thomson has demonstrated, in *The Young Charles Darwin*, how extremely improbable it is that Darwin could have felt 'astonishment' at Grant's Lamarckian views. 'His *Autobiography*'s misdirections and forgetfulness may be rather convenient. Even while he enjoyed a well-earned, heralded career in his old age, Darwin was incapable of sharing credit, of finding the shades and subtleties of intellectual debt that any creative person knows.'[45]

My own belief is that, by the time he wrote his *Autobiography*, Darwin – who had been failing to acknowledge intellectual debts and influences all his life, and who had his head turned further by the hero-worship of Huxley – was actually incapable of remembering, at some visceral level, that anyone beside himself had ever believed in evolution before 1859. In youth, however, his observant eyes were directed with equal beadiness upon the natural world and the main chance. Jameson and Grant could see in the young Darwin a genuinely brilliant naturalist in the making. In a paper to the Plinian Society, given when he was a mere eighteen years old, Darwin demonstrated that the little globular bodies (then sometimes called sea peppercorns) which had been supposed to be the young state of the (seaweed) *Fucus loreus* were in fact the egg-cases of the wormlike *Pontobdella muricata*, a marine leech. He also had discovered, in spite of being shy of the dissecting knife, that the so-called ova of *Flustra* had independent movement by means of cilia and were in fact larvae.[46] This was Darwin's first real scientific discovery, and it exploded one of Grant's pet theories, that these *flustrae* were midway between animals and plants. Grant felt some displeasure at being upstaged by his protégé, and so he tried to upstage Darwin, by giving the new information about 'flustrae' in a paper of his own, three days *before* Darwin's presentation to the Plinian. He gave no acknowledgement to Darwin for his discovery. Darwin boldly went ahead three days later and read his paper to the Plinian, making it clear that the discovery had been his own. This shabby behaviour by an older and well-established scientist, who even later was only grudging in half admitting Darwin's discovery, no doubt coloured Darwin's sour memories of Grant in the *Autobiography*. He had made his first genuine discoveries and Grant had been ungenerous. They were, however, small discoveries. Grant's opening Darwin's eyes to the truth of his grandfather's evolutionary theory was a huge thing, and, had Darwin been of a different character, he would have admitted that evolution was, as it were, the family business. As far as Western scientific thought was concerned, certainly from the late eighteenth century onward, Erasmus Darwin was its father. Yet Darwin asks us to believe that reading *Zoonomia*, the work which expounded the idea that species evolved into one another, was an experience he underwent without it 'producing any effect on me'. Any? Really?

The towering figure of natural history and life-science in the eighteenth century had, arguably, been the great Swedish naturalist Carolus Linnaeus (Carl von Linne, 1707–78). His attempt at a comprehensive plant-classification – *Genera Plantarum*, of 1737 – was the starting point of modern systematic botany. It was also, by extension, a starting point for looking at, and classifying, fauna as well as flora. Linnaeus's system implied a fixity of species: else how could it be possible to distinguish one from another? His main purpose, however, was to classify, not to theorize, and if the effect of his work was to strengthen, among many naturalists, the notion of fixity of species, this was not his primary intention. Whereas in the sphere of astronomy, from the seventeenth century onwards, Western humanity had been haunted by a sense of boundless space, following the divinely ordered laws laid bare by Isaac Newton, the earth seemed, by the science of Linnaeus and by the whole notion of taxonomy, or classification of species, to have been limited. It is worth reiterating, however, that belief in the fixity of homologues did not prevent Linnaeus or any taxonomist since from believing in a slow process of adaptive change within the species. The homologues are novelties of nature. They are features which appear to have no predecessors, no transitional forms leading up to them. They are *there*. It is from these fixed forms that fossil evidence, palaeontology, measures descent by modification or extinction.[47]

The taxonomists', however, was not the only perception of the world in the eighteenth century. There were the travellers, in fact and in imagination. Benoît de Maillet (1656–1738) was a traveller and French government official who was not a professional scientist. He was, however, one of the first to make observations which would change our perceptions of the world. He noted, for example, that fossil plants were of a species which 'exist no more'.[48] He understood that, just as the telescope had revealed the infinite dance of the planets, so we should eventually see that 'this whole System which we see, this fine Order which we admire, are subject to Changes'. And what held good for the spheres held good for the planet earth. Above all, he was bold enough to acknowledge that there were creatures in the Dutch Indies, orang-outangs, of whom 'it would be rashness to pronounce they were only brutes'. He quoted a Chinese author who had declared that 'men were only a species of Apes more perfect than those which did not speak'.[49]

The sense of variety in the world, which the travellers excited in their readers, led inevitably to a sense of adaptive change within species. This in turn led to the wild speculation that species might not be fixed, that one species might evolve into another, as Erasmus Darwin believed. James Burnett (1714–99), a Scottish judge who took the title Lord Monboddo, was an intellect of great originality. He studied humankind as if it were a branch of the animal kingdom, and collected information about primitive tribes for the light this might throw on more 'advanced' human groupings. His views, naturally, invited the derision of contemporaries. In Boswell's *Life of Johnson*, which Darwin read in his first year at Edinburgh, he would have found the friends laughing over Dr Johnson's quip that Rousseau 'knows he is talking nonsense . . . But I am *afraid* (chuckling and laughing) Monboddo does not know he is talking nonsense.'[50] On another occasion, Johnson said, 'Conjecture, as to things useful, is good; but conjecture as to what would be useless to know, such as whether men went upon all four, is very idle.'[51] When Boswell took Johnson to Edinburgh in 1773, they were able to meet lawyers who discussed Monboddo's belief that he could teach an orang-outang to speak. His view that men might once have had tails of course gave rise to the joke that he himself had a tail, Johnson opining on one occasion that Monboddo was 'as jealous of his tail as a squirrel'.[52]

These pleasantries however were the beginnings of an unease. What if Monboddo were right? What if the clear borderline between one Linnaean category and another were less firm than human beings might hope? The flickering of doubt about such matters is seen in the writings of the Comte de Buffon (1707–88). His survey of the natural world, *Histoire naturelle*, advanced a number of tentative hints in the direction of evolution, only to withdraw them. But he could see that species change or, as he would say, degenerate. Whereas the optimistic Charles Darwin, typically Victorian, saw evolution as a metaphor of progress, with each evolving species being better than the last, the pessimistic Buffon saw the changes in nature as a process of decline. Yet in many ways he had anticipated the 'uniformitarianism' of James Hutton – the belief that changes in nature were all part of the system, as it were, and not attributable in each case to external intervention by the Deity. Somewhat *malgré lui*, therefore, Buffon was a *kind of* evolutionist. Comparing species in the New World which are similar to those in Europe, he wrote, 'They have remote relations . . . which seem to indicate something common in their formation, and lead us to causes of degeneration (that is, evolution) more ancient, perhaps, than all the others.'[53]

Buffon was far from being alone in his hunch, not only that species – the human species included – were not fixed, any more than rocks and stones and trees were fixed, and also that life was in some impenetrable way interconnected. Darwin had questioned Grant's idea that plants and animals once had a common parentage, but such ideas were far from unusual in the time of the Enlightenment. Goethe's *Essay on the Metamorphosis of Plants* (*Versuch die Metamorphose der Pflanzen zu erklären*) of 1790 had been a deliberate riposte to Linnaeus, suggesting that all botany was in a state of flux, one species deriving from another.[54]

Goethe believed that all life – plant, animal and human – had a single source, and it was his distinction, in anatomy and osteology, to have discovered the intermaxillary bone, in 1784. Hitherto, it had been maintained that man was demonstrably different from the apes and other mammals because he had no intermaxillary bone (this is the bone in the upper jaw anterior to the maxilla bone). It was left to the scientist-poet to prove the professional anatomists wrong. 'Indeed, man is most intimately allied to animals. The coordination of the Whole makes every creature to be that which it is, and man is as much man through the form of his upper jaw, as through the form and nature of the last joint of his little toe. And thus is every creature, but *a note of the great harmony*, which must be studied in the Whole, or else it is nothing but a dead letter.'[55]

Goethe had certainly read Erasmus Darwin. Indeed, it would have been hard to find a cultivated reader, let alone one with scientific interests, in the period 1780–1830 who had *not* read Erasmus Darwin, which is what makes Charles Darwin's claim to be untouched by his grandfather's ideas all the more improbable.

Erasmus Darwin, in his books *Zoonomia* and *The Temple of Nature*, had proposed that all life originated from non-life in the ocean bed. All life, moreover, sprang from what he called a 'filament', an ur-life-form. Every form of life on earth, according to Dr Erasmus, had evolved from this original filament. The first volume of *Zoonomia* consisted of forty essays on a whole variety of scientific subjects, but all, basically, reverting to the original theory.

> Would it be too bold to imagine, that in the great length of time since the earth began to exist, perhaps millions of ages before the commencement of the history of mankind, would it be too bold to imagine, that all warm-blooded animals have arisen from one living filament, which THE FIRST GREAT CAUSE endued with animality, with the power of acquiring new parts, attended with new propensities, directed by irritations, sensations, volitions, and associations; and thus possessing the faculty of continuing to improve by its own inherent activity, and of delivering down these improvements by generation to its posterity, world without end.[56]

There, in essence, is the theory of evolution. Erasmus Darwin had worked it out in the 1770s, although *Zoonomia* was not published until 1794. In 1796, Tom Wedgwood, Sukey's brilliant, troubled elder brother, arranged for his friend, protégé and fellow junkie Samuel Taylor Coleridge to visit Erasmus Darwin in Derby. The philosopher-poet declared, 'Derby is full of curiosities, the cotton, the silk mills, Wright, the painter, and Dr Darwin, the everything, except the Christian! Dr Darwin possesses, perhaps, a greater range of knowledge than any other man in Europe, and is the most inventive of philosophical men. He thinks in a *new* train on all subjects except religion.'[57]

It is true that ideas emanate from individuals, and that Erasmus Darwin was an original. It is also true that ideas are 'in the air' – witness the fact that Alfred Russel Wallace would hit upon the idea of evolution by means of natural selection just as Charles Darwin was preparing to publish his thoughts on the subject in the late 1850s. Shortly after Erasmus Darwin published *Zoonomia*, a zoologist at the Muséum National d'Histoire Naturelle in Paris (the former Jardin du Roi) published his own version of some very similar ideas. This was the Chevalier Jean-Baptiste de la Marck. His *Systèmes des animaux sans vertèbres* appeared in 1801 and his *Philosophie zoologique* followed in 1809. De la Marck – Lamarck as he is invariably known – had read Erasmus Darwin, but it would seem that he had arrived at almost identical views of evolution independently, partly as a result of enormous knowledge and widespread research, partly after reading de Maillet, Buffon and others. Here we find the same essential story: that all natural life is one, deriving from a simple source. Even humanity is part of this chain of life. Lamarck proposed a constant and spontaneous generation of life – similar to the *scala naturae* of the ancients, often known as the great chain of being. Lamarck, like Robespierre and other French Revolutionists, and like Voltaire before them, was a Deist. He believed that the Creator had set this process in motion, and then, as it were, sat back. Although Dr Darwin writes of 'THE FIRST GREAT CAUSE' having started the evolutionary generative process, one suspects that if one hoisted his mighty form on to a truth machine one might have found him to be an atheist. Whichever side of the divide the two thinkers were to jump, there was no need, in their philosophical system, for an *interventionist* Deity. Nature was running itself, and one life-form was evolving into the next. Humanity derived from apes. 'It could easily be shown that his [man's] special characters are all due to long-standing changes in his activities and in the habits which he has adopted', as Erasmus wrote.[58]

Here we see the essence of the Lamarckian explanation of how generative evolution works. Characteristics acquired during a parent's lifetime can be handed down. Charles Darwin is credited with disproving this view of evolution, as when he rather loftily (and surely foolishly) remarked to Huxley, about the time of *The Origin of Species*, 'the history of error is quite unimportant, but it is curious to observe how exactly and accurately my grandfather gives Lamarck's theory'.[59] Foolishly, because the history of science is, surely, the 'history of error', one scientist building on the researches of predecessors and correcting their mistakes; foolishly, too, because, in the first edition of *The Origin of Species*, he would use four categories of evidence for biological evolution which Lamarck himself had used. Without acknowledging his debt to Lamarck, he would write, 'I think there can be little doubt that use in our domestic animals strengthens and enlarges certain parts, and disuse diminishes them: and that such modifications are inherited. Under free nature . . . many animals have structures which can be explained by the effects of disuse.' This is just a summary of the law by which Lamarck is best known.[60]

Back in Edinburgh, in April 1827, it is difficult to reconstruct Charles Darwin's state of mind, beyond knowing that his own account of it in the *Autobiography* must be false. His friendship with Grant, sharpened by the spat at the Plinian over the *flustrae*, had opened his mind to an entirely new way of viewing nature, deepened by reading his own grandfather's work. He had now become an accomplished young naturalist; indeed he was slightly more than that, a fledgling scientist, who had attended Jameson's boring but deeply informative series of lectures; had heard papers from many other young scientists at the Plinian; and had discussed evolution with a passionate Lamarckian, Grant, all the while continuing with his medical studies. His mind was teeming with knowledge, which we see him putting to good use four years later when he sailed with the *Beagle*, not as some ignorant bumbling amateur, but as one who was prepared for the journey. Edinburgh was the making of Darwin's young mind: Edinburgh, and reading old Dr Erasmus. Given his strange psyche, it was necessary to play these facts down, even to expunge them from the record.

'The history of error is quite unimportant.' In what did Lamarck's 'error' consist? In 1954 at the University of Melbourne in Australia, William Agar and colleagues reported on an experiment conducted over a period of twenty years. Laboratory rats were placed into a tank of water with two exits. The 'right' exit was in the dark; the 'wrong' exit was illuminated. Those who chose the 'wrong' exit received an electric shock. Those whose parents and forebears had been trained, by electric shock, to choose the darkened exit were shown to learn far more quickly than untrained rats how to escape the tank. This experiment suggests that Lamarck could have been right, and that in some circumstances acquired characteristics could be inherited.[61] Perhaps Lamarckianism (and Darwin's alternative theory of evolution by natural selection) belonged less to the history of error than to the category dismissed by Johnson when discussing Monboddo, 'conjecture as to what would be useless to know'. The *how* of evolution might indeed be unknowable (as opposed to the how of genetics, which is now demonstrable in all its wonderful intricacy and tininess and exactitude). Adaptive change within species, 'amidst the wreck of worlds',[62] however, is an overwhelming fact. Once a belief in a world aged only 5,000 years or so has been abandoned; once geology has been accepted as a science; once Hutton's uniformitarianism has displaced catastrophism in the mind; once the fossil evidence has been seen, and pondered, then it would not be merely perverse, it would be *impossible* to say that you did not believe in some form of evolution, however it takes place, whatever your theories respecting it. As a phenomenon, it is inescapable. Charles Darwin confronted and absorbed this scientific fact, this enormous phenomenon, this way of viewing the timelessness of the earth's past and the impersonality, so to say, of the natural process, as a late teenager in Edinburgh.

The objections to evolutionary theory at this stage of history were not scientific. They could not be, since there were no scientific facts which could, at this stage, even suggest what a disproof of evolution would look like. There existed the phenomena – some of them listed in the previous paragraph – and the Lamarckian/Erasmus-Darwinian theories were as plausible explanations as you were likely to get, *at that date*, from the geological, palaeontological and zoological facts and puzzles. The moral objections, which would be expanded when Charles Darwin linked the theory of evolution to the merciless ideas of Malthus, were that it appeared to promote an idea that selfish struggle was an underlying necessity of existence. The chief objection, therefore, if you put this purely moral objection to one side, was and is a religious one. Coleridge wrote to Wordsworth, 'I understood that you would take the Human Race in the concrete, having exploded the absurd notion of Pope's Essay on Man, [Erasmus] Darwin, and all the countless believers – even (strange to say) among Christians – of Man's having progressed from an Orang-Outang state – so contrary to all History, to all Religion, nay to all Possibility – to have affirmed a Fall in some sense.'[63]

Coleridge wrote that letter on 30 May 1815, just three weeks before the war of twenty-two years against France came to an end. Wordsworth relates in *The Prelude* how he and his friend Coleridge moved from being wild enthusiasts for the French Revolution –

> Bliss was it in that dawn to be alive,
> But to be young, was very Heaven!

– to being members of the Church of England who shrank from any of the ideology which had fuelled the Revolution. Lamarck's ideas about evolution were not frowned upon in England, and in the older universities of Oxford and Cambridge, because they were *scientifically* at fault. They were abhorred as being part and parcel of the Revolution which had brought to Europe the Terror and the Napoleonic Wars. It would need a generation for the English scientific establishment to come to terms with this, and to distinguish between Lamarck's politics and his science. Scotland was intellectually and spiritually detached from the Establishment. Its Church, Presbyterian, had nothing to do with the Church of England, with the bishops in the House of Lords, or with Oxford and Cambridge, most of whose academics were Church of England clergymen, and all of whose students at this date were obliged to subscribe to the Thirty-Nine Articles of Religion.

These matters would have been of purely academic interest had Charles Darwin pressed on with his medical studies and become a qualified physician. Word had reached The Mount, Shrewsbury, however, that Charles had cooled towards his medical career. With Ras having already made it clear that he was never really going to practise as a doctor, Dr Robert Waring Darwin's feelings can be imagined. The total absence of letters for the year 1827 in the diligently collected complete *Correspondence of Charles Darwin* must, surely, speak volumes. In the 'Chronology of his Own Life' which Darwin wrote as an appendage to his Journal, he recorded, having finished in Edinburgh, 'In Spring [1827] went tour. Dundee St Andrews Sterling. Afterwards Glasgow. Belfast Dublin. – Then London Paris. In Spring went to Dublin & Portran [Portraine, Co. Dublin], & then to London & Paris (in May) with Uncle Jos.' In other words, he was keeping out of his father's way. Uncle Jos was always the person to whom the young Darwin turned when Robert Waring was being especially difficult. 'He was very properly vehement', Darwin recalled in old age, 'against my turning into an idle sporting man which then seemed my probable destination.'[64] His fondness for sport was never in doubt, and he was always to be one with whom bursts of intellectual activity were punctuated by 'idleness', but this sentence, like so much in the *Autobiography*, is hugely misleading. Surely there was no danger of his becoming an idle sporting man, even though this is the version of events which Darwin was himself to give out. He wrote it in his *Autobiography* because he did not wish to advertise the fact that by the age of eighteen or nineteen he was well on the way to his inevitable destiny, which was to become an evolutionary scientist in the mould of Grant and Erasmus Darwin. Since his 'personal myth' depended upon the sense of his scientific life being *sui generis*, self-created, coming from nowhere and dependent upon no one, it was much more convenient to pretend that he was just an idle loafer with a shotgun under his arm who was forced to change career by his father.

Robert Waring decreed that, if Darwin would not become a doctor, he must go to Cambridge and prepare to become . . . a clergyman. Darwin admits that this idea was 'ludicrous'. At any stage in history, however, there must have been plenty of young men who, while not being especially religious, accept the role of a clergyman – especially in that agreeable period of the nineteenth century when a large parsonage house in a country parish, adjoining land where the shooting was good, and where natural history could be pursued freely, would have been so suitable for a man of Darwin's taste. But only to a man who had not read and absorbed *Zoonomia*; only to a man who had not had those long walks and conversations with Grant. Darwin's sisters and cousins were committed Christians, who were anxiously hoping that Charles would read the Bible. Dr Robert Waring was furiously impatient for his son to settle down. The young man was given a copy of John Pearson's *An Exposition of the Creed* and told to overcome his 'scruples'. Darwin had a lifelong horror of confrontation of any kind. He was temperamentally incapable of it. Like many a Darwin uncle, great-uncle and great-grandfather before him, he was lined up to become a clergyman of the Established Church.

The place to study for the priesthood of that Church was either Oxford or Cambridge, and given the many Darwins who had been at Cambridge, the choice was obvious. It must have been something of a shock to hear of this change of course, and to have it imposed from above by his tyrannical father. It is clear that Darwin skulked away from Shrewsbury as much as possible, either with the Wedgwoods at Maer Hall or with his Shropshire neighbours the Owens, whose estate, Woodhouse, was a happy refuge from the silences and moods of The Mount.

Woodhouse, a few miles from Shrewbury, was a fine brick residence, flanked with Ionic columns and pilasters, and surrounded by a park and a good shooting-estate. It was the seat of a convivial retired soldier, who enjoyed reminiscing about his campaigns in Holland. His flirtatious daughters were friends with Darwin's sisters. This summer they became, in Darwin's words, 'the idols of my adoration'. When he was not out shooting with their brothers, he loved flirting with Fanny or Sarah Owen. The Owens and the Darwins conversed in a private language. With allusion to Falstaff after the Battle of Shrewsbury (in *Henry IV Part Two*) they spoke of 'by Shrewsbury Clocks' as a synonym for time; 'broadcloth' was a man, 'black broadcloth' a clergyman. Getting married was 'being led to the halter'. Little was 'leetle'; a woman was a 'muslin'; a letter was a 'budget'; an invitation was an 'insinivation'; and rumours or bits of gossip were 'mysteries'.[65] Fanny Owen, in particular, caught his fancy – 'the prettiest, plumpest charmingest Personage that Shropshire possesses'.[66] Happy hours were spent with her, lying 'full length upon the Strawberry beds grazing by the hour'.[67]

Whatever else they spoke of, it was not of Charles Darwin's future, or anyway not with candour. When Darwin finally accepted the inevitable and went to Cambridge, he was followed by a reproachful letter. 'I never was so horror struck as to receive your leetle note the other day . . . I was very much surprised to hear from Sarah that you have decided to become a *DD* [Doctor of Divinity] instead of an *MD*. You never let *me* into the secret.'[68] The apparently amorous, frivolous young man, so addicted to shooting, seemed an improbable 'black broadcloth'. But in December 1827 (the academic year had begun in October) Darwin, having forgotten most of what Dr Butler had taught, had still not mastered enough Latin to pass into Cambridge; and he would postpone his cramming until the shooting season at Woodhouse was well under way. Ras, meanwhile, had been studying anatomy at the academy in Great Windmill Street – the theatre where John Hunter had dissected the corpse of Samuel Johnson in 1784. He had agreed to go back to Cambridge to read for the degree of Bachelor of Medicine.

The choice of college for Charles, therefore, was obvious. It would be Christ's, where Ras had already spent three years, and where cousin Hensleigh Wedgwood was a Fellow. Ras would follow, but it was alone that Darwin clambered aboard the coach, which took him through misty roads and ice-hard puddles to that fenland university.

4

Cambridge: Charles Darwin, Gent

OF THE TWO English universities, Cambridge was marginally more interested in science than Oxford. When Joseph Banks, a keen amateur botanist even at the age of eighteen, was entered as a gentleman commoner at Christ Church, Oxford in December 1760, he discovered that there were no lectures in botany. He promptly took the stagecoach to Cambridge and, at his own expense, returned with the astronomer and botanist Israel Lyons, who gave the lectures at Oxford to Banks's satisfaction. (Lyons would accompany Banks's friend Captain Phipps – later Lord Mulgrave – as the astronomer on his voyage towards the North Pole.) Banks, who came into a huge fortune in his first year at Oxford (on the death of his father, a Lincolnshire landowner), devoted his life to science. When he was twenty-three, he accompanied Phipps to Newfoundland to collect plants. He befriended Daniel Solander, one of Linnaeus's pet pupils, and procured him the appointment of assistant librarian at the British Museum. Together with Solander, he toured the world, collecting plants and animals and marine birds. He accompanied Captain Cook on his first great voyage of discovery in the *Endeavour*, exploring New Zealand and Australia and the islands of the Pacific. He was one of the great scientists of the eighteenth century. When Darwin became a world-voyager, he felt himself to be, almost literally, in the footsteps of Banks. 'I felt I was treading on ground, which to me was classic,'[1] he wrote to his sister on 30 March 1833. The example of Banks was an unforgettable one – President of the Royal Society, Director of the royal gardens at Kew, one of the great demonstrations of the truth that money could buy the independence to travel and to become a proper man of science.

One writes 'man of science' because, when Darwin went to

Cambridge, the word 'scientist' did not quite exist. In 1894, with absurd pomposity, Thomas Huxley wrote to the *Science-Gossip* magazine to protest against its use of a vulgar Americanism – the word 'scientist'. The English term, he insisted, was 'man of science'. He lumped the word 'scientist' with American coinages such as 'electrocution', a hybrid of 'electricity' and 'execution'. Other 'men of science' joined their voices to Huxley's over this trivial question: they included Darwin's old Kent neighbour Sir John Lubbock, his co-evolutionist Alfred Russel Wallace and the Duke of Argyll.[2]

In fact, the word 'scientist' was not an American neologism of the 1890s. It was first used in Cambridge, during the 1830s, not long after Darwin had left the place. It was a usage of William Whewell, Professor of Mineralogy, later Master of Trinity College, Cambridge. He had coined the word because of the thing. Before Whewell's generation, what we call scientists were called 'philosophers' or 'natural philosophers'. During Whewell's early lifetime, the phenomena which were under such free discussion when Darwin went to Edinburgh had become part of intellectual discourse throughout Europe. In 1818, when he was twenty-four, Whewell had been one of the founder-members of the Cambridge Philosophical Society (what we should call Scientific Society – in Cambridge they continue to speak of 'moral sciences' where others speak of 'philosophy'). Its aim was to keep abreast of developments in modern astronomy, anatomy, botany, taxonomy, mineralogy, mechanics, mathematics, but to keep these all within the focus of a metaphysical outlook. Whewell was of modest origins – his father was a master carpenter from Lancaster – whose cleverness had been fostered ever since he went as a young child to the 'Blue School', the town's grammar school. By the time he had been elected to his Fellowship at Trinity, he was not only an accomplished mathematician, but also a very good linguist who had become a disciple of Kant. He went to Potsdam and met the greatest scientist-explorer of the age, Alexander von Humboldt, who was later heard to complain that he had been expecting 'an English gentleman' to call on him. He had been happily discussing science with Whewell under the impression he was a German.

Darwin was not entering a university which was devoid of interest in those scientific subjects that were now so predominant in his mind.

Though Dr Erasmus Darwin had been a non-believer, and though Lamarck was far from being a Christian, was there necessarily a conflict between being a clergyman and taking the keenest interest in all the developments of modern science? Not at all, as was demonstrated by the career of Whewell, who became Professor of Mineralogy three months after Darwin arrived in Cambridge, and who had been ordained priest three years before that. Most of the dons at Cambridge were in holy orders. When Darwin was matriculated – that is, became a full member of the University, standing in line in the Senate House, vowing (in Latin) to keep all the rules of the University, and submitting to the Articles – he did so before the Senior Proctor, Adam Sedgwick, who was to become not merely an academic mentor, but a friend.

Sedgwick, like Whewell, was a Fellow of Trinity; like Whewell, a priest; like Whewell, a scientist – Professor of Geology since 1818. Although he had been given this job purely because he was well liked in Cambridge, he recognized that there was something ludicrous about accepting the post with an almost complete ignorance of the subject. By the time Darwin knew Sedgwick, he had made himself an expert in the field, though his knowledge was chiefly of British geology. More parochial in every sense than Whewell, Sedgwick – a tall, smooth-shaven man, with a fiery temper and a Yorkshire voice – was another reminder that Cambridge was a place which now took science seriously.

But, of course, you could not graduate in science at Cambridge. Ras followed his younger brother back to Christ's. He was going to finish his medical studies and read for the degree of MB – Bachelor of Medicine. Darwin was preparing to take holy orders, at least notionally. The required syllabus for undergraduates was by modern standards limited, remarkably so. Mathematics and classics were studied – mathematics at a more advanced level than classics. Darwin later regretted not having grasped 'something of the great leading principles of mathematics, for men thus endowed seem to have an extra sense'.[3] In his first year, he was idle and needed coaching in the vacation to reach the required level. In the second year, he continued to read mathematics and classical texts. The final year would be devoted to a few classical texts, to algebra, to Euclid's geometry, which he enjoyed, and to some elementary theology – a book called *Evidences of Christianity* by William Paley. In order to proceed to holy orders, he would have needed to show competence in New Testament Greek, though this would have been examined not by the University, but by whichever bishop was to admit him to the order of deacons.

The work levels, in other words, were minimal. A student of the twenty-first century could easily master, within a few months, everything which had been studied in an entire university course in 1828.

'The three years which I spent at Cambridge were the most joyful in my happy life,' wrote Darwin in his *Autobiography*, and it is not difficult to see why. Although notionally engaged in the study of mathematics and a few books of Homer, Darwin was in fact continuing to study science, at the feet of the Cambridge scientists. He enjoyed college life, and the company of other young men. And he formed at least one friendship – with a cousin, inevitably – which would last through life.

The friend was Darwin's second cousin William Darwin Fox (1805–80). Being four years Darwin's senior, Fox took him in hand and became his 'entomological tutor'.[4] Fox had a high forehead, lustrous dark-brown hair, very blue eyes, a sharp nose and a humorous mouth. He was mad about insects. They called one another 'Fox' and 'Darwin', just as Robert Waring called Uncle Jos 'Wedgwood', and Uncle Jos called Robert 'Darwin' – even though all were kinsmen. We get a taste of the quality of their friendship when we read a letter Darwin wrote to Fox at the end of the first year, shortly after they had gone their separate ways to spend the Long Vacation with their families. 'My dear Fox, I am dying by inches, from not having any body to talk to about insects.'[5] There followed what was in effect a journal of the beetles and creepy-crawlies he had encountered since returning to Shrewsbury, with illustrations.

Fox would continue, throughout his seventy-five years of life, to be an avid amateur naturalist. Like Darwin at Cambridge, he was preparing to take orders. For thirty-five years, he would be the Rector of Delamere in Cheshire, where he would follow the pattern of life enjoyed by dozens of Victorian parsons, tending a rural parish and learning more each year about the natural history around him. Darwin's own father, though a doctor, not a clergyman, modelled himself on Gilbert White,[6] the parson at Selbourne in Dorset, whose account of the birds, flora, insects and animals of his country parish was recorded in round-the-year diaries.

'I had scruples about declaring my belief in all the dogmas of the Church of England,' Darwin would write, 'though otherwise I liked the thought of being a country clergyman.'[7] Many readers of the *Autobiography* take this at face value, and imagine that Darwin's life of seclusion at Down House, after he married, was a repetition of the life of a naturalist parson such as Gilbert White, only without the religion. Such a reading overlooks two, possibly related, facets of the story. One is Darwin's immense wealth. Dons love money. They can smell it. And the arrival in the then tiny community of Cambridge of another Darwin boy had not gone unnoticed. Professors at Cambridge are not required to teach, or even to notice, the undergraduates. That is the job of the college Fellows who act as supervisors. There is something very conspicuous about Adam Sedgwick, for example, taking great note of Darwin, and we may question whether the Woodwardian Professor of Geology would have befriended an undergraduate of humbler means. The gentle and apparently unworldly Professor of Botany, John Stevens Henslow, was another who became Darwin's friend even while he was just an undergraduate. (It was a friendship with momentous consequences.)

Henslow was Darwin's senior by thirteen years. He at first held the chair of mineralogy and then, after 1825, that of botany, which was his real passion. His predecessor, the Revd Thomas Martyn, had created the first Botanic Garden in Cambridge, on a site now occupied by the Cavendish and other laboratories, but it had fallen into decay in Martyn's latter years, and there had not been any botanical lecturers in the University for a long time when Henslow took over. It was in 1831, in Darwin's last year, that the University, under Henslow's direction, acquired the forty acres off the Trumpington Road and, after an exchange of lands between the University and Trinity Hall, effected by Act of Parliament, created the new Botanic Gardens modelled on the gardens at Kew. (It was not until 1851 that the Natural Science Tripos was established at Cambridge.)

As well as being an inspiring botanist, and geologist, Henslow had another characteristic which was a crucial factor in his friendship with Charles Darwin. It had always been part of Henslow's fantasy-life that he should become one of the great explorer-botanists, like Bougainville or Joseph Banks. Fired by reading Levaillant's *Travels* he had imagined himself discovering the plant life of Africa.[8] His grandfather was Sir John Henslow, Chief Surveyor of the Royal Navy during Nelson's time, and Henslow, who grew up on the Medway in Kent, was accustomed to the sight of tall ships and those who exercise their business in great waters. His father, however, a solicitor in Rochester,[9] who found himself the parent of eleven children, had no money to finance such an adventure. You had to be rich to be an explorer. So Henslow was sent to St John's College, Cambridge, where his cleverness brought the success described.

Darwin was a rich young man through whom the innocent Henslow could live vicariously. Darwin became known in Cambridge as 'the man who walks with Henslow', and he often dined with Henslow's family in the evenings. These were the days before mere Fellows of colleges were permitted to marry. Professors could marry and have family life. So once again, here is Darwin enjoying an unusual privilege. Most undergraduates had no taste of family life during their three Cambridge years, but by 'walking with Henslow' Darwin could enjoy a hearth, children's voices, firelight flickering on silver – all things denied the average undergraduate with his pewter mug of ale and his college dinner served at a long refectory table. Everyone who knew Henslow remarked upon his benevolence. After Darwin had come to know him well, he acquired the living of Hitcham in Hertfordshire, and took his duties seriously as a parish priest, with 'excellent schemes for his poor parishioners'. He was unaffected in his Christian faith, and 'so orthodox', recorded Darwin, 'that he told me one day he should be grieved if a single word of the Thirty-nine Articles were altered'.[10]

Darwin's wealth would not have impeded, but nor did it facilitate, comparable simplicities. When he was dying he became aware that he had never lived for others, and his granddaughter Gwen Raverat noted that the poor did not really exist for him.

The Darwin and Wedgwood money, however, meant that there was no danger of his being neglected as an undergraduate, and he was taken up, not only by the professors but by his fellow students. This is the only time in Darwin's life when we see him being touched by the arts. His Shrewsbury schoolfriend Charles Whitley, who had followed him to Cambridge and was at St John's, took him to the Fitzwilliam, the University museum, and showed him pictures. The taste lasted for a few years, so that he was able to experience, for a while, a fondness for pictures in the National Gallery in London. (Whitley was later a canon of Durham, and reader in natural philosophy – that is, science – at Durham University.) His friend John Maurice Herbert, later a county court judge in Cardiff, nurtured Darwin's musical tastes. Darwin listened to Herbert and his friends play chamber music, and Herbert steered him in the direction of King's College Chapel, so that weekday walks in pursuit of beetles would now conclude, for Darwin, with taking in the anthems sung by the choir there. He loved the sound so much – 'intense pleasure, so that my backbone would sometimes shiver' – that he actually used to hire the King's choristers to sing in his rooms at Christ's. These were the happy days that he recalled from an arid old age, by which time he had lost all taste for art and music and declared, 'I am so utterly destitute of an ear, that I cannot perceive a discord, or keep time and hum a tune correctly; and it is mystery how I could possibly have derived pleasure from music.'[11] The plaintive words suggest bafflement at himself, unable in old age to see how he could have lost such capacities for inward joy.

For there was something else in Darwin, apart from his great wealth, which made it unlikely that he would ever happily settle down, as Fox did, and become one of hundreds of naturalist-clergymen, riding round their parish in early days on horseback and in later times on pennyfarthing bicycles, butterfly net over their shoulder, shovel hat on head. That is, Darwin always wanted to cut a dash; his eye for the main chance was unfaltering. The diffident manners concealed this, sometimes even from himself, but it was there from an early stage. It had certainly begun to stir in Edinburgh, which perhaps made Grant uneasy when Darwin, with a little too much vigour and a little bit too much pleasure, decided to challenge Grant at the meeting of the Plinian Society. The same ardent ambition surfaced as Henslow spoke of his own youthful desire to become a great explorer, a naturalist on the world stage. Had Dr Robert Darwin seen the latent, highly aggressive ambition which lay buried beneath the surface of the shy, moody, shooting-obsessed younger son, he might have been less worried. All he could see, however, after a lifetime of hard work himself, building up a distinguished medical practice and managing what was in effect a highly successful small private bank, was a pair of loafers: Ras, while finishing his medical studies in Vienna, was becoming increasingly dependent upon opium and unable to settle to anything, Charles an idler.

The Long Vacation of 1828 was long indeed, and you sense its seeming endlessness in Darwin's letters to Fox, which were largely about beetle-collecting. 'My head is quite full of Entomology. I long to empty some information out of it into Yours';[12] 'My dear old Fox I long to see you again, but I suppose it will not be long before Term begins.'[13] Those words were written in July, and he did not return to Cambridge until the very end of October. Dr Robert was neither in the best of health (gout) nor in the best of tempers. Darwin was dispatched for two months to Barmouth where he endeavoured to revise mathematics, but concentrated rather upon the local insects. Barmouth is a delightful little resort on the mouth of the River Mawddach in north-west Wales, an easy journey from Shropshire. After two months there, lodging with a private tutor, with whom he did not get along, and in whose company his mathematics did not improve, he longed for congenial companionship, and for shooting. He went to stay with Fox at his family home, Osmaston Hall near Derby, and, as Darwin most revealingly remarked to Fox, 'Formerly I used to have two places, Maer & Woodhouse, about which, like a wheel on a pivot, I used to revolve. Now I am luckier in having a third, & I hope I need not say that third is Osmaston.'[14] The Mount was no longer pivotal. Perhaps it had stopped being pivotal when Sukey died. In September at Maer, he managed to kill '75 head of game; a very contemptible number, but there are very few birds. I killed however a brace of Black Game [black grouse] . . .'[15] By the beginning of October, his sister Caroline drew up the following 'petition' directed at the Darwin family from The Mount:

To the

Charitable & Humane The Case of a Distressed Sportsman – 1828. Oct 3 –

Charles Darwin gent – humbly petitions all benevolently disposed persons to pay attention to his case –

Whereas he the aforesaid formerly gained a respectable livelihood by destroying hares, pheasants, partridges & woodcocks, with the aid of a double barrelled gun, & the said gun becoming dangerous & liable to destroy the aforesaid Charles Darwin's legs, arms, body & brains & consequently unfit for use – he is reduced to lay his deplorable case before the charitable & humane being utterly unable to raise the sum requisite for the purchase of new Double barrd Gun – Value £20 –

	£.	s.	d.
R. W. Darwin	5	0	0
Miss Darwin	5	0	0
Miss Susan Darwin	5	0	0
Miss Cath Darwin	5	0	0[16]

The gun was duly purchased. (Six years later, in Dorset, George Loveless and five other farm labourers were transported to Australia for daring to protest that they were unable to feed their families on the wages of six shillings per week, that is, £15 twelve shillings per annum. They were named the Tolpuddle Martyrs, and the trade union movement was born.) Darwin clearly felt that his father kept him on a tight leash financially.

Just before the end of this first Long Vac, Darwin, at a dinner at Netley Hall, five miles south of Shrewsbury, met Frederick William Hope, a clergyman-entomologist who would found the chair of zoology at Oxford in 1849. Hope was friends with Darwin's Edinburgh mentors Jameson, Grant and others, and he possessed one of the largest collections of insects in Britain, which he gave to his old University when he founded the Oxford chair. He promised Darwin 300 or 400 insect specimens before Christmas. It was in many ways the high point of his vacation.[17]

The next term was dominated by his friendship with Fox and

his passion for insects. 'I live almost entirely with Fox and Entomology goes on most surprisingly.' So much did the scientific interest predominate over the theological that Fox began to fear that he would be 'plucked'[18] – refused a degree – which might have spoilt his chances of being ordained and getting a curacy.

He went home to The Mount in late December, giving his father a death's-head hawkmoth for his Christmas present. 'You cannot imagine how pleased my Father was with the Death's-Head's; to use his own words "if he himself had thought for a week he could not picked out [*sic*] a present so acceptable".'[19] Sometimes one suspects that Dr Robert's ironies were lost on his children. In January 1829, stricken with fever and in an agony of gout, the Doctor took to his bed while Darwin returned to Cambridge.

In those days, the first examination sat by a Cambridge undergraduate was in the Lent Term of his second year. Called 'Little-Go', it consisted of a fairly rudimentary set of mathematical problems, and a few classical set texts. 'For a serious reading man, the "Little-Go", was an irritant rather than a worry.'[20] For Darwin, it was a worry. It was one thing to irritate his father, from whom he was increasingly estranged. Quite another was to disappoint Henslow, his new father-substitute. Darwin had started his university career a term late – January 1828 – whereas most undergraduates began the academic year in October. His tutor at Christ's advised him to put off Little-Go for a year, rather than taking it in the spring of 1829 after only four terms. The mathematics, in particular, was a worry to him, and it was with exultation that he was able, in March 1830, to write to Fox, 'I am through my little Go! I am too much exalted to humble myself by apologising for not having written before . . . I went in yesterday, & have just heard the joyful news.'[21]

He celebrated by going up to London, where Erasmus had already set up house in Great Marlborough Street, increasingly drug-dependent, but quietly social and eccentric. He also saw Hope, with whom he planned a summer expedition of insect-collecting. Much of that summer was spent with Hope – and with Thomas Campbell Eyton, a Shropshire coeval and fellow entomologist who was also at Cambridge with him. Visits to Wales deepened Darwin's insect-knowledge, while keeping him safely out of the influence of home. It is striking just how little time Darwin spent in his father's company when an undergraduate; and for the rest of his life he kept his distance from the increasingly sick and lonely parent. Christmas 1830, for example, was spent, not at The Mount, but in his college rooms at Christ's, preparing for the final heave of exams.

Although it fitted the image he was later to promote of himself as an 'idler', who acquired his scientific knowledge all by himself and by observation when on board the *Beagle*, Darwin was always a voracious reader and a careful student, even if he was not always reading the books set by the Cambridge examination syllabus.

Conversations with Henslow, the world-traveller manqué, had whetted Darwin's appetite for reading the great explorers. Above all, he read Alexander von Humboldt. After Humboldt's lifetime (1769–1859), world literature was so vast, scientific knowledge so advanced and multifarious, technological progress so prodigious, that there could be no universal genius who knew all that there was to be known. But in his own time Humboldt almost was that man. There had been many helps along the way. One was the death of his father, a Prussian army major, when Humboldt was just ten. His formidable mother asked no less a personage than Goethe to advise on the education of her two sons. Goethe, himself the 'universal genius' of his own generation and, in the eyes of nearly all Germans, for all time, suggested that the older boy, Karl Wilhelm, should be reared as a philologist and man of letters, while the younger, Alexander, should be a scientist. Both Humboldts were very fortunate in their friends. Their mother engaged the greatest scholars in their various fields to teach them, but they also simply happened to meet the right people at the right time. (Darwin had this gift.) Alexander von Humboldt also had the great advantage of being gay, but without, so far as one can tell, any neurosis about it, or the tendency to make himself unhappy in the manner of Michelangelo's or Shakespeare's sonnets about unrequited love for boys. So he was detached, and in no need of the consolations and distractions of home. He was therefore an ideal scientific explorer. It was George Forster, himself a heterosexual, the boy who had travelled as a naturalist-illustrator on Captain Cook's second voyage – aboard the *Resolution* – who had introduced Humboldt to travel. Forster took young Alexander on his first serious journey, a mineralogical excursion up the Rhine, which enabled Humboldt to write his first scientific book, when he was twenty-one, on basaltic rock formations. Forster's magnetic personality and legendary conversations persuaded Humboldt that he should become a traveller, and he set about preparing himself for the task by learning languages (something Darwin never did) and by attaching himself to a proposed voyage of circumnavigation with a French sea-captain called Baudin. This expedition was frustrated by the wars (1796). There was an unsuccessful attempt to reach Egypt in time to coincide with Napoleon's occupation. By the time he was thirty, however, Humboldt, by now the master of many languages and as well informed about science as any European of his age, was in a strong position to set off, with Aimé Bonpland (who had been going with Baudin as botanist to the failed voyage round the world). The two friends sailed from Corunna in June 1799, crossed the Atlantic, stopping in Tenerife to see the remarkable meteor-shower of 12–13 November, reached Caracas by the end of the year, spent four months following the course of the Orinoco, and then made extensive journeys, both in the Caribbean (memorably to Cuba) and in South America, exploring the Amazon and Peru. Humboldt formed an encyclopaedic collection of minerals and archaeological relics. All the time he was travelling, he was engaged in scientific investigations, for example into the geographical distribution of plants, the origins of tropical storms, the relationship between the decrease in mean temperature and the increase of elevation beyond sea-level. It was he who discovered the decrease of intensity of the earth's magnetic force from the poles to the Equator. In 1808, when he had returned to Europe, he wrote up the travels in a monumental *Personal Narrative*, and it has been rightly observed, that 'with the exception of Napoleon Bonaparte, he was the most famous man in Europe'.[22]

It was all a far cry from Gilbert White's country garden at Selborne, or, come to that, the life of Darwin's cousin Fox and the hundreds of others of naturalist-clergymen whose ranks Darwin was preparing to join. As he read Humboldt, his mind was filled with wider horizons, more exotic shores. There were three ways, in 1830, to excel as an English man of science. One was to read medicine, and to use the study of anatomy as a way into the study of other sciences. The other was Sedgwick's way, Henslow's way, to obtain a Cambridge or Oxford Fellowship and hope to be appointed to one of the very few chairs of botany, geology or mineralogy. The third was to be rich enough to become, like Joseph Banks, an independent traveller and collector. Darwin was unlikely to have done so well in the Cambridge examinations as to be elected to a Fellowship. The third course, that of a rich, independent man of science, was his only option. Meanwhile, in obedience to his father's wishes, he prepared for the humbler option of becoming a clergyman.

The final part of his Cambridge exams, in May 1831, would require him to display knowledge of divinity, in particular of the work of a former fellow of Christ's called William Paley. The set book was Paley's *Evidences of Christianity*. It remains to this day a powerful piece of work. Paley, who lived from 1743 to 1805, saw during those years the greatest assaults on Christian belief which that religion had endured since the times of the Roman persecutions of the third century. Mention has already been made of the French Revolution, whose fervent attempts to root out Christianity were viewed with appalled astonishment by most Britons, though not all. Old Dr Erasmus Darwin was among those who openly welcomed the Revolution, and who undoubtedly sympathized, if not with the murderous methods of Robespierre, at least with his desire to worship Reason rather than the – as they would see it – irrational faith of revealed Christianity.

Moreover, it was not possible to ignore the fact that the infidels of Revolutionary France drew much of their inspiration from British philosophy. Voltaire, whose bust and whose portrait medallions were among the proudest productions of old Josiah Wedgwood's manufactory at Etruria, openly acknowledged his debts to John Locke and the English empiricists. And in Paley's lifetime lived the most sceptical British philosopher of them all, David Hume. The most devastating of his analyses was to be found in his *Dialogues Concerning Natural Religion*, which would have been dynamite had he published it in his lifetime. This posthumous work, which appeared in 1776, had probably been completed in its first version by 1755, but he revised it at least twice, in 1761 and shortly before his death in 1776.[23] Although notionally, in his early work as a philosopher, Hume appeared to go along with the argument from design – that is, that the existence of the laws of nature and the intricacy of nature support the idea of a lawgiver and a Creator – this opinion did not survive. Huxley, in his book on Hume for the English Men of Letters series in 1887, wrote that 'if we turn from the *Natural History of Religion* to the *Treatise*, the *Enquiry*, and the *Dialogues*, the story of what happened to the ass laden with salt, who took to the water, irresistibly suggests itself. Hume's theism, such as it is, dissolves away in the dialectic river, until nothing is left but the verbal sack in which it was contained.'[24] Clearly, by 1887, this was Huxley's position, and Darwin, who had first seriously studied science in Edinburgh, the city of David Hume, with men, particularly Grant, who had soaked themselves in Hume's philosophy, had begun to imbibe the Humeian scepticism much earlier than is sometimes supposed.

Hume had two arguments with which he countered 'natural theology', or anyway Christianity's version of it. One is that when Christians, from the earliest times to the present, attempt to impress the incredulous, it is with appeals not to the created order itself, but to departures from it: in short, to miracles. This argument, it might be thought, is a bit of a smokescreen. Although the argument from design leads you nowhere – or not very far – when you are considering the veracity of miracle-stories, this does not in itself invalidate the believers' conviction that there *is* a natural order, from which it pleases God to depart.

Much more undermining to faith was Hume's attack on the argument from design itself. The *Dialogues* were, according to Leslie Stephen – the clergyman-father of Virginia Woolf, who lost his faith so wholeheartedly – 'the first work of our literature to subject the argument from design to a passionless and searching criticism'.[25] Hume does this partly by pointing out how often the argument from design resorts to metaphor, to anthropomorphic projection, to 'a great analogy to the productions of art'.[26] The basic analogy is between the cause or causes of order in the universe and human intelligence. But this is all it is – an analogy, says Hume.[27] For our purposes, the most destructive and perhaps the most 'Darwinian' moment of the *Dialogues* bears the mark of having been composed at quite a late stage.

> Look around this universe. What an immense profusion of beings, animated and organized, sensible and active! You admire this prodigious variety and fecundity. But inspect a little more narrowly these living existences, the only beings worth regarding. How hostile and destructive to each other! How insufficient all of them for their own happiness! How contemptible or odious to the spectator! The whole presents nothing but the idea of a blind nature, impregnated by a great vivifying principle, and pouring forth from her lap, without discernment or parental care, her maimed and abortive children.[28]

It is impossible to believe that this paragraph, and the arguments implicit throughout the *Dialogues Concerning Natural Religion*, made no impression upon Darwin, the strength of the point being that there is no need for the analogy with human intelligence or art. *Nature is blind.* Its processes, and the great vivifying principle which animates it, do not need to be given names, though it is the task of science to understand those processes. Supposing it were possible for science to demonstrate just such a 'blind' process as Hume posits . . .

It was as a counterblast to such scepticism that Paley wrote his own apologetics – the most celebrated two books of which are *View of the Evidences of Christianity* (1794) and, his last work, *Natural Theology* (1802). In *Evidences*, the book on which Darwin was examined for his Cambridge degree, Paley has Hume in his sights all the time, particularly Hume's attacks on miracles. Paley's argument is from evidence, and, for him, the strongest argument for the truth of his religion is that there is overwhelming historical evidence that, from the very earliest times – certainly from the time of Nero in the AD 60s – those who claimed to be witnesses of the events described in the New Testament were prepared to undergo torture and death rather than deny what they had experienced. Paley, in other words, had an empirical defence for the likelihood of Christianity being true based upon actual historical truths.

The only way in which one could refute Paley's *Evidences* – since Juvenal, Pliny and many other witnesses attest to the Christians' willingness to die for their claim to be witnesses to Christ – is to say that they were liars, or innocently deluded. Darwin did not have to read Paley's *Natural Theology* for his examination, but he did so anyway, and he records in his *Autobiography* that the logic of Paley's *Evidences* gave him as much delight as did Euclid. So too, he adds, did his *Natural Theology*.

Whereas Hume saw the various species of nature as 'hostile and destructive to each other', Paley wondered in awestruck delight at the sheer variety of nature, and at its multiplicity of forms, its thousands of species of flies, its hundreds of different types of butterfly. As for the destructive and predatory aspects of animal life, 'What one nature rejects another delights in. Food which is nauseous to one tribe of animals becomes, by that very property which makes it nauseous, an alluring dainty to another tribe. Carrion is a treat to dogs, ravens, vultures, fish. The exhalations of corrupted substances attract flies by crowds. Maggots revel in putrefaction.'[29]

Paley matters, because he was almost the only theologian (apart from Pearson on the Creed) whom Darwin ever read. The opening chapter of Paley's *Natural Theology* remained in Darwin's head for the rest of his life. This is the celebrated analogy:

> In crossing a heath, suppose I pitched my foot against a *stone*, and were asked how the stone came to be there: I might possibly answer, it had lain there forever; nor would it perhaps be very easy to show the absurdity of this answer. But suppose I had found a *watch* upon the ground, and it should be inquired how the watch happened to be in that place . . .[30]

Paley's geology, incidentally, is here shown to be up to date: he casually points out that it would be absurd to suppose that the stone had been there 'forever'. As an arresting analogy, the discovery of the watch reminds any reader not merely of nature's intricacy, but of its orderliness. All language is metaphor, a fact of which Darwin sometimes, in his writing, appeared heedless; but scientists, regardless of their religious opinions, find it useful to apply the word 'law' to natural phenomena. This is because, as with Paley's analogy with the watch, nature does follow what we call laws. Whether you are studying thermodynamics, astronomy, quantum physics or chemistry, you will not progress far until you have mastered and learnt the laws which are observable in all these branches of knowledge.

Darwin in his later career would claim that 'the theory of Evolution is quite compatible with the belief in a God',[31] though in his posthumously published *Autobiography* he would be somewhat more candid, admitting, 'I had gradually come by this time [January 1839, thirty-three years before the final edition of *The Origin of Species*], to see that the Old Testament from its manifestly false history of the world . . . was no more to be trusted than the sacred books of the Hindoos, or the beliefs of any barbarian.'[32]

Plainly, this is empirically the case, since there are millions of people in the world who believe in both evolution and God. With his distinctive way of arguing, however, it is plain that Darwin, anyway as an old man, believed that his own particular theory of how evolution worked – the theory of the evolution of species by natural selection – removed the need for believing in God.

> Although I did not think much about the existence of a personal God until a considerably later period of my life, I will here give the vague conclusions to which I have been driven. The old argument from design in Nature, as given by Paley, which formerly seemed to me so conclusive, fails, now that the law of natural selection has been discovered. We can no longer argue that, for instance, the beautiful hinge of a bivalve shell must have been made by an intelligent being, like the hinge of a door by man. There seems to be no more design in the variability of organic beings, and in the action of natural selection, than in the course which the wind blows.[33]

There is a certain amount of muddle here – not least in the suggestion that wind speeds and directions occur randomly, rather than being attributable to causes. If Darwin's theory of natural selection is true, then this would, for the theist, be the manifestation of how the 'first great cause' set evolving life in motion. For the atheist, there would be many other reasons for unbelief – the unfairness of things, the suffering of the innocent, the unlikelihood of life after death, and so forth – to be going on with, quite apart from adding the Darwinian natural selection to the list. What Darwin's linking of Paley to his theory reveals is that he saw his *Origin of Species* idea as theological, or anti-theological: an answer, not to other scientists, but to the theology he had read at Cambridge. The more ardent neo-Darwinians of our own times retain this line, as demonstrated by Richard Dawkins's wonderfully lucid exposition of these ideas in his 1979 book *The Blind Watchmaker.*

Paley's *Natural Theology* did not end with its first paragraph. One can see why the young Darwin appreciated the book, since it revels in the plenitude and variety of plants, insects, birds and beasts. Natural theology, however, that is to say, the deduction of the probability that there is a God from a study of the laws of nature, is for Paley the Christian theologian only the *beginning* of the true faith, which is the acceptance of a personal God, and the acceptance of His revelation of Himself in the person of Jesus. 'Under this stupendous Being we live. Our happiness, our existence, is in his hands . . . The hinges of the wings of an *earwig*, and the joints of its antennae, are as highly wrought, as if the Creator had nothing else to finish . . .'[34]

But, of course, He did have other things to finish, whereas the Darwinian 'answer' to Paley stops with a consideration of the *Just So* story – how the earwig got its wings. For Paley, this is only the beginning. 'The works of nature want only to be contemplated,' he wrote. But such a contemplation can only go so far. It was for this reason that William Blake, for example, in 1788 wrote that 'there is no natural religion'.[35] W. H. Auden, in his Commonplace Book, copied some words of Ferdinand Ebner with which Paley would have agreed: 'To talk *about* God except in the context of prayer, is to take His name in vain.'[36] Darwin was to be, for many people, during and after his lifetime, the embodiment of the essentially Victorian myth that science had somehow disproved, or invalidated, religion. In fact, of course, rather than disproving religion, science had become their religion, but neither as a young man nor as an old one would Darwin have agreed with that statement or have understood it. One learns, however, to tread carefully when he makes religious statements. Which came first? His decision that scientific facts made belief in a personal God impossible? Or his lack of belief in a personal God, and his inward unhappiness, making Hume's bleak view of the universe more plausible to him than that of Paley, Henslow or Sedgwick?

Darwin mastered enough of Paley's *Evidences*, of Euclid and of his smattering of classical texts to sit his 'Mays', his final exams at Cambridge, four days sitting papers, with two papers each day lasting two hours each. In fact, having come up to Cambridge late, he sat his 'Mays' not in May but in January 1831. He did not put in for honours, and he came tenth on the list of what he called 'hoi polloi'. It was a creditable result, but it scarcely put him in the running for a college Fellowship. The unpalatable prospect of a curacy beckoned, a possibly unsuitable role for one who 'did not think much about the existence of a personal God'.[37]

It was not until this late stage of his Cambridge career that he actually met the Professor of Geology, Sedgwick, an encounter which, like so much that was momentous in Darwin's story, was engineered by Henslow. It was Henslow who proposed that Darwin should branch out into geology, perhaps aware that this was the area of science in which the great advances were being made. In 1830, Charles Lyell (1797–1875) had published the first volume of his *Principles of Geology*. Lyell, a Scotsman from a naval family, had studied geology at Oxford under the formidable William Buckland. He had absorbed the ideas of Hutton, that, far from being of recent origin, planet earth was very old, perhaps almost infinitely old. He had read Lamarck. Henslow, despite his remark to Darwin about being grieved if a single one of the Thirty-Nine Articles were altered, was aware of the changes in the air, and he could see what a groundbreaking work Lyell's was. After Lyell, it was no longer going to be possible to do science in the old way. And botany and entomology and zoology and all the subjects which so obviously interested Darwin were going to be affected by the knowledge which Lyell and his fellow geologists imparted.

Whereas Henslow was gentle and fatherly to Darwin, Adam Sedgwick, an extrovert Yorkshireman (1785–1873), was an amusing companion for the introvert Charles Darwin. Any scientist would have been interested to stay in the Darwin family residence, whatever their views of *Zoonomia* (and Sedgwick's views of Erasmus Darwin Senior would have been robust). Moreover, Shrewsbury was en route to Wales, and Sedgwick had planned a summer of geological exploration in Anglesey, Caernarvonshire, Merioneth and Cardiganshire. Incidentally, that Sedgwick asked Darwin to accompany him on this trip gives the lie to Darwin's implication that, until he sailed with the *Beagle*, he had scarcely studied geology. Sedgwick would not have wanted to be accompanied by an ignoramus, and it was obvious from Darwin's conversation that while in Edinburgh he had learnt much from Grant, Jameson and others.

They set off together in July. Sedgwick was a restless companion. One morning, having left their inn a mile or two behind them, Sedgwick began to speculate about a sixpence which he had put into the hands of the waiter, telling him to give it to their chambermaid. What if that 'damned scoundrel' had failed to give the girl the money and pocketed it himself? It was all Darwin could do to persuade Sedgwick that there were no reasons for his suspicions and that a tiring walk back to the inn to confront the waiter would have been a waste of time.[38]

From the first, Sedgwick treated Darwin, not as some ignorant student brought along for the fun of a walking trip, but as a serious geologist who was knowledgeable enough to help him in his searches – he was looking for the reported presence of an extended layer of Old Red Sandstone (Devonian) in contact with Carboniferous Limestone in the Vale of Clwyd – and also debate with him. Darwin had attended Sedgwick's geology lectures at Cambridge – something which he would later deny having done,[39] even though two of his contemporaries remembered seeing him there. When Sedgwick and Darwin separated on their Welsh journey, Sedgwick sent Darwin a detailed letter about the organic remains of slate in Moel Shabod in Snowdonia. Darwin had meanwhile encountered an Oxford undergraduate hiking in the mountains, the future Lord Sherbrooke, Chancellor of the Exchequer in Gladstone's first administration, who remembered:

> He was making a geological tour in Wales, and carried with him, in addition to his other burdens, a hammer of 14 lbs weight. I remember he was full of modesty, and was always lamenting his bad memory for languages . . . I saw something in him which marked him out as superior to anyone I had ever met: the proof which I have of this was somewhat canine in nature, I followed him. I walked twenty-two miles with him.[40]

Darwin, however, had ambitions to travel further than Wales. With Henslow, he had hatched some as yet inarticulate plans to study tropical vegetation in Tenerife, the Darwin money effecting, in the younger man's life, the fulfilment of ambitions which Henslow had never managed to accomplish in his own. Whether it would have been possible to persuade Dr Robert Darwin to let him go to Tenerife, we do not know. It would seem likely that, without his quite knowing it, Darwin, when he set forth with Sedgwick, was on probation. Surely one Cambridge professor would report back to another how the budding young geologist had acquitted himself. And, as Sherbrooke's memory attests, there was something intensely motivated and immediately impressive about Darwin. While Darwin tapped and hammered the rocks of Upper Carboniferous deposits in North Wales, however, Henslow was preparing him for a life-changing opportunity.

5

The Voyage of the *Beagle*

HMS *Beagle* was built as a ten-gun brig-sloop of the Royal Navy – in other words, a small fighting vessel, originally 235 tons, just over 90 feet in length, 24–25 foot in the beam and designed to hold 120 men as a ship-of-war. In fact, she was never used for hostilities, and five years after she was launched she was refitted, in 1825, as a survey ship. Ever since the great voyages of Captain Cook in the 1770s, it had been realized that the sloop, rather than an enormous ship, was the ideal vessel of survey, because it could be taken inshore easily, either for cartography or for observation of natural phenomena. The *Beagle*'s first voyage of discovery had been in 1826, under the command of Captain Pringle Stokes, and its task was a hydrographic survey of Patagonia and Tierra del Fuego. It is a desolate place, the weather is terrible, the semi-naked inhabitants presented a woeful appearance, and Captain Pringle Stokes fell into a dangerous depression. He shot himself on 2 August 1828, and died ten days later. The First Lieutenant, W. G. Skyring, was appointed commander and the poor little *Beagle* made her miserable way to Rio de Janeiro for refitting and provisions. The Commander-in-Chief of the South American station was Admiral Sir Robert Otway, and he ordered the *Beagle* to Montevideo for repairs. It was there that he made the bold decision to place her under the command of a twenty-three-year-old aristocrat named Robert FitzRoy, son of Lord Charles FitzRoy, grandson of the Duke of Grafton, and the nephew of Lord Castlereagh who had been Foreign Secretary at the time of the Congress of Vienna. It was a highly unbalanced family – Castlereagh committed suicide in 1822, fearing his unmasking as a homosexual, and he was one of the most unpopular politicians in British history. (Although he had cut his own throat, which would have disqualified lesser mortals from Christian burial at that date, Castlereagh was given a funeral in Westminster Abbey and his coffin was greeted by whoops of merriment when it came out into the street.)

Captain FitzRoy was, however, a largely competent officer. Passing back through Tierra del Fuego, the *Beagle* suffered the loss of one of her jolly-boats, stolen by natives. In retaliation, FitzRoy took four Fuegian hostages with him back to London, naming them Jemmy Button, York Minster and – this was a young girl – Fuegia Basket. The fourth died when they reached England.

FitzRoy had them educated at his own expense, but it was obvious that they should be returned to their own people, and a supplementary reason for the second voyage of the *Beagle* was that Jemmy Button, York Minster and Fuegia Basket, now accompanied by a missionary called Richard Matthews, should return to their desolate homeland and win the natives for the Church of England.

FitzRoy was a scientifically interested man, and, apart from giving the Fuegians safe transport home, he planned to make a survey of South America. For this, he wanted to take the most up-to-date navigational instruments. On the first voyage, he had felt keenly the lack of a good geologist, and his aim, should the Admiralty agree, was to set out again in the *Beagle* with a geologist 'qualified to examine the land; while the officers, and myself, would attend to hydrography'.[1] FitzRoy was destined to become a distinguished meteorologist who invented the term 'synoptic chart', by which he pioneered a method of giving reliable gale warnings to ships. The daily recitation of weather conditions in British coastal waters which is heard on British radio – a secular equivalent to the Angelus in other cultures – is one of the direct consequences of FitzRoy's work.

The successor of a captain who had taken his own life in the gloomy, windy, rain-tossed Tierra del Fuego, the nephew of a famous suicide and, as it happens, a man who was doomed himself, aged sixty, to take his own life, FitzRoy was a vulnerable man, and he knew that Captain Pringle Stokes would probably still have been alive if he had enjoyed the comradeship of some congenial fellow human beings during the darkest days at sea.

When FitzRoy eventually received permission from the Admiralty, therefore, to go ahead with his voyage of survey, he began to ask around, to see if a companion of about his age could be found.

While Darwin was in Wales with Sedgwick, in August 1831, George Peacock, tutor in mathematics at Trinity College, Cambridge – future Lowndean Professor of Astronomy and Geometry at Cambridge, and Dean of Ely – wrote to Henslow. 'Captain Fitz Roy [*sic*] is going out to survey the southern coast of Terra [*sic*] del Fuego & afterwards to visit many of the South Sea Islands & to return by the Indian Archipelago: the vessel is fitted out expressly for scientific purposes, combined with the survey: it will furnish therefore a rare opportunity for a naturalist & it would be a great misfortune that it should be lost . . .'[2] Peacock floated the name of Leonard Jenyns, a naturalist-clergyman, who was a friend of Darwin's. Henslow himself would have loved to realize a lifetime's ambition in making such a voyage, but by now he was married with a child.

It is not clear who, precisely, was using Peacock as the go-between to find a suitable naturalist. Sir Francis Beaufort, Hydrographer of the Navy and inventor of the Beaufort scale of wind-measurement, had told him about the voyage, and it may well have been Beaufort, rather than FitzRoy,[3] who first began the search via the Cambridge network. Jenyns was approached, but decided that he could not leave his parish (Swaffham Bulbeck). It was then that Peacock wrote to Darwin, who found the letter at The Mount awaiting his return from North Wales. Peacock emphasized that there was no time to lose, and that Darwin should make up his mind immediately, since the *Beagle* was to sail at the end of September. After a day's lengthy discussion with his father, Darwin wrote to Henslow:

> My Father, although he does not decidedly refuse me, gives such strong advice against going. – that I should not be comfortable, if I did not follow it. – My Fathers objections are these; the unfitting me to settle down as a clergyman. – my little habit of seafaring. – the *shortness of the time* & the chance of my not suiting Captain Fitzroy [*sic*] . . . But if it had not been for my Father, I would have taken all risks.[4]

Darwin then turned to his Uncle Jos at Maer Hall, who on 31 August wrote to 'My dear Doctor' the letter which sealed Darwin's destiny. He itemized the objections and answered them. The journey would *not* be 'in any degree disreputable to his character as a clergyman'. It was inconceivable that the Admiralty would 'send out a bad vessel on such a service'. 'The undertaking would be useless as regards his profession, but looking upon him as a man of enlarged curiosity, it affords him such an opportunity of seeing men and things as happens to few.'[5] This letter won Dr Darwin round.

The objections raised by Dr Darwin about the comfort and safety of a small sailing ship were very reasonable ones. The Cherokee-class ships of which the *Beagle* was an example were known in the Royal Navy as 'coffin brigs'. They were frequently top-heavy and it required great skill to sail them without capsizing. The *Beagle* had been extensively rebuilt since she first saw the light of day in 1820. She was adapted as a barque not a brig when her function changed from ship-of-war to sailing vessel. This involved reducing her ten cannon to six, and adding a mizzen mast to improve manoeuvrability.[6] In readiness for this new voyage, FitzRoy had her taken to Devonport for refitting. A new deck was added, but, with the constant danger of being top-heavy, this involved raising the upper deck by a mere eight inches aft and twelve inches forward. FitzRoy's aim was to make the vessel less likely to topple, the extra deck-space allowing the boat drainage. (One of the reasons they toppled was that water collected in the gunwales.)[7]

So Dr Darwin's fear that the voyage would not be safe was no mere landlubber's fantasy. These boats were highly dangerous. The six cannon were necessary. The purpose of FitzRoy and his crew was peaceful – charting the coast of South America, and to fix a more accurate measurement of longitude, hence FitzRoy's purchase of the most modern and accurate chronometers which had ever set sail. These peaceful and scientific aims would not be known, or necessarily believed, in every South American port where they tried to dock. The *Beagle* would witness military action in Montevideo. She would be caught up in a major naval blockade in Buenos Aires. Moreover, many of her stopping places would be controversial: for example, it was intended that they call at the Falkland Islands,[8] then as now disputed territory with Argentina. Dr Johnson's words might well have come to Dr Darwin's lips as he tried to dissuade his son from the journey: 'No man will be a sailor who has contrivance enough to get himself into a jail; for being in a ship is being in a jail, with the chance of being drowned.'[9]

Darwin was not a fat man, like his father, but he was, like Dr Robert, tall. When not on deck, or in the very centre of the Captain's cabin, he would be obliged to stoop all the time. The ninety-foot barque would contain, as well as York Minster, Jemmy Button and Fuegia Basket, over seventy other people: the master, two mates, the boatswain, the carpenter, clerks, eight marines, thirty-four seamen and six boys. Then there were the officers – John Wickham, First Lieutenant, James Sulivan, the Second, and John Lort Stokes, who would assist FitzRoy with the surveying. Then there was Robert McCormick, the ship's surgeon. Here was a thorny problem. By convention, the ship's surgeon was the person who also doubled as ship's naturalist. It would normally be the surgeon who made the special collections of flora and fauna and fossils, so that the addition of a special naturalist, to keep the Captain company, was controversial from the start, and whoever went aboard the *Beagle* as the naturalist was going to incur McCormick's hatred. In addition to McCormick, there was his assistant Benjamin Bynoe the purser George Rowlett, a midshipman – King – and an artist, Augustus Earle,[10] a vital figure in those pre-photographic days, to record what they saw. This was a lot of people to squeeze into a very small space, and it was essential that they should coexist in reasonable harmony. Was Darwin, used to large houses and large rooms, prepared for the expedition? Was he prepared to be cooped up in this ship, not for a few weeks but for – as they then supposed – *two years*? And, a vitally important question, was he temperamentally suited to be the companion of Captain FitzRoy?

It was to answer this final question that Darwin set off for London at the beginning of September. On the first of the month, Sir Francis Beaufort had written to the Captain, 'I believe my friend Mr Peacock of Trinity College Camb has succeeded in getting a "Savant" for you – A Mr Darwin grandson of the well known philosopher and poet – full of zeal and enterprize and having contemplated a voyage of his own to S. America.'[11]

This letter makes clear that FitzRoy was on the look-out, not merely for a collector, such as the ship's surgeon might be, but for a real man of science, such as he, FitzRoy, considered *himself* to be. By the time he reached London, Darwin had been warned by Henslow that he was not, in fact, the only man being considered for the post. A naturalist called Chester was another candidate, and much would depend upon the impression Darwin made upon FitzRoy.

FitzRoy was twenty-six, Darwin twenty-two. Both men were depressives – FitzRoy was what would now be called bipolar. Both had uncles who had committed suicide. Both had lost their mothers at an early age. Both were obsessed by their health. Both looked upon themselves as scientists, though pursuing different callings. Both were highly intelligent and well read.[12] It was hugely cheering to Darwin to hear that FitzRoy was taking a large library on the voyage.

Although after their first meeting Darwin was euphoric – 'You cannot imagine anything more pleasant, kind & open than Cap. FitzRoy's manners were to me'[13] – FitzRoy was still keeping his options open. He confided in a friend that he had developed cold feet about taking on board 'someone he should not like'.[14] The two men met again the next day, and it appeared that they were to be congenial companions. The Captain's bipolar character was quite unsuspected by Darwin. He had no idea of what lay in store – FitzRoy's captaincy was that of a martinet; as a companion, he could be cordial and friendly one moment, and cold and distant the next. He was full of aristocratic hauteur and made it clear that he was Darwin's social superior. And if, in leaving The Mount, Darwin thought he was escaping outbursts of rage, he would be proven wrong. FitzRoy was more wildly irascible even than Dr Robert.

But on that September day in London FitzRoy was all cordiality. The sailing had been delayed until 10 October, so there would be time for Darwin to kit himself out. Seasickness? Stormy seas had been much exaggerated, said FitzRoy. If Darwin ever decided he had had enough of life at sea, he could leave. As for their getting along together, FitzRoy 'asked me at once. – "shall you bear being told that I want the cabin to myself? When I want to be alone. – if we treat each other this way, I hope we shall suit, if not, probably we shall wish each other at the Devil" . . .'[15]

Darwin wrote home to his sister Susan for stuff – 'Tell Nancy to make me soon 12 instead of 8 shirts,' he wrote, with only a month to go before he sailed. In addition, he asked her to get arsenic from his father to treat the eczema on his hands. And he wanted his carpet bag, his Spanish books, his slippers, his walking shoes. He also used Susan as the go-between over the delicate question of money. Would she please tell their father – interestingly, Darwin refers to him as 'my Father' when writing to his siblings – that Captain FitzRoy 'is all for Economy excepting on one point, viz fire arms he recommends me strongly to get a case of pistols like his which cost £60!! & never to go on shore anywhere without loaded ones'.[16] They were not very reassuring words for a father to hear, particularly at second hand. In fact FitzRoy had spent over £400 on his own firearms, as Darwin discovered when the pair set out to buy guns on the following day.

They set off round town in a gig. London was filled with crowds, swarming in to witness the coronation procession for William IV, so that Regent Street was a solid jam of coaches. Darwin bought a ticket to watch the procession on 8 September, and to see the Sailor King trot past in his gold coach with his German wife Queen Adelaide. It was a punctual ceremony, the coach arriving at Westminster Abbey at exactly the moment the clock struck eleven. Only six months before, the first Reform Bill had been read in the House of Commons, and the King had been hooted and pelted with stones when he went to the theatre.[17] In May, when Reform was being debated in Parliament, the Queen's carriage had been surrounded by the mob.[18]

On 11 September, as a little trial of their compatibility as shipmates, FitzRoy took Darwin from London to Plymouth by ship to see the *Beagle* in dock. Darwin's praise for the Captain was ecstatic. To a male friend, Charles Whitley, he wrote that 'Cap FitzRoy is very scientific.'[19] To Susan, 'Perhaps you thought I admired my beau ideal of a Captain in my former letters: all that is quite a joke to what I now feel. – Everybody praises him . . .'[20]

It was something of a shock to see how *very* small the ship was, but he consoled himself by saying that, although 'the want of room is very bad', his cabin was the second largest in the ship, next to FitzRoy's. The date of their departure had now been put back to 20 October, and, as he wrote to Susan from Devonport, 'I found the money at the Bank, & I am much obliged to my Father for it.'[21] Again – 'my' Father, not 'papa' or 'Father', as would be natural when addressing a sibling; and no direct thanks to the father himself.

Back in London, Darwin wrote to thank FitzRoy. Already, the sore question of the ship's surgeon had arisen. Darwin clearly wanted no ambiguity, so put in writing that 'I mentioned [to Beaufort] that I believed the Surgeons collection would be at the disposal of Government, and this he thought would make it much easier for me to retain the disposal of my collection among the different bodies in London.'[22] The specimens collected by Dr McCormick would therefore belong to the government; Darwin's collection – and how vast this would turn out to be no one could at this date have predicted – was his. He generously would distribute this among the London museums, but in doing so made abundantly clear his foremost position in this sphere. It is interesting to note that, *even before he set sail*, this was part of his ambition.

Now all that remained to Darwin was to make his farewells, which he intended to squeeze into the final week of September. He went to Cambridge to take leave of Henslow and Sedgwick and other friends and mentors. Then there was time for only a few days in Shrewsbury before he returned to Plymouth. In Shropshire, Fanny Mostyn Owen, with whom he had lain in the strawberry beds, asked if he could persuade FitzRoy to take on her brother as a midshipman, but this request was turned down. The number of 'mids' had been determined when the vessel was commissioned by the Admiralty. In conveying this brush-off, FitzRoy also warned Darwin that the dockyard in Devonport was making very slow progress in repairing and refurbishing the ship and that their departure would be yet further delayed.

It was a frustrating time for all the Beagles, but especially for Darwin, who had never been to sea before, and whose head was already filled with 'date & cocoa trees, the palms & ferns so lofty & beautiful – everything new, everything sublime',[23] as he rhapsodized to Fox. In Plymouth, things were the reverse of sublime. Darwin went on board the *Beagle* in dock, and even slept on board, finding the cabin so restricted that he had to remove a drawer from his locker to make room for his feet when he lay on the truckle bed.

FitzRoy had crammed the ship with no fewer than twenty-two chronometers, packed in sawdust on the shelves.

And there was an ominous portent of things to come when Darwin accompanied the Captain into a Plymouth china shop. A piece of crockery had been bought for the ship, but FitzRoy wanted to exchange it. When the shop man refused, the hitherto charming FitzRoy flew into a rage. Then he asked the price of the most expensive item in the shop and said, 'I should have purchased that if you had not been so disobliging.' The two young men left the shop in silence. Apart from the disconcerting rage, the incident had revealed the class abyss between them. Uncle Jos, now a member of the landed gentry, had once served in his father's London's show-rooms and had begged his father's leave to discontinue, for dread of the embarrassment of serving his grand friends across the counter should they come into the store.

After some awkward silence, FitzRoy said, 'You didn't believe what I said' (about having intended to buy the china). Darwin answered, 'No, I did not.' FitzRoy burst out, 'You are right. I acted wrongly in my anger at the blackguard.'

The delays were agonizing. The *Beagle* was at last ready for sea by the beginning of December, but the sea was not ready for her. On 10 December, and again on 21 December, the ship set out into choppy waters, and was driven back to Plymouth by violent turbulence. They set out again on 27 December. They escaped a heavy gale in the Channel, but felt its consequence – a 'heavy' sea. The little barque was tossed on vast, heaving waves. Darwin wrote to his father:

> In the Bay of Biscay there was a long & continued swell & the misery I endured from sea-sickness is far far beyond what I ever guessed at. Nobody who has been to sea for 24 hours has a right to say, that sea-sickness is even uncomfortable. – The real misery only begins when you are so exhausted – that a little exertion makes a feeling of faintness come on. – I found nothing but lying in my hammock did me any good. – I must especially except your receipt of raisins, which is the only food the stomach will bear.[24]

Added to the wretchedness of seasickness was his sense of failure, the thought that FitzRoy would find him too 'soft' to undertake the voyage at all. FitzRoy, for all his faults of character which were beginning to surface, had become Darwin's mentor, the parent-substitute which throughout his young manhood (Jameson, Grant, Henslow) he had needed. Though so little older than Darwin in years, the fact that FitzRoy was his Captain meant much; Darwin revelled in acceptance, and in the nickname given him by FitzRoy – 'Philos' – short for Philosopher, or Scientist. He need not have worried. 'Darwin is a very sensible hard-working man, and a very pleasant mess-mate,' FitzRoy wrote at an early stage of the journey. 'I never saw a "shore-going fellow" come into the ways of a ship so soon and so thoroughly as Darwin.' And again, to Beaufort, FitzRoy wrote, 'He was terribly sick until we passed Teneriffe, and I sometimes doubted his fortitude holding out against such a beginning of the campaign. However, he was no sooner on his legs than anxious to get to work, and as a child with a new toy could not have been more delighted than he was with St Jago [in the Cape Verde islands, now Santiago].'[25] The other officers also got on well with Philos and picked up the collecting habit, though none was so thorough as he. (Wickham, First Lieutenant, affectionately called Philos 'that damned flycatcher'.)[26]

After the Cape Verde islands, the routines of sea life resumed. Breakfast was at eight. Darwin ate it with FitzRoy in the Captain's cabin. FitzRoy then went off on his morning inspection of the decks and Darwin absorbed himself in marine animals, dissecting, classifying and noting. Dinner was at 1 p.m., a vegetarian meal consisting of rice, peas, bread and water. They drank no alcohol. Supper was at 5 p.m., with such anti-scorbutics as pickles and dried apples eaten with a little meat.

Darwin's attitude to FitzRoy slowly changed. 'Never before have I come across a man whom I could fancy being a Napoleon or a Nelson. I should not call him clever; yet I feel nothing is too great or too high for him. His ascendancy over everybody is quite curious . . . Altogether he is the strongest marked character I ever fell in with.'[27]

Junior officers would ask among themselves, 'Has much hot coffee been spilled this morning?' It was code for asking about FitzRoy's mood. His tours of inspection made the sailors mending sails or heaving on ropes fling themselves into the work with near-frenzy. The men did not hate FitzRoy, they admired his seamanship. When the outbursts of wrath got too much for Darwin, he could retreat to the company of the young officers. He had acclimatized himself to shipboard life, and, providing the sea was relatively calm, he worked and read with concentration.

After breakfast on 17 January 1832, Darwin accompanied by Captain FitzRoy, went to Quail Island – 'a miserable, desolate spot, less than a mile in circumference'.[28] They intended to set up an observatory here under canvas. Darwin spent the morning gazing into rock pools, and collecting geological specimens and shells. What chiefly struck him, however, about this outcrop of volcanic rock was the island itself, its geological essence:

> a stream of lava formerly flowed over the bed of the sea, formed of triturated recent shells and corals, which has baked hard into a hard white rock. Since then the whole island has been upheaved. But the line of white rock revealed to me a new and important fact, namely that there had afterwards been subsidence around the craters, which had since been in action and had poured forth lava. It then first dawned on me that I might perhaps write a book on the geology of the various countries visited, and this made me thrill with delight.[29]

These reflections passed through Darwin's mind as he sat, in intense heat, on a rock eating his luncheon of ripe tamarinds and biscuit. It was a moment of enormous significance for him. He had moved, already, a mere three weeks after leaving England, from being an amateur naturalist to being a scientist.

What had happened, in the intervals of seasickness and abject self-pity, in his cabin on board, was that Darwin had begun to read the first volume of Charles Lyell's *Principles of Geology*. 'This book was of the highest service to me in many ways. The very first place which I examined, namely St Jago in the Cape Verde Islands, showed me clearly the wonderful superiority of Lyell's manner of treating geology, compared with that of any other author, whose works I had with me or ever afterwards read.'[30]

Henslow had recommended Darwin to take the first volume of

Lyell on his journey, 'but on no account to accept the views therein advocated'.[31] Darwin's own copy of the book, in the University Library at Cambridge, is inscribed, 'From Capt FitzRoy'.

Lyell was thirty-three years old when his revolutionary work was published by John Murray (in July 1830). One of his primary aims had been 'to free the science [of geology] from Moses'.[32] It was while working as a barrister on the western circuit in 1827 that he had first read the speculations of Jean-Baptiste de Lamarck (1744–1829) and realized that if species were mutable, as Lamarck proposed, this would have profound implications for the science of geology. Lyell at this stage recoiled from the idea of mutable species: he believed that it demeaned the dignity of man. He believed (then) that the fossil record was too sketchy to provide proof of an evolving life. He believed, rather, that fossil evidence suggested that species became extinct, to be replaced by new ones, and this was a view he would maintain (with some modifications) until he had become convinced by the Darwinian theory of evolution thirty years later.

Lyell's progress in geology, anyway as far as the English-speaking world was concerned, was to explain the changes of the earth's surface by reference to 'causes now in operation'. His master – he learnt German in order to read him – was Karl von Hoff (1771–1837), who had compiled an immense volume of data culled from accounts of voyages and expeditions round the world. Lyell explained the volcanic eruptions, earthquakes, sedimentation and erosion which were the story of geology in terms of long-term climate-change. He had two broad viewpoints to confront and refute. One was that of previous geologists, who had clung to the notion that the history of geology was one of inactivity interrupted by 'catastrophes' – such as the biblical Flood. The other was the belief, all but universal, in the 'biblical' dating of the earth's age. When Lyell spoke of freeing the science from Moses, he would have been more accurate to say he was freeing it from Ussher (see p. 47). He feared that the first volume of *Principles* would 'irritate . . . If we didn't irritate, which I fear that we may . . . we shall carry all with us.'[33]

When Lyell, in 1831, was offered the chair of geology at the newly founded King's College, London – his appointment was queried by only one of the clergy on the Governing Body – the Revd Edward Copleston, who was afraid that Lyell's views on creation and the Flood were incompatible with orthodoxy. The rest of the Governing Body, who included the Archbishop of Canterbury and the Bishop of Llandaff, accepted Lyell as the innovative and brilliant scholar that he was. His resignation from King's was not because it was a religious foundation, but because he could not live – after marriage – on the paltry salary they gave him. While Darwin sat on the *Beagle* reading the first volume of *Principles*, Lyell was expounding his geological theories, not only to the young men at King's, but to mixed audiences at the Royal Institution.[34]

The nineteenth-century public were being introduced to one of the fundamental truths which would eventually enable them to understand Darwin's theory of evolution: that the world was older, much, much older, than had ever been conceived. That so many other aspects of Lyell's *Geology* were false – his concept of a self-balancing non-progressive earth in which species do not evolve – and that many of his ideas were shared by other geologists and were not, in themselves, original does not diminish the revolutionary impact that Lyell had upon his readers. And in Darwin he had found 'his first and, at that time, only scientific disciple'.[35]

By the time the *Beagle* was once again at sea, FitzRoy could note, in his official report to Beaufort,

> Mr Darwin has found abundant occupation already, both at sea and on shore; he has obtained numbers of curious though small inhabitants of the ocean, by means of a Net made of Bunting, which might be called a floating or surface trawl, as well as by searching the shores and the Land. In Geology he has met with far much more [*sic*] interesting employment in Porto Praya than he had at all anticipated. From the manner in which he pursues his occupation, his good sense, inquiring disposition, and regular habits, I am certain that you will have good reason to feel much satisfaction in the reflection that such a person is on board the *Beagle* . . .[36]

★

FitzRoy's favouritism towards his young protégé did not go unnoticed by the ship's surgeon, Robert McCormick. As we have observed, it was taken as read that when a naturalist was required on board a Royal Navy vessel, the role was traditionally filled by the surgeon. This was, indeed, McCormick's brief, and it was agreed that specimens collected by McCormick in the course of the voyage would be 'at the disposal of Government'.[37] It was with understandable envy and chagrin that McCormick saw Darwin's nets and trawls hanging over the sides of the ship. It was Darwin, the rich young man, who dined alone with the Captain on a daily basis, while McCormick ate in the gunroom with the young officers.

Like his rival, McCormick was a doctor's son, but not the son of a prosperous medic such as Robert Darwin. McCormick's father had been a naval surgeon. The boy grew up in Norfolk. No Shrewsbury School or Cambridge for him. He had been taught at home by his mother and sisters and had then studied at Guy's and St Thomas's hospitals in London before entering the Royal Navy as an assistant surgeon aged twenty-three. It had become his ambition to be an explorer-naturalist and he had volunteered for an Arctic voyage, north of Spitzbergen, in 1827. Sickness interrupted many of his voyages – yellow fever in the Spanish main in 1825 had been followed up by two occasions in the West Indies when he was invalided home, in 1827 and 1828. In spite of these setbacks he had become a keen geologist and natural historian.

As the *Beagle* made her way south towards the Equator, they approached St Paul's rocks, which Darwin 'considered as the top of a submarine mountain. – It is not above 40 feet above the sea, & about ½ a mile in circumference.'[38] Two boats were lowered so that the rocks could be examined in detail. The first, containing Wickham (First Lieutenant, soon to become Commander of the *Beagle*) and Darwin, landed on the rocky island. They were surrounded by birds, and were able to collect eggs and geological specimens. McCormick, in the other boat, did not so much as get the chance to land.

Two days later, they crossed the Equator, and the tradition was followed of the crew being allowed to torment their superiors, as the world turned literally topsy-turvy. 'For the first time in my life I saw the sun at noon to the North,' Darwin noted on 26 February.

'The disagreeable practice', wrote FitzRoy, 'has been permitted in most ships, because sanctioned by time; and although many condemn it as an absurd and dangerous piece of folly, it has also many advocates. Perhaps it is one of those amusements, of which the omission might be regretted. Its effects on the minds of those engaged in preparing for its mummeries, who enjoy it at the time, and talk of it long afterwards, cannot easily be judged of without being an eye-witness.'[39]

Darwin and thirty-one other 'griffins' were locked in their cabins, until four of 'Neptune's constables' came and led them up on deck, blindfolded. They were then soused by a bathtub of cold water. Darwin's face was next lathered with pitch and paint. Some of this was scraped off with an iron hoop. Then he was ducked in the bathtub again. 'At last, glad enough, I escaped – most of the others were treated much worse, dirty mixtures being put in their mouths & rubbed on their faces.'[40]

This ritualized reversal of the hierarchies, paralleling their entry into an up-ended hemisphere, was a raucous way of acknowledging that, as far as victims and perpetrators were concerned, the social structures were as immovable as the laws of nature.

Between the officer class and the ratings was a great gulf. Only political dreamers felt pain as they stared across it. The hidden valleys within the middle class of money-dominated England were more painful to contemplate, for in some cases they might be traversed with ease, and in others, where luck was absent, they were insurmountable. As well as being much richer than the Irish doctor, Darwin, having read Lyell's *Principles*, or having begun to read it, could remark, when they had geologized together in St Jago, that McCormick 'was a philosopher of rather antient [*sic*] date; at St Jago by his own account he made *general* remarks during the first fortnight & collected particular facts during the last'.[41]

The ship sailed quietly, in calm warm weather, until the coast of Brazil came into sight and they glimpsed Bahia, or San Salvador, 'embosomed in a luxuriant wood & situated on a steep bank'.

After a fortnight in this paradise, in which Darwin felt ashamed of his idleness ('a few insects & plants make up the total'),[42] the *Beagle* weighed anchor and slowly made her way along the coast to Rio de Janeiro. The weather was perfect and the sea calm. Just after midnight on April Fool's Day, Sulivan, Second Lieutenant, called out, 'Darwin, did you ever see a Grampus [killer whale]: Bear a hand then.' The naturalist rushed out from his hammock 'in a transport of Enthusiasm, & was received by a roar of laughter from the whole watch'.[43] Porpoises, sharks and turtles all were seen in the blue sea during the ensuing days, and on 5 April the ship moored at Rio. Darwin, to whom expense seemed little object, commissioned carpenters to encase the collections he had already catalogued. The sight was too much for McCormick who complained to FitzRoy that Darwin had been given preferential treatment.

McCormick's complaint was not to be tolerated. In later life, the surgeon expressed 'unavailing regret, as so much time, health, and energies utterly wasted'; he regretted having gone on the voyage in the first place.[44] He applied to be 'invalided' out of the expedition. Benjamin Bynoe, the assistant surgeon, served as surgeon for the rest of the voyage.[45] There is something chilling about Darwin's scorn for McCormick, whom he saw as having 'chose[n] to make himself disagreeable to the Captain & to Wickham'.[46] Darwin used McCormick as a postman to take a letter to his sister Caroline back to England in HMS *Tyne*; in the letter which the poor doctor carried, his rival said bluntly, 'He is no loss.' The two naturalists had locked antlers and in his passive, gentlemanly way Darwin had ruthlessly demonstrated the Survival of the Richest.

While the *Beagle* sailed back down to Bahia, Darwin stayed behind in Rio, taking a cottage in nearby Botafogo. They were a revelatory few weeks. On 6 April he 'frittered away' a day obtaining passports to explore the interior: 'It is never very pleasant to submit to the insolence of men in office; but to the Brazilians who are as contemptible in their minds as their persons are miserable it is nearly intolerable. But the prospect of wild forests tenanted by beautiful birds, Monkeys & Sloths, & Lakes by Cavies and Alligators, will make any naturalist lick the dust even from the foot of a Brazilian.'[47]

Darwin was not alone in the cottage. He had as a companion Augustus Earle, the thirty-nine-year-old official artist on the *Beagle*. Although in the early stages of his time ashore, Earle was bed-bound with rheumatism, he was a likeable and useful companion. He had been to Rio before and was able to show the sights to Darwin and to one of the midshipmen, Philip Gidley King (aged fifteen), whom Darwin found to be 'the most perfect, pleasant boy I ever met with'.[48]

Earle took them to dine at a *table d'hôte*, where they met several English officers serving under Brazilian colours: but 'it is calamitous how short & uncertain life is in these countries: to Earle's enquiries about the number of young men whom he left in health and prosperity, the most frequent answer is he is dead and gone – the deaths are generally to be attributed to drinking'.[49]

When the old conquistador Tomé de Souza landed on this spot in 1552, he wrote, 'Tudo e graça que se dele pode decir': 'Everything here is of a beauty which can hardly be described.'[50] They are words which are echoed in Darwin's diaries nearly three centuries later. In Darwin's time, the city had a population of some 229,600.[51]

Rio was a South American embodiment of Grandfather Wedgwood's mercantile values. It had become the capital of the Brazilian Empire because of the proximity of the gold- and diamond-mining areas of Minas Gerais. It was also a centre of trade and commerce. 'The whole day has been disagreeably frittered away in shopping' is one diary-entry,[52] while Darwin was impressed by 'the gay colours of the houses, ornamented by balconys, from the numerous Churches & Convents, & from the numbers hurrying along the streets' – all of which 'bespeaks the commercial capital of Southern America'.[53]

Most of the splendid buildings were but a few decades old. When Napoleon had invaded Portugal in 1807, the royal family relocated to Brazil, and in the next eight years many of the most magnificent streets and buildings in Rio were constructed around the royal court.[54] As well as buildings, the arrival of the Portuguese government in exile brought institutions – the Treasury, the High Court, the Court of Appeal. The 'royal hand' had 'regenerated America'.[55] A new political order had come into being, and after 1821, when the King returned to Lisbon, Brazil became an independent kingdom – as it remained at the time of Darwin's residence in Botafogo, then a district much favoured by English merchants and bankers.[56] There was a great appropriateness about Darwin, so much of whose money derived from trade, settling in this spot. 'With the claws of a lion', it has been aptly said, British trade, soon after the establishment of the independent kingdom, 'controlled almost the entire Brazilian market'.[57] Britain took three-quarters of the cotton exported by Pernambuco, half of its sugar, cacao, coffee, rubber, timber, as well as gold and diamonds from the mines. All this helped boost the British economy: it was the ore from which Conrad's *Nostromo* would be fashioned.

For Darwin, however, the primary excitement of Rio and its environs was to be found in natural history. In the bay itself, 'on the sandy plain, which skirts the sea at the back of the Sugar Loaf',[58] he collected 'numerous animals'. The sheer abundance of vegetation was a source of boundless rapture. 'The number of oranges which the trees in the orchards here bear is quite astonishing. I saw one today where I am sure there were lying on the ground sufficient to load several carts, besides which the boughs were almost cracking with the burthen of the remaining fruit.'[59]

If the coastline provided him with an abundance to admire, how much more was this the case as he made his way into the interior. 'If we rank scenery according to the astonishment it produces, this most assuredly occupies the highest place.'[60]

Through May and June he divided his time between encounters with English exiles – 'called on a Mr Roberts, one of the endless nondescript characters of which the Brazils are full – broken down agents to speculation companies; officers who have served under more flags than one: &c &c'[61] – and natural history.

In the month of May, concentrating upon land animals, he collected 'an host of undescribed species' for dispatch to England. Together with its cotton, cacao and diamonds, Brazil was to give up its beetles, its spiders, its very rocks, for 'geology carries the day'.[62] On the coast he gathered 'several specimens of an Octopus, which possessed a most marvellous power of changing colours, equalling any chamaelion & evidently accommodating the changes to the colour of the ground which it passed over'.[63]

The tropical forest 'in all its sublime grandeur' yielded a 'host of parasitical plants'.[64] Darwin could scarcely contain his excitement as he wrote back to Henslow and Fox, cataloguing his discoveries.[65]

He saw it all with fresh eyes. It was indeed epiphanic. At the same time, he saw it through the eyes, and through the pages, of his role model Alexander von Humboldt, the European traveller in South America who had gone on, in *Kosmos*, his *magnum opus*, to write a book which tried to explain, or describe, everything that is. 'I formerly admired Humboldt,' Darwin told Henslow, 'I now almost adore him; he alone gives any notion, of the feelings which are raised in the mind on first entering the Tropics.'[66] In an earlier part of that letter, Darwin revealed to Henslow that he actually had passages of Humboldt by heart: he speaks of himself 'at Santa Cruz, whilst looking amongst the clouds for the Peak & repeating to myself Humboldt's sublime descriptions'.[67]

It is striking that, with the spoilt languor of the rich, Darwin should have written to Fox in May 1832, 'I suppose I shall remain through the whole voyage, but it is a sorrowful long fraction of one's life.'[68] Only a few months later, he had changed. There had always been a contrast in Darwin's nature between the amiable loafer, who preferred shooting and country-house life to any commitment to work, and the insatiably curious naturalist. During the first few months in Brazil, Darwin quite consciously walked in the footsteps of Humboldt.

'We have been 3 months here,' he wrote to his sister Catherine,

> & most undoubtedly I well know the glories of a Brazilian forest. Commonly I ride some few miles, quit my horse & start by some track into the impenetrable – mass of vegetation. Whilst seated on a tree, & eating my luncheon in the sublime solitude of the forest, the pleasure I experience is unspeakable. The number of undescribed animals I have taken is very great – & some to Naturalists, I am sure, very interesting. I attempt class after class of animals, so that before long I shall have a notion of all – so that if I gain no other end I shall never want an object of employment & amusement for the rest of my life.[69]

He was happier systematizing Brazilian fauna in the forests than contemplating the Brazilians, who were from his perspective 'ignorant, cowardly, & indolent in the extreme'. As for the monks 'it requires little physiognomy to see plainly stamped preserving cunning, sensuality and pride'.[70]

Early in his stay at Rio, trying to make himself understood while crossing in a ferry and talking to 'a negro who was uncommonly stupid', Darwin had talked in a loud voice and made manual gestures which passed near the man's face. The Brazilian had flinched, making it clear that he expected his face to be struck. 'This man had been trained to a degradation lower than the slavery of the most helpless animal.'[71] Earle told Darwin that it was usual for Brazilian gentlewomen to keep thumbscrews to punish their slaves and that he had 'seen the stump of the joint, which was wrenched off in the thumbscrew'.[72] He felt bound in honesty to record that 'the slaves are happier than what they themselves expected to be or than people in England think they are', though his diary does not make it clear how he arrived at this conclusion. The grandson of two conspicuous abolitionists, Erasmus and Josiah, could not fail to 'hope the day will come when they will assert their own rights & forget to avenge their wrongs'.[73]

The Brazilian slave trade was not abolished until 1850, partly as the result of an outbreak of yellow fever, brought from Africa, which killed some 16,000 Brazilians; partly because of Lord Palmerston's policy, when British Foreign Secretary, of sending gunboats to the Brazilian ports where the vile trade was conducted. 'These half-civilised governments', he opined, 'all require a dressing down every eight or ten years to keep them in order.'[74]

Yet while the cruelty of slavery shocked Darwin, there is no evidence that he believed, either as a young man or as a mature one, in the equality of the human race, whether as a political ideal to be hoped for or as a scientific fact. His cool appraisal, in *The Descent of Man*, of the inferiority of 'savages' to 'civilised people' shows that if he were forced to answer the enchained man's question on old Wedgwood's medallion – 'Am I not a Man and a Brother?' – the answer could have been negative. He believed, what has now been roundly refuted by medical science, that the size of people's skulls and their intellectual faculties were correlative.[75] He was to express the view that the nineteenth century was one in which 'civilised nations are everywhere supplanting barbarous nations',[76] and attributed the 'remarkable success of the English as colonists over other European nations' to their 'energy', implying at least that the English were more 'civilised' than, say, the Italians or the Germans. They were clearly in every way superior to the 'immense mongrel population of Negroes and Portuguese' to be encountered in Brazil.[77] When he had developed the central idea of *The Origin of Species*, it was essential for him to believe that 'the American aborigines, Negroes and Europeans' stemmed from 'a common progenitor'. This did not commit him, however, to the view that they had progressed at the same pace, and the doctrine of European superiority to other peoples of the planet underlay all his later work on the descent of the human animal. Even if this had not been the case, Darwin was in a political sense cautious, shy of anarchy. Though a supporter of Whigs versus Tories in domestic politics, he was no radical and would probably have shared Goethe's view: 'ich will lieber eine Ungerechtigkeit begehen, als Unordnung ertragen' – 'I would rather perpetrate an injustice than put up with anarchy.'[78]

So, when he rejoined the *Beagle* and they cruised south from Rio to Montevideo in July, his diary betrayed no sense of angst at Captain FitzRoy's gung-ho approach, either to the volatile politics of the Latin colonists or to the plight of their conscripted slave soldiers and sailors. The aim of the voyage, after all, from FitzRoy's perspective, was primarily cartographical, and from Darwin's it was natural history. Why should they waste their time attempting to understand these obviously laughable foreigners? 'Everybody is full of expectation & interest about the undescribed coast of Patagonia,' Darwin wrote on 13 July.[79] There was a good mood aboard, though Darwin suffered from the falling temperatures. 'Everything shows we are steering for barbarous regions, all the officers have stowed away their razors, & intend allowing their beards to grow in a truly patriarchal fashion.'[80] As the ship got up speed, eight or nine knots, Darwin delighted in the sight of hundreds of porpoises crossing their bows, and of flying fish leaping from the water.

When they reached harbour at Montevideo, however, they learnt ('to our utter astonishment and amusement') that the 'present Government is a military usurpation': 'The revolutions in these countries are quite laughable.'[81] The Captain of an English frigate, the *Druid*, told them his men had not been allowed to go on shore for several weeks. When they went to Buenos Aires they were forbidden to land because of a cholera scare. And when they tacked back to Montevideo, they found themselves being asked by 'the Minister for the present military government' for assistance against an insurrection of 'some black troops'.[82]

Old Samuel Johnson, on a visit to Oxford, proposed as a toast 'Here's to the next insurrection of the negroes in the West Indies!'[83]

His reason was simple: that 'the laws of Jamaica allow a Negro no redress'.[84] It is a measure of how the British capacity for sympathy had been coarsened by colonialism, in the fifty years since Johnson died, that no one aboard the *Beagle*, least of all Darwin, appeared to side with the insurgents in Montevideo. Captain FitzRoy sent fifty-two men ashore 'heavily armed with Muskets, Cutlasses & Pistols', but withdrew them a day later. His evident reason was cowardice, but the reason he gave was that if he gained possession of the citadel, which the mutineers had occupied, it would impugn his 'character of neutrality'. Had he wished to preserve this 'character', FitzRoy would never have sent his men ashore in the first instance. 'There certainly is a great deal of pleasure in the excitement of this sort of work,' commented Darwin, who, with some of the young officers, had 'amused ourselves by cooking Beefsteaks in the courtyard' of the fortress of St Lucia while the working-class sailors and the besieged rebel slaves threatened one another's insignificant lives. On 9 August, Captain FitzRoy thought to ask what exactly was going on to cause the disturbance. He found out that there 'are actually 5 contending parties for supremacy'. Darwin inevitably added, 'It makes one ask oneself whether Despotism is not better than such uncontrolled anarchy.'[85]

They sped southwards, sometimes through heavy rain. Sometimes seasickness overwhelmed Darwin ('two of the days I was on my "beam ends"').[86] Sometimes, being at sea, enclosed in his small cabin, enabled him to catch up on correspondence, and to catalogue what was already becoming a prodigious collection of specimens. Off Baia Bianca they encountered a sealer, a small schooner hunting seal. Landing, they visited the settlement of 'wild Gaucho cavalry', wearing brightly coloured shawls around the waist, fringed drawers and ponchos. The 'old Spaniards' of 'pure blood' impressed Darwin very much, with their sabres, and powerful horses and warlike talk. The Indians, by contrast, 'whilst gnawing bones of beef, looked, as they are, half recalled wild beasts'.[87] The Commandante had the previous week led an expedition to seize horses and cattle from the Indians. His son had been captured and tortured with nails and small knives by Indian children. 'The Indians torture all their prisoners & the

Spaniards shoot theirs.'[88] As well as the excitement of this encounter, Darwin was pleased to be given an ostrich egg by the Spaniards, and subsequently to discover a nest containing twenty-four more. FitzRoy and Harris, a British trader they had met in Patagonia (he owned two sealers), were engaged in surveying the coast while Darwin went ashore and saw deer, ostrich and a warren of Agouti – 'or hare of the Pampas'.[89]

The Spaniards offered Darwin some hunting, and he was able to accompany nine men and one woman ('the woman was a perfect nondescript: she dressed & rode like a man')[90] pursuing an ostrich with two or three balls fastened to leather thongs; these were whirled above the gaucho's head before being flung at the bird and instantly lashing its legs together. Having chased an ostrich and some cavies, the gauchos then plundered an ostrich nest and caught a number of armadillos, which they roasted, in their hard cases, over a camp fire. Some days later they caught a large puma which they sold for its skin.[91] In addition to this sport, Darwin found time for geologizing and at Punta Alta chipped out a fossilized jawbone and a tooth belonging to 'the great ante-diluvial animal the Megatherium': a giant sloth. In common with other contemporary naturalists Darwin wrongly believed this creature to be like an armadillo with an osseous coat – he had confused it with the *Glyptodon*. Punta Alta, both on this visit and on their return a year later, was one of the richest sources of palaeontological discovery for Darwin: a low bank on the shore about twenty feet in height, made up of shingle and gravel with a stratum of muddy red clay running through it.

South American palaeontology was in its infancy. Humboldt had, in the course of his travels, found a few mastodon teeth. And in the 1780s a *Megatherium* had been discovered in Argentina and sent back to Madrid.[92] But at Punta Alta, Darwin was in effect a pioneer. Tapping with their pick-axes, he and his personal servant, Syms Covington, were revealing a plenitude of extinct species. FitzRoy wrote, 'My friend's attention was soon attracted to some low cliffs near Punta Alta, where he found some of those huge fossil bones, described in his work; and notwithstanding our smiles at the cargoes of apparent rubbish which he frequently brought on board, he and his servant used their pick-axe in earnest and brought away what have since proved to be most interesting and valuable remains of extinct animals.'[93]

Darwin was gazing at creatures, most of them, which were unknown to modern zoology: a giant hippopotamus, a creature like a llama, big as a camel, packed in a matrix of seashells. Almost the most arresting discovery were the bones of what was evidently a horse. The conquistadors in the sixteenth century had found a horseless South America. Yet here, embedded in the shells of Punta Alta, was clear evidence that an animal very like a modern European horse had once inhabited Patagonia. Moreover, the resemblance which these fossils bore to their modern equivalents provoked inevitable questions. These animals were similar in all but size to their contemporary counterparts. The giant hippo had become extinct, to be replaced at a later juncture by a more moderate-sized hippo. It was inevitable, when he had brought all these specimens home, that he should ask: what if the ur-horse to be seen in fossilized form in Argentina was not a species separate from the modern horse, but . . . its ancestor?

When Darwin had attended Robert Jameson's lectures in Edinburgh on the origin of species, he had heard of the vital importance of studying fossils. Jameson had emphasized the importance of studying the higher plants and mammals – where fossil evidence was available – because here you could actually observe the process of evolution at work. You could see new faunas and floras arising. In some cases, they were replacing old taxa. In other cases, old forms were morphing into new ones. This was to be the central puzzle of Darwin's intellectual life. Georges Cuvier and the French natural historians, about whom Darwin learnt from Jameson's lectures, had provided, in their museums of palaeontological specimens, abundant evidence of species becoming extinct. They also believed that, somewhere, if the right interpretation could be found of all this evidence, the fossils would provide, not just the history of *what* happened, but the explanation of *how* it happened.

Although Darwin was not destined to be a geologist, his geological investigations, while a young man on HMS *Beagle*, were absolutely central to the later pursuit of the origin of species. Only relatively recently, in the last decade or so, have Darwin scholars realized the importance of his geological notes. The geological journals were for a long time unread, lying with the other abundant manuscript material in the Cambridge University Library. Eventually, in this century, they were put online. Paul Barrett's monumental edition of Darwin's works included a volume on Darwin's geological diaries, published in 2016. And scholars such as Sandra Herbert (*Charles Darwin: Geologist*, 2005), George Chancellor, who edited and introduced the Darwin Online geological journals, M. D. Brasier (*Darwin's Lost World: The Hidden History of Animal Life*, 2009) and Niles Eldredge (*Eternal Ephemera*, 2015), elucidated their significance for us. The geological journals which he kept on the voyage show us the burgeoning of Darwin's mind. The notes he kept on the fossils of Bahia Blanca are especially detailed. He found numerous bones of quadrupeds, five or six different species, which he identified as *Megalonyx* – a genus of North American ground sloth, originally named by President Thomas Jefferson – as well as extinct rodents which had not been previously identified. He also later identified a species of as-yet-unknown endemic South American mammals, the Notoungulata, later classified by Owen, in the Natural History Museum in London, as *Toxodon*. We see, in every such fossil-gathering expedition Darwin made in South America, how closely the geological work he had undertaken since attending lectures in Edinburgh, which was stimulated by reading Lyell, would feed into his biological theories – how and when the species became extinct; whether (though this thought had not yet occurred to him) the species might not be as distinct as Linnaeus had supposed; that they were fluid, morphing and blending into one another.[94]

And on they sailed, to Tierra del Fuego.

6

'Blackbirds . . . gross-beaks . . . wren'

As we have seen, on a previous voyage to South America, a year or so earlier, FitzRoy had adopted four Fuegians, one of whom had died of a smallpox vaccination. The rest, a young man whom FitzRoy named York Minster, a boy who had been purchased for a few buttons and so was called Jemmy Button, and a girl named Fuegia Basket, were put in the care of the Church Missionary Society. The aim was, when they had been taught the rudiments of how to use common tools and how to speak English, that they should also have accepted the Christian Gospel. They were now passengers on the *Beagle* – York Minster aged twenty-seven, Jemmy Button fifteen and Fuegia Basket ten. York Minster and Jemmy in their nankeen trousers and frock coats, Fuegia in her bonnet, were like parody-people, something waiting (as we might think, for it was four years before *Sketches by Boz*) to come to life in the mind of Dickens. They were accompanied by an earnest trainee missionary called Richard Matthews. They were now returning to Tierra del Fuego as to the Vineyard of the Lord, and as to a home in which, perhaps, York Minster, supplied by the London Missionary Society with a beaver hat, wine glasses, cutlery, soup tureens and white tablecloths,[1] would probably never again feel quite at home.

It had been in 1520 when Magellan was making his way in the *Trinidada* through the Straits which he had discovered that the Portuguese navigator had seen beacons, lit by the natives in the southern hills, and named it the Land of Fire. Tierra del Fuego, the territory around Cape Horn, has about the harshest climate in the world. Icy squalls hurl albatrosses and blue petrels through the relentless air. The *Beagle* rolled and turned like a cork. The sky was gloomy. The coast was made up of fifty miles of low, horizontal rock strata. In the distance could been seen the glowering, snowy mountains. On 17 December

1832 they anchored in the very place – the co-called Harbour of Good Success – where Captain Cook and Joseph Banks had landed in January 1769.[2] On 18 December, FitzRoy sent out a party to communicate with the natives.[3]

A group of them had assembled on the shore, and when FitzRoy, with an advanced party, had come to land in the jolly-boat, the Fuegian men had sent the women and children into the safety of their huts while they greeted the Beagles with shouts and gesticulations.

In the choppy sea, the rowing-boats borne up and down on waves the height of a London house, the men of the *Beagle*, sometimes with laughter, had rowed ashore the supplies deemed necessary, by the London Missionary Society, for the implanting of civilization in this inhospitable terrain. There were especial whoops of amusement as they unpacked the chamber-pots.[4] For now, the men who were to be persuaded to make use of these commodities stood before the Europeans: huge individuals, with long matted hair and mahogany-coloured faces daubed with stripes of black and red, with white stripes around their eyes. Darwin, when he first set eyes upon them, thought of the devils 'which come on the stage in plays like *Der Freischütz*'.[5] He told his diary, 'I would not have believed how entire the difference between savage and civilized man is. It is greater than between a wild and domesticated animal.'[6]

Although Darwin reassured himself (or tried to do so) by noting that their diet chiefly consisted of limpets, mussels and the occasional seabird, Jemmy Button had told him that in the winter the Fuegians ate their womenfolk. A sealing Captain with whom Darwin later discussed the matter said he had heard the same, and had asked his informant 'why not eat dogs?' The reply had possessed an inexorable logic: 'Dog eat otter; woman good for nothing: man very hungry.'[7] 'Viewing such men', Darwin confided in his diary,

> one can hardly make oneself believe that they are fellow creatures placed in the same world . . . What a scale of improvement is comprehended between the faculties of a Fuegian savage & a Sir Isaac Newton – Whence have these people come? Have they remained in the same state since the creation of the world? What could have tempted a tribe of men leaving the fine regions of the North to

> travel down the Cordilleras the backbone of America, to invent & build canoes, & then to enter upon one of the most inhospitable countries in the world?[8]

The allegation that the Fuegians were sometimes cannibals, repeated by FitzRoy in his *Narrative* of the voyage, and Darwin in his *Journal of Researches*, was later seen to be based on misleading interviews, and that Jemmy Button was an unreliable witness.

They spent a couple of months, bobbing around the coast, battling against the westerly winds, soaking themselves in rain – 'I have scarcely for an hour been quite free from seasickness' was one January diary entry. When it was possible to land, Darwin did so, climbing the mountains, enraptured by superb views, collecting tiny alpine flowers and noting birds – the steamer, a large variety of goose, being especially remarkable. But the weather was inconducive to study, and on a disastrous 14 January seawater splashed and destroyed all his drying-paper and plant collection.

Every so often, contact would be renewed with one Fuegian group or another. They managed to build some primitive houses and to set up quarters near Jemmy's people. They were swift. They ran so fast their noses bled, spoke so fast their mouths frothed. Jemmy seemed to forget his English as soon as he was reunited with his mother, his uncle and his brother. It seemed impossible to prevent the Fuegians from stealing. The Beagles felt sad to be leaving 'their' Fuegians behind when eventually, in late February, they weighed anchor; at the beginning of March, they reached the Falkland Islands.

They found these islands, on which there was such an abundance of wild oxen (5,000), horses, pigs, wild fowl and rabbits, to be inhabited only by one Englishman, Mr Dixon, twenty Spaniards and 'three women, two of whom are negresses'. These were guarding the British flag.[9] Darwin, reflecting upon Mr Dixon's life, mused, 'It is surprising to see how Englishmen find their way to every corner of the globe. I do not suppose there is an inhabited & civilized place where they are not to be found.'[10]

One disaster which occurred here was the death of Captain FitzRoy's clerk, Edward Hellyer. He had shot a seabird, swum out to retrieve it and become entangled in kelp. A melancholy funeral followed, the coffin covered by the British ensign, and a Union Jack borne aloft at half-mast. Darwin spent several days walking, shooting snipe and collecting 'the few living productions which this island has to boast of'.[11]

Then it was back to the ship, which sailed to the coast of Patagonia.

For much of the year 1833, Darwin was on land, while the *Beagle* surveyed the coastline. Several weeks, from April to May, were spent in Maldonado, packing up bones, rocks, plants, skins of birds, fish, fungi, to be taken back to England in spirits of wine. He had already, in a little over a year, assembled a formidable collection – he had catalogued 1,529 specimens.

From April to July Darwin lived at Maldonado. Already – in June – some of the fossil bones which he had dispatched had been exhibited at Cambridge. Henslow wrote in August, from Cambridge, that although a few items in the cask of spirits had been spoiled, 'owing to the spirit having escaped thro' the bung-hole', the majority of specimens were safe.[12] Senior geologists in Britain were already taking serious notice of Darwin's work. He had now passed from being the keen amateur naturalist to being a respected scientist. He had joined the club. In December, he was inscribed as a founding member of the Entomological Society by F. W. Hope, who wished him 'all success in collecting & Health to enjoy yourself & a safe return to Old England with 1,0000,000,0000 Insects'.[13]

Meanwhile, between August and September Darwin had made an overland trip from Rio Negro to Bahia Blanca and Buenos Aires. He had made an expedition to Santa Fé, returning from there to Montevideo. Darwin became acquainted with the rhea, the South American ostrich, smaller than the African one. Later on the Santa Cruz river, Darwin twice saw the ostriches swimming. He caught one example of a particular species, very rare, and sent it to the Zoological Society; later it was named after him: *Rhea darwinii*.[14] He came to know the appalling smell of the zorillo or skunk. Very little missed his notice – birds, armadillos, hunting dogs, sheepdogs, he noted them all. Darwin was one of the greatest cataloguers of the world, perhaps *the* greatest, a taxonomist in the Linnaean league. He also developed a great fondness for the gauchos, in their scarlet ponchos and white riding-drawers. Superb horsemen, they forced their animals into great rivers: the naked rider, once the horse was out of its depth, would slip off its back and take hold of its tail and drive it forward.

Darwin stayed at the encampment of General Rosas, a cattle rancher who ruled Argentina as Governor of Buenos Aires in the years 1829–32 and 1835–52. When Darwin encountered him he was leading a campaign 'to exterminate the Indians'.[15] Rosas was 'a perfect Gaucho'. His popularity was attested by an anecdote in the diary. A man was arrested for murder, but when he explained what had happened – 'the man spoke disrespectfully of General Rosas & I killed him'[16] – he was instantly released. 'I never saw anything like the enthusiasm for Rosas & for the success of this most just of all wars against Barbarians.'[17] The enthusiasm for Rosas was 'universal'.[18] The Argentinians accepted without question that the Indians (who resented the Argentinians overrunning their hunting ground) must be eliminated. Rosas eventually also overturned the Argentinian government and ruled as a dictator for seventeen years (when he was ultimately deposed he went to live in England, in Hampshire, where Darwin met him again). 'It is clear to me', the naturalist wrote with some approval in his diary, 'that Rosas ultimately must be absolute Dictator.'[19] It would be evident to most political thinkers and historians that Rosas's activities – a campaign of violence in which he eliminated a weaker group of individuals for the sake of their hunting-prey, followed by a strutting, macho seizure of political power – were not a recipe either for progress or for stability. Darwin, however, was not a political philosopher, and there can be little doubt that the career of Rosas, observed through the naturalist's lens, became for him a template of human behaviour. While Britain advanced in political power – and far greater influence than nineteenth-century Argentina – by its lack of 'Darwinian' savagery in the public sphere, its gradual extension of the franchise, its increasing attempts to look after the 'weaker' elements in society in schools, hospitals and the like, the young Darwin was able to see the South American dictator as an example of 'natural' aggression, of the inevitability that 'struggle' and violent conflict were necessary ingredients in the development of human societies.

At the beginning of the next year – 1834 – they once again negotiated the Straits of Magellan, and in February they made a detailed survey of the east coast of Tierra del Fuego. During this surveying work, they landed only once, at what had been called St Sebastian's Channel – closer investigation showed it to be a large wild bay. When James Cook and Joseph Banks had tried to draw an accurate map of 'Terra del Fuego' as Cook called it, in January 1769, the weather had made the task impossible: 'Foggy weather, and the westerly winds which carried us from the land prevented me from satisfying my curiosity.'[20] In 1834, the world – every inch of which, as moderns, we reckon on being able to find in the atlas – was still an incomplete puzzle, an unfinished chart.

In the bay, they watched a multitude of spermaceti whales, some of them leaping from the water. Darwin's delight in, fascination with, the abundance and variety of species he encountered and noted never wavered. Constantly, he reread *Paradise Lost*; his pocket edition made it the ideal companion not only on board ship, but on his many land expeditions. He responded with particular sympathy to the beautiful passage in Book VII, in which the Archangel Raphael told Adam the story of creation:

> Forthwith the Sounds and Seas, each Creek and Bay
> With Frie innumerable swarme, and Shoales
> Of Fish that with their Finns and shining Scales
> Glide under the green Wave . . .[21]

and Darwin, as he looked at the spermaceti whales disporting themselves in Tierra del Fuego, had some of the excitement experienced on the day such creatures were first created:

> there Leviathan
> Hugest of living Creatures, on the Deep
> Stretcht like a Promontorie sleeps or swimmes,
> And seems a moving Land, and at his Gilles
> Draws in, and at his Trunck spouts out a Sea.[22]

Milton, of course, was only writing a poem, as were the authors of Genesis who compiled the creation myths in the Bible. It is probably safe to say that of the two texts, Genesis and *Paradise Lost*, Milton's was the more influential upon the British nineteenth-century mind. While the Bible, with its broad creation myths, by no means rules out an evolutionary idea of nature, Milton, for poetic effect, specifically envisaged the arrival on the earth, ready made, of all the species in their immutable form. When God called for earth to bring forth the animal kingdom,

> The Earth obeyd', and strait
> Op'ning her fertil Woomb, teem'd at a Birth
> Innumerous living Creatures, perfect formes
> Limb'd and full grown . . .[23]

Many of the Victorians who agonized about the specific question of the mutability of species, and about the more general idea of creation, did so as men and women, such as Charles Darwin, who had had these passages of Milton instilled in their minds since childhood. Wordsworth, the Poet Laureate until 1850, had apostrophized the blind poet, 'Milton! thou shouldst be living at this hour . . .' In a sense, he was – Lord Macaulay had a better memory than most, but he would probably not have been alone, among the educated classes of the nineteenth century, when he said that if 'by some miracle of Vandalism all copies of *Paradise Lost* . . . were destroyed off the face of the earth, he would undertake to reproduce it from recollection'.[24]

If the poetry of Milton fed the nineteenth century's belief-systems about science (though Milton himself in his own day had been a progressivist who admired Galileo's new astronomy), another of the Victorian certainties, which seems so strange to later generations, is the belief in colonialism, and the confidence that they could impose their own brand of civilization upon others. Captain FitzRoy, supported by the London Missionary Society, had generously sent Bibles, a white missionary, a collection of white table-linen, ceramic dinner services and chamber-pots to Tierra del Fuego, but upon the *Beagle*'s return, in spring 1834, they were dismayed to find that the good seed had fallen on stony ground and failed to take root.

When they came ashore on 25 February, they pulled alongside six Fuegians, 'stunted in their growth, their hideous faces bedaubed with white paint and quite naked'. Another canoe-load, encountered on 1 March, struck Darwin as 'more amusing than any Monkeys'. The Beagles hoped for better things when they moved round the coast, anchoring in Ponsonby Sound, to find themselves in 'Jemmy

Button's country'.[25] Alas, when they came across him – 'without a remnant of clothes, excepting a bit of blanket round his waist . . . it was quite painful to behold him'.[26] Far from wearing his well-tailored English clothes, and persuading his fellow Fuegians to read the Authorized Version of the Bible, and to eat their food off dishes and plates made in Stoke-on-Trent, Jemmy had reverted to his Fuegian ways. He had taught some of his friends a little English, including a woman who was identified as his wife. His own grasp of the language, however, seemed to have deserted him. As for York Minster, he had persuaded Jemmy, and his mother, to come to his own country several months previously and had robbed them of everything.

> Every soul on board was as sorry to shake hands with poor Jemmy for the last time, as we were glad to have seen him. I hope and have little doubt he will be as happy as if he had never left his country; which is much more than I formerly thought. He lighted a farewell signal fire as the ship stood out of Ponsonby Sound, on her course to East Falkland Island.[27]

Jemmy perhaps did not retain quite such happy memories of the British as they did of him. When, in November 1859, the Patagonian Missionary Society sent a ship, the *Allen Gardiner*, to establish a mission at Wulaia, their reception was unfriendly. Jemmy and his brother, assisted by fellow Fuegians, massacred the missionaries and the entire crew of the *Allen Gardiner*, excepting the cook.

Fuegia Basket married York Minster, in 1833. He was subsequently killed in retaliation for a murder, and she married a much younger man. E. L. Bridges, author of *Uttermost Part of the Earth*, met her as late as 1883, aged sixty-two and near her death, at London Island, to the extreme west of Tierra del Fuego.

The violence of the Fuegians might have provided a good example of the 'struggle for existence' which Darwinists would come to believe was the underlying secret of how sustainable life moves on from one generation to the next. If so, it would question rather than confirm the belief that struggle was conducive to evolutionary progress. By the end of the nineteenth century, the three Fuegian tribes were almost extinct. The Alacalufes, the canoe people of the western channels, numbered 10,000 at the time of Darwin's visit. By 1960 there were hardly a hundred.[28]

As soon as Button appeared to them semi-naked, Darwin saw the folly of the attempt to westernize the Fuegian people. For Captain FitzRoy, the failure of the Fuegian experiment was a personal tragedy. As they sailed away from Jemmy, who stood beside his fire on the shore, the Captain could hardly restrain his feelings, and was on the verge of tears.

After they had left Tierra del Fuego, FitzRoy hardened. For the Captain, it had been a difficult year. While his young naturalist-companion Darwin had been free to go ashore, to idle away a week or two if the mood took him in Buenos Aires or to make exciting expeditions into the rainforests, FitzRoy had responsibility for everyone on board the *Beagle* and, moreover, he had been entrusted by the Admiralty with a task which had, in part at least, defeated one of the greatest cartographers in history: namely to make accurate maps of the entire coast of South America.

Cook had sailed, on each of his great voyages, with another ship to chaperone him. A year into the voyage of the *Beagle*, FitzRoy had come to see that this had been not a luxury but a necessity. Not only had the weather been consistently against them, but there had been much ill health on board, and life would clearly be safer for everyone were a second vessel to be acquired, which could be surveying other stretches of coast, or allowing half the crew to rest up. At Maldonado in the summer of 1833 FitzRoy had chartered two small boats and purchased an American 170-ton sealing vessel without getting clearance for the purchase from the Admiralty. He spent £1,300 buying the vessel and a considerable sum in reconditioning her. To make clear her function and his purpose he renamed her the *Adventure*, the name of the consort-ship (commanded by Captain Furneaux) in Cook's epic second voyage in which he had made three valiant efforts to discover the Southern Continent, in which he had charted nearly all the islands of the South Pacific and in which he had discovered South Georgia, Norfolk Island and others. The Admiralty refused to reimburse FitzRoy for his purpose, but he knew his business: 'I had often anxiously longed for a consort, adapted for carrying cargoes, rigged up so as to be easily worked with few hands, and able to keep company with the

Beagle.'[29] Since the *Adventure* had come into service, with Lieutenant John Wickham as its highly competent commander, life had been busy. FitzRoy had run the *Beagle* constantly back and forth between Montevideo and Patagonia, usually in terrible weather, completing his cartographical work as assiduously as any man could. He worked long hours. On many days he dined alone in his cabin, bent over his charts.

When the time approached for them to round Cape Horn and head up the western coast of Chile, they dropped anchor in the mouth of the river of Santa Cruz. The *Beagle* was found to have lost some of her false (protective) keel. She had to be beached. All but her mainmast had to be taken down. Guns, anchors and heavy gear had to be brought ashore while the carpenters got to work. It was a minor repair, but it needed doing, and while the *Beagle* lay ashore like a beached whale, the Captain took three whaleboats to explore up the Santa Cruz river. They moved at about ten miles per day. While the crews slept, two men and an officer kept watch, maintaining the campfires and keeping a look-out for Indians.

It was a strenuous trip. After two days, oars and sails were abandoned, with every man, FitzRoy included, taking turns to tow the heavy boats upstream. It was also bitterly cold and Darwin at first had to record that 'the country remains the same, & terribly uninteresting'.[30] But there was wildlife to observe – ostriches, huge herds of guanaco (one with as many as 500 of these camelids), pumas, lions and, hovering over their heads, the huge-winged condor. Little by little, Chile began to work its magic on Darwin. His attentive eye made out fields of lava which must have flowed down from the Andes. Then there was sport – he shot a condor, whose wingspan was over eight feet. At the end of a long hike, 'we hailed with joy the snowy summits of the *Cordilleras* as they were seen occasionally peeping through their dusky envelope of clouds'.[31] Within a week they reckoned they were 140 miles from the Atlantic and 60 from the nearest inlet of the Pacific.

Throughout the next year, Darwin frequently complained, in his journal, of tiredness, and there were occasional illnesses: this was to be expected. Given the man he would become, relatively soon after his return to England, however, we cannot but note his strength and vigour. As homesickness grew sweet, rather than acute, and as the letters from his sisters grew ever more parochial and inane, the space and the scope of South America were especially intoxicating. To Fox, back in November 1832, he had written from Rio Plata, 'Poor dear old England. I hope my wanderings will not unfit me for a quiet life, & that in some future day I may be fortunate enough to be qualified to become like you a country clergyman.'[32] It became a less inviting prospect with every month which passed. The letters from his sisters and female cousins read like the outpourings of a Jane Austen heroine. It is a comparison made by Charlotte Wedgwood, when a Maer neighbour entreated the girls to entertain a party of officers. No one would do so but Emma Wedgwood, who when she could not persuade her sister Fanny 'to go & keep her in countenance had great scruples lest she should appear too Lydiaish':[33] that is, like Lydia Bennet in *Pride and Prejudice* whose flirtations with soldiers had almost disastrous consequences. Fanny Wedgwood was one of the partners whom his unstoppably matchmaking sisters had in mind for Darwin. When news reached him that she had died aged twenty-six, after a few days' illness from some inflammatory attack, he does not appear to have responded in any way. Fox, writing from the Isle of Wight, mentioned that he had heard from the Darwin family: 'Poor Fanny Wedgwood's death had just been a great shock to them, as it will be to you I am sure,'[34] again, as far as we can see, eliciting no response at all from Darwin.

It would be preposterous to suggest that Darwin was other than devoted to his family. He lived his whole life in their bosom and when he chose to marry, inevitably, he married a cousin – Fanny's sister Emma Wedgwood. We can but note that when he was thousands of miles away from them, he was usually fit, strong and frequently bursting with joy – particularly during the first few months in Chile. The years of voyaging with the *Beagle* were laying the seeds of a long life of intellectual inquiry. They were also years of liberation, the only five years in which he was an individual rather than a member of a family. 'I use the term Struggle for Existence', he wrote in his most famous work, 'in a large and metaphorical sense, including dependence of one being on another and including (which is more important) not only the life of the individual, but success in leaving progeny.'[35]

Darwin was always affectionate in his letters home, and no doubt sincere when he wrote to his sister Catherine, 'continue in your good custom of writing plenty of gossip'. Yet, to have all his family life encased by paper envelopes and red sealing wax, to be free, for five years, from their actual company, was a palpable liberation. 'Tell Charlotte . . . I should like to have written to her; and to have told her how well everything is going on – But it would only have been a transcript of this letter, & I have a host of animals, at this minute, surrounding me, which all require embalming & Numbering.' And besides, 'I long to be at work in the Cordilleras, the geology of this side, which I understand pretty well is so intimately connected with periods of violence in that great chain of mountains.'[36]

It was that sublime double mountain chain the Andes which now, gleaming and snow-capped, was their alluring goal. It was desperately hard making their way up the Pacific coast in unstoppable northerly gales in June 1834. The purser on the *Beagle*, George Rowlett, the oldest officer aboard at the age of thirty-eight, died of exhaustion and illness. On 28 June, Captain FitzRoy read the funeral service and Darwin noted, 'it is an aweful & solemn sound, that splash of the waters over the body of an old ship-mate'.[37]

Not long afterwards, they landed, though thick mists made the longed-for Cordilleras invisible. Then they took to the sea again and sailed on to Valparaiso, and the two ships, the *Beagle* and the *Adventure*, anchored together in the beautiful harbour. The sky was bright blue. A range of hills, 1,600 feet in height, surrounded the straggling but pretty little town. 'When in T. del Fuego I began to think the superiority of Welsh mountain scenery only existed in my imagination. Now that I have again seen in the Andes a grand edition of such beauties, I feel sure of their existence.'[38] He threw himself at once into strenuous mountain walks, from the luxuriant valleys up into the rough mass of greenstone from which he could turn and watch the glories of the setting sun. The sublimity of the mountain scenery brought forth from Darwin an excitement, and awe, which recall the Romantic poets Goethe and Wordsworth in Alpine or Lakeland settings. 'Who can avoid admiring the wonderful force which has upheaved these mountains, & even more so the countless ages which it must have required to have broken through, removed & levelled whole masses of them?'[39]

Moreover, in Valparaiso, there were intelligent English-speakers with whom he could share his enthusiasms. 'We all, on board, have been much struck by the great superiority in the English residents over other towns in S. America.'[40]

Lyell's first volume of *Principles of Geology* had been Darwin's companion during the earlier part of his voyage, and even in St Jago, when he had read only a few chapters, it had begun to change him – in its exposition of the age of the planet, in its puzzlement (even though Lyell at this stage rejected Lamarckian or any other notions of transmutation) about the origin of species, and about the fossil evidence for creatures which had ceased to be. In the course of the voyage, he had Lyell's subsequent volumes sent out to South America. By April 1833 he had read volume two.[41] By November, his sister Catherine was writing to him of volume three, and asking if it had reached him.[42] It had certainly done so by July 1834.[43] (The inscribed copy, with no date, is in the University Library at Cambridge.) He told Fox, 'I am become a zealous disciple of Mr Lyell's views,' and by 1836 he was offering to Lyell his paper 'Observations of proofs of recent elevation on the coast of Chili'.[44] Darwin's notion was that 'the Andes, at the period the Ammonites lived (which corresponds to the secondary rocks), must have been [a] chain of Volcanic Islands, from which copious stream[s of] lava were poured forth & subsequently covered with Conglomerates. Such beds form the Cordilleras of Chili.'[45]

As Darwin observed in Valparaiso, Lyell's *Geology* was not read by the scientific academy alone. 'Already I have met several [English] people who have read works on geology and other branches of science . . . It was as surprising as pleasant to be asked, what I thought of Lyells [*sic*] Geology.'[46] The generality were immediately able to grasp the significance of what Lyell was making clear. Tennyson's *In Memoriam*, as well as being a record of bereavement, is a poetic diary of an intelligent man, brought up in a parsonage to conventional Christian belief, and coming to terms with the implications of Lyell's book. No longer living on a planet which was a few thousand years old, that generation of men and women now looked back into a past which was unimaginably distant. Their childhood faith in a heavenly Father who did not allow a single sparrow to fall to the ground without his notice was confronted by fossil evidence of whole species 'cast as rubbish to the void'.[47] Lyell's harsh refusal to believe in the transmutation of species made the extinction of the fossilized fish, birds, reptiles and mammals appear to be the work of a relentlessly indifferent and impersonal force. Evidence, however, was evidence: and though Lyell's rigidity about the transmutation or development of species would soften, and though scientific opinion would eventually embrace the idea of species modifying, rather than being destroyed outright and replaced by new 'creations', the basic premises of Lyell's *Geology* had laid the foundations of the modern scientific study of the subject. The human race was beginning to realize that the planet on which it lived was almost unimaginably old – thousands of millions years old, as we now know.[48] Just as James Cook in the 1770s had established the existence of unknown parts of the world, and thereby made everyone's world a larger place, so Charles Lyell in the 1830s made the world not only older but more palpable. The fathers of modern science had been the chemists who, while Cook sailed the seas, had begun to demonstrate the properties of matter itself and, after the groundbreaking work of Priestley and Lavoisier, to see the very nature of elements and particles. It led, almost at once, to prodigious technological change, especially the discovery that water was H2O which allowed James Watt and others to pioneer the use of steam in industrial technology.

In the post-Napoleonic times, Lyell was one of the great pioneers not only of geology but of the wider ramifications of geological discovery. It began to seem tantalizingly possible that, by investigating rocks and fossils, volcanoes and ancient lava, that generation could see into the heart of things, comprehend the origins of life on earth. Historians of ideas sometimes represent this as a conflict between a 'scientific' outlook and an older 'religious' outlook, and this is how it sometimes seemed at the time. The reality is, however, that it was a new beginning. Not since the Arab mathematicians and astronomers of the middle ages imported their discoveries to Spain, and in some cases not since the ancient Greeks, had human beings in the West asked, or answered, many of the scientific assumptions which in the twenty-first century we take for granted. It is easy to blame 'religion' for this, but such a simple view overlooks the fact that most 'modern' scientists, from Galileo to Boyle, from Newton to Lyell himself, had some religion: many were pious, as Newton had been. The lazy mindset which assumed all scientific truth, together with moral and spiritual truth, was contained in Scripture, belonged largely to those who had no interest in science. If one had to find an underlying cause it is more attributable to the Latinization of the West than to its Christianization. The Romans were technological wizards when it came to building roads or under-floor heating, but they had no interest in mathematics, and from their conquest of the Eastern Mediterranean until the Islamization of Spain there was not a single scientific development. Modern science coincided with the revival of Greek learning. Lyell's generation was more or less coeval with the arrival of the Parthenon marbles in London.

Even Captain FitzRoy, who in later life backtracked and became what might be termed a biblical fundamentalist, read Lyell and discussed him eagerly with Darwin. Devout Protestant though he was, FitzRoy could see that Lyell drew a very different picture of the origin of the earth from that in Genesis, or, more accurately, Genesis as interpreted by the Protestant seventeenth century. Lyell wrote, as was customary, of Moses as the author of the first five books of the Bible: and, looking back to the time of the voyage when he was convinced by Lyell, the penitent FitzRoy could not imagine how he could have overlooked the fact that 'the knowledge of Moses was super-human'.[49] He regretted that while Captain of the *Beagle* he had 'suffered much anxiety . . . from a disposition to doubt, if not disbelieve, the inspired History written by Moses'.[50] This essay, which FitzRoy attached to his own account of the voyage, was published in 1839. If, to a callous modern eye, it verges upon the ludicrous, it is also a good illustration of the mentality which was not untypical. Loyalty to scientific truth was challenged by a higher loyalty, or what seemed a higher loyalty: faithfulness to the inspired word of God.

In Valparaiso, Darwin had met a school contemporary, now a merchant, and taken up residence in his house. It was doubtless in R. H. Corfield's company that Darwin met the intelligent English who had kept up with the publication of Lyell's *Geology* – in particular another person by the name of Robert Edward Alison. Corfield would return to England where he worked as a sugar merchant in Liverpool, retiring at the age of seventy-six, and dying aged ninety-three in 1897. Alison, who wrote about South American affairs, shared Darwin's interest in geology. He became the director of a Chilean mining company and died a year after Darwin in 1883.

Geology was Darwin's uttermost passion in these months, and he made expeditions into the mountains. The 'chaos of mountains' inspired inevitable wonder – it was 'like hearing a chorus of the *Messiah* in full orchestra. I felt glad I was alone.'[51] Yet the solitary Darwin was not only the awestruck wanderer: he was also the geologist – 'I had some more hammering at the Andes,'[52] he told his sister – and, as the fascinated grandson of a great industrialist, he could not fail to notice the mines. 'Almost every part of this mountain has been drilled by attempts to open gold mines . . . The rage for mining has left scarcely a spot in Chili unexamined, even to the regions of eternal snow.'[53] In the copper mines beyond Quillota he met the manager, 'a shrewd but ignorant Cornishman', married to a Spaniard. The copper ore was shipped to Swansea to be smelted. The Cornishman's skill as a mining engineer was matched by lack of other sophistications. Now that George Rex was dead, he asked Darwin, 'how many of the family of Rex's' were yet alive?[54] (George Rex was, of course, George IV in 1830.) Darwin speculated that 'This Rex certainly is a relation of Finis who wrote all the books.' The loftiness behind this joke is very 'Darwin', just as the flintheartedness with which he observed the working conditions of the miners is very 'Wedgwood'. Old Josiah had been a good-natured paternalist, but as he accumulated his tens of thousands he had begrudged his workforce every shilling. It is almost with approval that his grandson noted, 'Poverty is very common with all the labouring classes. One of the rules of this mine sounds very harsh, but answers pretty well – the method of stealing gold is to secrete pieces of the metal & take them out as occasion may offer. Whenever the major-domo finds a lump of ore thus hidden, its full value is stopped out of the wages of all the men, who are thus obliged to keep watch on each other.'[55] Nothing about this arrangement appears to have shocked Darwin, still less touched his heart.

He travelled on horseback to St Jago ('enjoyed myself very much')[56] and for much of September the mountains continued to delight him. Then, as he made his way back to Valparaiso, illness struck. At first he attributed an upset stomach to 'sour new made wine' which he had drunk while at the gold mines. In a later account, however, in his diary for 26 March 1835, he described sleeping in a village five leagues south of Mendoza, where he was attacked by 'the Benchuca, the great black bugs of the Pampas. It is most disgusting to feel soft wingless insects, about an inch long, cawling [*sic*] over one's body; before sucking they are quite thin, but afterwards round & bloated with blood.'[57]

Speculation naturally arose afterwards whether the illness of September 1834 or the 'attack, & it deserves no less a name'[58] could have been the origin of Darwin's lifelong mysterious illness – palpitations, vomiting and so forth. The great black bug of the pampas – not Benchuca as Darwin called them but more properly *Triatoma infestans* – are the principal vectors of South American trypanosomiasis or Chagas's disease. They spread the disease not directly through their bites, but by defecating in the contaminated wounds which they inflict. Many of Darwin's later symptoms – fevers, such as those suffered in autumn 1834, vomiting, palpitations – do correspond to those of Chagas's disease. It has been convincingly argued, however, that Darwin suffered from many of these symptoms before he even left England for South America and that 'it is beyond credibility that Chagas's disease could produce symptoms of cardiac insufficiency for between 40 and 50 years and not produce some physical signs'.[59] No other Beagle suffered from the infection and moreover, in 1838, two years after his return to England, Darwin undertook a strenuous mountain tour in Scotland, which would scarcely have been possible for a sufferer from Chagas's disease. The simpler explanation, that Darwin was suffering from a bad case of poisoning – either from the filthy wine or from something he ate – is more plausible. He just managed to get back to Corfield's house in Valparaiso where he languished in bed for a fortnight. 'Capt. FitzRoy very kindly delayed the sailing of the Ship till the 10th of November, by which time I was quite well again.'[60]

In Darwin's sickly absence, FitzRoy had had more than his share of problems. The Admiralty in London persisted in its refusal to pay for the *Adventure* and FitzRoy came to the reluctant conclusion that the second ship must be sold for 7,500 dollars – nearly £1,400 in the money of that time. Darwin believed that the Admiralty had treated FitzRoy badly ('solely . . . because he is a Tory').[61] He noted with alarm that the hard work, the distress of discharging the crew of the *Adventure* and the difficulty of equipping and arranging the now grievously overcrowded *Beagle* had brought on 'a morbid depression of spirits & loss of all decision & resolution. The Captain was afraid that his mind was becoming deranged (being aware of his hereditary predisposition)'[62] – an allusion to the suicide of FitzRoy's kinsman Lord Castlereagh.

Before the *Beagle* at last set sail in November 1834, Darwin arranged for two more boxes of specimens to be conveyed back to England in HMS *Challenger*. This sadly fated vessel was wrecked in a storm at Aranco. It would seem that, for some reason, Darwin's cases were transferred, at Rio, to another vessel and so did not go down with *Challenger*.

The Captain's temper was frayed to snapping point. When Darwin eventually went aboard the *Beagle* (and one can imagine that, even having disposed of two large packing cases, he was taking up a lot of space with all his equipment and his specimens), he found FitzRoy in a foul mood. The obligation to give a party on board ship to thank all the local residents who had been helpful to him caused an agony of irritation and, upon Darwin's appearance, FitzRoy burst out into a fury, accusing the naturalist of being one who 'would receive any favours and make no return'.[63] Darwin was stung. His quiet but intensely ambitious pursuit not only of scientific knowledge but also of fame was marked by, precisely, receiving favours and making no return. He went back to Corfield's house, momentarily wondering whether he should not take another ship direct for home. Two days later, however, he returned to the *Beagle* and was welcomed back by a cordial Captain. Wickham, however, serving once more as First Lieutenant having lost the *Adventure*, said: 'Confound you, philosopher, I wish you would not quarrel with the skipper; the day you left the ship I was dead-tired (the ship was refitting) and he kept me walking the deck till midnight abusing you all the time.'[64]

On 10 November, they got under way and southerly winds bore the *Beagle* up the Chilean coast again. They anchored at the island of Chiloe and tried (unsuccessfully) to penetrate the lush forests, drenched with rain. It was a tangle of bamboo creepers which enmeshed them like fish in a net. For two months of mostly bad weather they sailed up and down surveying the mainland coast. On 18 January 1835, they anchored for a second time in the bay of San Carlos on the island of Chiloe. The next day the volcano of Osorno was in great activity.

> At 12 o'clock the sentry observed something like a large star from which state it gradually increased in size till three o'clock when most of the officers were on deck watching it – It was a very magnificent sight; by the aid of a glass, in the midst of the great red glare of light, dark objects in a constant succession might be seen to be thrown up & fall down. The light was sufficient to cast on the water a long bright shadow. By the morning the Volcano seemed to have regained its composure.[65]

Afterwards they heard that the volcanoes of Aconcagua in Chile, 480 miles to the north, and Cosegüina in Nicaragua, 2,700 miles north again, had also erupted that same night. Darwin, by now steeped in the works of Lyell, had travelled in regions which most European geologists could only dream of. His palaeontological finds placed him at the forefront of fossil research. (F. W. Hope had told him that sending home 'the much desired bones of Megatherium' had ensured that his name 'is likely to be immortalized'.)[66] With his eyes fixed on erupting volcanoes he was now seeing geology in action, as it were.

Four weeks later, when the *Beagle* was anchored off the town of Valdivia on the south Chilean coast, Darwin went ashore with Covington in search of specimens. After a walk through an apple orchard they lay down to rest and felt the ground beneath them tremble. When they sailed north, to the port of Talcahuano, they realized the extent of the earthquake. The shore was strewn with debris 'as if a thousand great ships had been wrecked. Besides chairs, tables, bookshelves &c &c in great numbers there were several roofs of cottages almost entire . . .'[67]

By 4 March, when they landed at the island of Quiriquina, they learnt that the town of Concepción, and the port of Talcahuano, had been devastated by the earthquake, with not a house left standing. Even on the island, Darwin noticed great cracks in the rocks on the beach. Surface slabs of slate had been smashed to smithereens. 'For the future when I see a geological section traversed by any number of fissures I shall well understand the reason.'[68]

Riding through the ruins of Concepción with FitzRoy at his side was an awful, and awe-inspiring, experience. Within six seconds the town had been wrecked. Had the quake happened at night, three-quarters of the population would have perished. The accounts of it by 'Mr Rous the English Consul' – a thunderous noise, a whole side of his house disappearing while he ate his breakfast, the sky dark with dense dustclouds, the air unbreathable – were not to be forgotten. Darwin, the banker's son and industrialist's grandson, thought immediately of the economic implications of living in a volcanic environment. Imagine 'if beneath England a volcanic focus should reassume its power . . . England would become bankrupt; all papers, accounts, records as here would be lost; & Government could not collect the taxes.'[69] Pliny the Younger, who left us his immortal description of the eruption of Vesuvius and the destruction of Pompeii, watched the disaster with cinematic eye and, after all the flame and dust and mayhem, focused on one death, that of his beloved uncle. Voltaire, after the Lisbon earthquake of 1755, was a changed man, his capacity for optimism destroyed, his world-outlook, and that of a generation thanks to what he wrote about it, disrupted, shaken, changed. Darwin saw the ruins of Concepción with distress; but his immediate thoughts, financial and parochial, were of the Treasury in London if such a calamity were to occur at home.

Two adventures lay ahead: one of great excitement to Captain FitzRoy and his men would be a piece of gallantry and superb seamanship; the other, for Darwin, in his mental journeyings, an event of the greatest importance.

On the *Beagle*'s last visit back to Valparaiso, after a second trip to Concepción, the news was waiting from the Admiralty in London that FitzRoy, hitherto only a lieutenant in rank, had been promoted to full captain. No doubt this was intended to soothe his bruised ego after the *Adventure* fiasco. The strategy worked and it put not only FitzRoy but all the Beagles in a happy frame of mind and in a mood for a proper naval adventure. When FitzRoy heard of the plight of the *Challenger*, the man-of-war wrecked in a storm at

Arauco (an effect of the earthquake which destroyed nearby Concepción) he determined to go in search of the ship's Captain, Michael Seymour, and such of his crew as had survived the wreck. Seymour was an old friend of FitzRoy's.

The Admiralty had entrusted the task of rescue to HMS *Blonde*, the British man-of-war on the station, but the *Blonde*, under the command of an elderly commodore, was reluctant to get on to a lee shore in winter. FitzRoy placed the *Beagle* under Wickham's command, and set off on the *Blonde* having volunteered his service as a pilot. They anchored in the Bay of Concepción and FitzRoy, having hired horses, made a 100-mile trek overland, braving all the hazards caused by currents, streams, hostile Indians and exhausted pack animals, to find his British comrades. An encounter with a party of Chileans made the expedition all the more urgent, for these men told FitzRoy that 3,000 hostile Indians had assembled on the Chilean frontier and that, until driven off by some other Indians, they had been on the verge of plundering the meagre supplies of the marooned Challengers. When FitzRoy finally found the shipwrecked crew, they were hungry, semi-mutinous and dead tired, but astonishingly, given all they had been through, only two had died. Thanks to FitzRoy they all got back to Coquimbo where a rescue-ship awaited them. Captain Seymour had to answer for the loss of HMS *Challenger* before a court martial in Greenwich when he reached home. It was largely because of FitzRoy, who told the court of the tsunami which had accompanied the earthquake, that he was acquitted.

FitzRoy rejoined the *Beagle* at Callas, in Peru. By now, Darwin, together with the rest of the crew, was yearning for home. They had been absent from England for four years. FitzRoy caused them all an agony of frustration when he insisted on one last trip to Lima, where he consulted maps and charts. He was determined to do as thorough a job as possible of charting the coastline before they left South America, heading out north-west into the Pacific Ocean past the little archipelago known as the Galápagos Islands (off the coast of Ecuador) before heading down towards the Society Islands and the Friendly Islands, following the route taken by Captain Cook in 1775 on his sea-path to New Zealand.

'My dear Henslow,' Darwin wrote from Lima on 12 July, 'This is the last letter which I shall ever write to you from the shores of America, and for this reason I send it. In a few days time the *Beagle* will sail for the Galapagos Isds. I look forward with joy & interest to this, both as being somewhat nearer to England, & for the sake of having a good look at an active volcano. Although we have seen lava in abundance, I have never yet beheld the crater.'[70]

His mind still, primarily, absorbed in geology, he did not realize that when he reached the little cluster of Pacific islands, he would behold biodiversity on a scale unmatched on the planet.

When he first went ashore at Chatham Island, the barren volcanic rock was uninviting; and the small black cones, former craters, resembled nothing so much as the foundries at Coalbrookdale in Shropshire, believed by some to be the cradle of the Industrial Revolution. He saw them as 'the ancient chimneys for the subterranean melted fluids'. Yet the grandson of the Industrial Revolution was soon superseded by the naturalist who rejoiced to see that 'The Bay swarmed with animals; Fish, Sharks & Turtles were popping their heads up in all parts.'[71] There was good fishing, and the first hunting party on the island brought back fifteen tortoises which were good to eat. Too good. On Charles Island, they met the acting Governor, Nicholas Lawson, an Englishman working for the government of Ecuador. Lawson told them that the American whalers regularly hunted the tortoises and turtles here. One tortoise provided them with upwards of 200 pounds of meat. Lawson believed they would last twenty years before becoming extinct.

Today, the giant Galápagos tortoises are all but extinct, as is the land lizard – 'hideous animals; but . . . considered good food'.[72] Even some of the finches, which supposedly first set Darwin's mind moving towards an explanation of the evolutionary process, have become extinct.

The *Beagle* spent a month exploring the Galápagos Islands. While the ship cruised, boatloads of men could be sent ashore to investigate. The birds, of which there was a prodigious variety, were quite tame. Darwin pushed a hawk off a branch with the end of his gun.[73] On Charles Island he saw a boy sitting beside a well with a long stick in his hand. As the doves came to drink he casually killed as many as would feed his family, and had collected easily enough within half an hour. The sheer variety of birds was intoxicating. 'Doves and finches swarmed,'[74] Darwin excitedly told the diary. The days were cloudless, overpoweringly hot. They slept on the soft, warm sand.

Darwin's collecting-jars filled up with plants, seashells, insects and reptiles. It was as they moved from island to island in the archipelago that he saw the truth of what Lawson had told him: you could tell from the look of a tortoise which island it came from. Captain FitzRoy was to write, 'All the small birds that live on these lava-covered islands have short beaks, very thick at the base, like that of a bull-finch. This appears to be one of those admirable provisions of Infinite Wisdom by which each created thing is adapted to the place for which it was intended.'[75] The key word here is 'adapted'. Though it is well known that FitzRoy became a diehard opponent of evolutionary theory, the use of the word 'adapted' implies that, in conversation with Philos, even on the archipelago, some discussion took place, not merely about the variety of the birds, tortoises and other creatures, from island to island – as pointed out by Lawson – but also about the underlying cause.

The Darwinian mythology suggests that it was on the Galápagos Islands that the naturalist first began to observe the different characteristics of species from one island to the next, and that, in particular, the beaks of the finches, differing from one island to the next, suggested to him the phenomenon of descent by modification. Darwin's granddaughter, Nora Barlow, published extracts from his *Beagle* notebooks in 1945 and propounded the notion that already on the islands 'a revolution was taking place in his views on the immutability of species'. Anticipating the obvious question – why, were this the case, her grandfather had not expressed such thoughts at the time – she hazarded, 'Probably, there was some deference to FitzRoy's emphatically creationist opinions in this delay; but mostly the need in his own mind to marshal all the facts in logical sequence.'[76]

Today, the mythology has become so fixed in the public mind that when, for example, the *Encarta World Dictionary* turns to the finches of the Galápagos Islands it speaks of 'the birds . . . on which Charles Darwin based his theory of natural selection through observation of their feeding habits and corresponding differences in beak structure. Subfamily Geospizinae.' A typical tutorial, at Palomar College in California, teaches that:

> Darwin identified 13 species of finches in the Galápagos Islands. This was puzzling since he knew of only one species of this bird on the mainland of South America, nearly 600 miles to the east, where they had all presumably originated . . . Each of the various islands had its own species of finch, differing in various ways, and in particular in beak shape. Darwin's idea was that these finches were all descendants of a single kind of ancestral finch, but that the different environments of the different islands had given advantages to different characteristics in its finch population . . . Eventually the separated populations would become too different to interbreed, and would be separate species.[77]

Although this is what modern biologists believe, such thoughts did not occur to Darwin at the time. As a matter of fact, he failed even to identify most of the finch specimens which he collected on the Galápagos as finches. Some he labelled blackbirds, others 'gross beaks' and one a wren. He gave them to the Ornithological Society of London, which in turn gave them to John Gould, an ornithologist, to be identified, and it was Gould, not Darwin, who recognized that they were all distinct species of finch.[78] It was FitzRoy, not Darwin, who made collections of finches and labelled them correctly, and, as Harvard University's Frank Sulloway demonstrated in 1982, it was FitzRoy's identification of the differences between the finches which enabled Gould to make his remarkable observations.[79] Darwin never in fact mentioned the differences between the finches in *The Origin of Species*, even though, during the celebrations of the 150th anniversary of publication, Gould's drawings of the Galápagos finches were reproduced again and again as if they were Darwin's 'discovery'.

Moreoever, Peter and Rosemary Grant, evolutionary biologists from Harvard University, spent over twenty-five summers studying these birds, mainly on the island of Daphne Major. They revealed that the beak changes were reversible – this is hardly 'evolution'. Beaks adapted from season to season depending upon whether droughts left large, tough seeds, or heavy rainfall resulted in smaller, softer seeds. Even had Darwin noticed the supposed evolution of finches' beaks on the Galápagos Islands and had thereby become an instantaneous convert to his famous theory, the epiphany would have been wrong.[80]

Nora Barlow expounded the classic confrontation which the thirteen varieties of Galápagos finch supposedly force upon the observer, the choice between 'creationism' as represented by FitzRoy and 'evolution' as represented by her grandfather. She overlooked, as Gould did not, the fact that the variations in the finches are in most cases minute, just as she overlooked the fact that, when he was labelling his finches as blackbirds and wrens, her grandfather was, in any case, a rather simple creationist with *Paradise Lost* in his pocket. She also overlooked another possible explanation for different varieties flourishing on different islands, namely that they had migrated from the South American mainland to environments which they found congenial, and that their evolution, if it occurred, might have owed nothing to the climate and conditions of the Galápagos at all. If further proof were needed that Darwin did not, at the time, see the significance of the Galápagos Islands' biodiversity, one has only to think of the fate of the giant tortoises which he collected. Here indeed was evidence that, from island to island, there were differences between these fascinating and amiable creatures. Few as their descendants are today, we know that, as Nicholas Lawson observed, there were significant variations between the tortoises from one island and the next. This much Darwin noted, as the *Beagle* sailed away from the archipelago towards Tahiti, but he regarded the anomalies as 'insignificant' and, together with the officers at the Captain's table, ate all the evidence and allowed the cook to dump even their carapaces overboard.[81] This hardly suggests a scientific epiphany of the kind invented in the pious pages of Lady Barlow over a hundred years later.

If Darwin had not yet developed his theory of the origin of species by means of natural selection, this was not to say that no one else had done so. On 19 July, when the ship had anchored 'in the outer part of the harbour of Callao'[82] (the port for Lima), Darwin jotted in his notebook, 'Smelling properties discussed of Carrion Crows, Hawks. Magazine of Natural History'.[83] Clearly he had picked up a copy of the *Magazine of Natural History* which had been sent out by mail for Darwin from England. The reference to the 'smelling properties of carrion crows, hawks' refers to an article on this subject by Charles Waterton (and on the faculty of scent in the vulture) in the issues of January and May 1833.[84]

We do not know the exact point at which Darwin got hold of the issue of the *Magazine of Natural History* for January 1835, for this was the one which contained an article by Edward Blyth, expounding an evolutionary theory which Darwin would eventually appropriate and make famous.

> It is a general law of nature for all creatures to propagate the like of themselves: and this extends even to the most trivial minutiae, to the slightest peculiarities; and thus, among ourselves, we see a family likeness transmitted from generation to generation. When two animals are matched together, each remarkable for a certain peculiarity, no matter how trivial, there is also a decided tendency in nature for that peculiarity to *increase*; and if the produce of these animals be set apart, and only those in which the same peculiarity is most apparent, be selected to breed from, the next generation will possess it in a still *more* remarkable degree; and so on, till at length the variety I designate a *breed* is formed, which may be very unlike the original type.

Just as Darwin would experiment with pigeons to see the ways in which species could adapt, from one sexual conjunction to the next, breeding characteristics which would be useful or strengthening, so Blyth asked whether the process we can watch at work in artificial breeding was not also at work in nature.

> A variety of important considerations here crowd upon the mind, foremost of which is the enquiry, that, as man, by removing species from their appropriate haunts, superinduces changes on their physical constitution and adaptations, to what extent may not the same take place in wild nature, so that, in a few generations, distinctive characters may be acquired, such as are recognized as indicative of specific diversity . . . May not, then, a large proportion of what are considered species have descended from a common parentage?[85]

Having raised the idea, however, which would eventually be coterminous with Darwin's name, Blyth rejected it, on the grounds that, 'were this self-adapting system to prevail . . . we should seek in vain for those constant and invariable distinctions which are found to obtain'. In other words, he thought, if the theory were true, living species would blend into each other, which we can see is not the case in nature. Edward Blyth, a twenty-five-year-old pharmacist from the Surrey village of Tooting (now absorbed into the suburban sprawl of south London) was a poor man of no reputation.

In the ornithological notebooks which Darwin kept at the time, however, Barlow found reflections upon 'Thenca (Mimus Thenca). These birds are closely allied in appearance to the Thenca of Chile . . .' He recorded that he had collected different specimens from different islands, noting similarities and differences. 'The only fact of a similar kind of which I am aware is the constant asserted difference between the wolf-like Fox of East and West Falkland Isds. – If there is the slightest foundation for these remarks, the Zoology of Archipelagoes will be well worth examining, for such facts would undermine the stability of species.'[86]

Barlow added, 'It is astonishing, when we consider that more than twenty years were to elapse before this "undermining" of the stability of species was sufficiently documented to be given to the world in *The Origin of Species*.'[87]

A biography of Darwin must, chiefly, be the biography of an idea. This is Darwin's idea, first published to the world in 1859, of what we now call evolution. He took some time to arrive at his conclusions, and he would then spend the rest of his life wrestling with them, and sometimes changing his mind about them. Some Darwin scholars question whether he had in fact read Blyth's article in the *Magazine of Natural History*, even though he had plainly read another article in the same issue. This is in some senses of less significance than Darwin's eventual conclusion about evolution.

Before proceeding, we should be clear in our minds what were the issues at stake for science here. There are three crucial points to master. First, is Blyth's 'were this self-adapting system to prevail . . . we should seek in vain for those constant and invariable distinctions which are found to obtain'. Biology proceeds on the basis that taxa are distinct. It is not simply a question of convenience to say this. Since Linnaeus – you could say since Aristotle and Pliny – it is a matter of actual observation, of what can be seen and tested. Comparative anatomy leads us to observe the phenomenon of what is called homology. A homologue is a constant. Take as an example the forelimbs of vertebrates. In *The Origin of Species*, Darwin would write,

> What can be more curious than that the hand of man, formed for grasping, that of a mole for digging, the leg of a horse, the paddle of the porpoise, and the wing of the bat, should all be constructed on the same pattern, and should include similar bones, in the same relative positions?
>
> We may call this conformity to type, without getting much nearer to an explanation of the phenomenon . . . but is it not powerfully suggestive of true relationships, of inheritance from a common ancestor?[88]

Here you have the essence of the puzzle, laid out in Darwin's clear, good prose. In each case – horse, bat, man, mole, porpoise, etc. – we find the same basic building-blocks in the forelimbs – humerus, radius, ulna, carpals, metacarpals and phalanges. You find, to use anthropomorphic language, that they all, basically, have something equivalent to our upper arm, our forearms, with its two bones radius and humerus, and they all have the five-fingered 'hand', which, in the case of the horse – for example – you can watch, through the gradations of fossil evidence, changing and adapting, never using its 'thumb' even when it was a little foxlike creature, and eventually encasing its 'hand' in a hoof.

That is the phenomenon. Blyth was not the first to see it. Cuvier had seen it. Lamarck had seen it. Goethe had seen it. Erasmus Darwin had seen it. Blyth was perhaps the first to spell out what it implied for taxonomy (that is, species classification) if you thought that one species evolved into another.

No serious observer of the fossil evidence, and of the bones dug up and studied by the palaeontologists, would question that we can observe a process of adaptation at work in nature. The basic building-blocks, the homologues, adapt themselves according to need, circumstance, environment.

The further question, however – and this is where Darwin would eventually become revolutionary – is whether these basic building-blocks themselves evolve, or whether they were simply given. We watch the five-fingered 'hand' changing in the evolutionary cycle of the mole, the porpoise, the bat, the horse, the man. Do we find any evidence that the hand itself, the five-fingered entity, had evolved?

That is a question which Darwin has to answer if we are to consider him the greatest revolutionary scientist of the nineteenth century, or merely a naturalist of genius, whose theory of how things came into being remains questionable. There is plenty of evidence for the building-blocks adapting themselves. But such things as hair, feathers, eyes or even hands – is there any evidence for these forms, often in themselves extremely complex, having 'evolved'? Darwin had cottoned on to the process of micro-evolution (as with the finch-beaks) in some species and had made it into a universal. The same principle which accounts for the shape of a finch's beak also accounts for the difference between a bee and an elephant.

And then again – and this is fundamental to the whole story – what do we mean by using this word 'evolution', which Darwin himself was slow to use, and which was certainly not current when he was aboard the *Beagle*? The word carries two connotations which we must hold in our heads for the rest of this story: what can be termed micro-evolution, and macro-evolution.

When John Gould, of the Ornithological Society of London, began to examine the finches which Darwin had sent home, he saw a process at work which we could usefully describe as micro-evolution. Little changes were occurring which made the beaks of the finches more useful for them in various particular environments. Now, Darwin is famous for the conclusion he reached, in *The Origin of Species*, that ultimately all life derives from a common source. Micro-evolution cannot be denied. Ever since the development of palaeontology and fossil study in the early nineteenth century, it has been possible to see, within species, the same basic homologues or building-blocks being adapted for different purposes.

Darwin was to develop the theory (with which Blyth in the *Magazine of Natural History* had only doodled) that all life-forms evolve in this gradual way. Just as the finches' beaks changed little by little, so too all forms – including the basic building-blocks – came about by these infinitesimally gradual processes. Nature does not make leaps. This was to be Darwin's central contention. Even forms which look as if they are highly complex – such as eyes, or hair, or feathers – slowly evolved. This we can call macro-evolution.

Some people, both during Darwin's lifetime and since – have seen his ideas in a religious light, as he did himself. They think that if this version of evolution could be demonstrated, it would remove the need for belief in a Creator. To use the words of Darwin's wife Emma Wedgwood, his ideas seemed to be 'putting God further and further off'.[89]

If one leaves the religious question on one side, however, it might help to clear the mind for a consideration of the factual evidence. Evolution clearly takes place. The question which science would love to answer is – how? When Darwin had developed his theory, would nature provide him with evidence to prove it? Or would there remain an everlasting puzzle, at the apparent 'leaps' in the living world and its development? This really is the key question. From the beginnings of the idea that all life derives from one, or a few, simple sources, there have been religious believers who espoused it, and scientists who refused to believe it until the evidence was provided. This is our story, Darwin's story. The biography of this idea is, in the end the story, not simply of one man, but of the Western world which came to believe that 'what are considered species have descended from a common parentage. The microevolution of the finches was observable. The common ancestry of mammals is not. It is hardly surprising that Darwin, who always guarded his public reputation so carefully, should have waited before committing himself to the view that giraffes, elephants, blue whales and Queen Victoria herself all descended from a tiny animal a bit like a shrew.'[90] It is indeed necessary to believe this if you accept Blyth's 'common parentage' theory of the origin of species. Darwin would not go public with such a view until circumstances forced him into it.

When they left the Galápagos Islands, in October 1835, Darwin, in common with the majority of the officers and men, was ready for home. In fact, a year would pass before their Odyssey was done. They sailed through the South Pacific, as Cook had done in the 1770s, spending ten glorious days on Tahiti – 'the Queen of the Islands',[91] Darwin called it. Darwin shared FitzRoy's admiration for the missionaries there, who 'have done much in improving' the moral character of the natives. 'It is something to boast of, that Europeans may here, amongst men who so lately were the most ferocious savages probably on the face of the earth, walk with as much safety as in England.' George Forster's descriptions of Tahiti, in his account of Cook's visits, dwelt lovingly upon the beauties of the all but naked women who had swum out to the *Resolution* and offered themselves to the English sailors (a passage which James Boswell particularly savoured, to Johnson's annoyance). Darwin, over sixty years later, by contrast, 'was much disappointed in the personal appearance of the women; they are far inferior in every respect to the men'.[92] He disliked their style of cutting their hair, 'or rather shaving it from the upper part of the head in a circular manner so as only to leave an outer ring of hair', and he found their tattoos unseemly. 'They are in greater want of some becoming costume even than the men.'[93]

Were this entry written by a strict Christian like FitzRoy, or by a much older man, it might pass without remark. The man who wrote about the Tahitian women was not yet twenty-six years old. The sentence prompts the inevitable – if unanswerable – question of Darwin's sexuality, and how this side of life was tackled during his years in South America. The diary gives no clue, and one gets no sense, when he visited Rio de Janeiro, Buenos Aires, Valparaiso or Lima, that he pursued women. The lack of evidence suggests either very great discretion or very great restraint; if the latter, it was perhaps more easily achieved in one with very low libido.[94] Certainly, in Tahiti, where there would undoubtedly have been the opportunity to enjoy the women, we find him merely basking in the sunshine, enjoying the fruit ('I do not know anything more delicious than the milk of a young Cocoa Nut').[95] And of course, short as his time was on Tahiti – just eleven days – he explored and botanized and mused on the origins of coral. The splendid Queen Pomare of Tahiti came on board the *Beagle* – 'an awkquard [*sic*] large woman without any beauty, gracefulness or dignity of manners'.[96]

They got under way on 25 November and sailed to New Zealand, which they reached in three weeks. As they explored the Maori villages, Darwin was as shocked as FitzRoy by the heathenism. 'This little village is the very stronghold of vice; although many tribes, in other parts, have embraced Christianity, here the greater part are yet remain [*sic*] in Heathenism.' One can only conclude that the 'vice' here referred to is sexual. Certainly, on all Cook's visits, the openness with which the Maori women greeted his crews was remarked upon in the normally broad-minded Captain's journals. Darwin also found the Maoris dirty – 'the idea of washing either their Persons or clothes never seems to have entered their heads'.[97]

Darwin attended divine service on Christmas morning 'in the chapel of Pahia . . . So excellent is the Christian faith, that the outward conduct of the believers is said most decidedly to have been improved by its doctrines, which are to a certain extent generally known.'[98]

The New Zealanders in general failed to impress. Darwin came to the conclusion that 'the whole population is addicted to drunkenness & all kinds of vice'.[99] The native New Zealander compared ill with the Tahitian. 'One glance at their respective expressions, brings conviction to the mind that one is a savage, the other a civilised man.'[100] The English residents were, in Darwin's view, 'of the most worthless character . . . many of them run away convicts from New South Wales'.[101] And it was to New South Wales, after Christmas, that the *Beagle* sailed.

They anchored within Sydney Cove, and he walked the new-built streets. Gigs, phaetons and carriages whizzed about this rapidly expanding city, whose population was 23,000. There was something in the air which was instantly familiar, and reassuring, to the banker-doctor's son of Shrewsbury: money. Whereas in the South American cities the landed and propertied classes were 'known', here the new rich swanned about in their carriages, careless of whether or not they were of 'known family'. He was in the land of *Great Expectations*. 'An auctioneer who was a convict, it is said intends to return home & will take with him 100,000 pounds. Another who is always driving about in his carriage, has an income so large that scarcely anybody ventures to guess at it, the least assigned being fifteen thousand a year. But the two crowning facts are, first that the public revenue has increased 60,000 £ during this last year, & secondly that less than an acre of land within the town of Sydney sold for 8000 pounds sterling.'[102] (Darwin's father, when he died in 1848, left £223,759,[103] and was one of the richest men in Shropshire.) Only fifteen years had elapsed, when Darwin reached Sydney, since the retirement of Lachlan Macquarie, the 'father of Australia', as Governor of New South Wales. It was Macquarie who 'found a gaol, he left a burgeoning colony'.[104] It was he – the child of a tenant farmer from the tiny Hebridean island of Ulva, who rose to be lieutenant-colonel in the 77th Highland Regiment – who had decided to transform Australia from a place which was primarily a penal colony into an incipient modern nation. Darwin saw an infant nation: the Australian Cricket Club had been formed in 1827. Each Sunday, the uniformed Governor would appear at St James's Church, an emblematic assembly in which the left and centre aisles were packed with uniformed officers and rich merchants, and the right-hand gallery was full of convicts.[105] It was Macquarie's genius to pursue an 'emancipationist' policy, allowing the convicts in the right gallery to transmute, eventually, into the rich merchants in the centre aisle. Darwin, whose honest antecedents in Staffordshire had made comparable stratospheric leaps through the ranks, could not fail to be impressed.

In his short time in New South Wales, he wanted to see as much as possible. He rode out to Bathurst, 'the centre of a great pastoral district'.[106] He saw gum trees against bright-blue skies. He met a party of 'Aboriginal Blacks, who, in exchange for a shilling, "threw their spears for my amusement"'. He went out to the Blue Mountains. He went kangaroo-hunting with the superintendent of a farm called Walerawang – 'but had very bad sport, not seeing a kangaroo or even a wild dog'.

Lying in the sunshine after this unsatisfying experience, Darwin had some metaphysical reflections:

> on the strange character of the Animals of this country as compared to the rest of the World. An unbeliever in everything beyond his own reason, might exclaim, 'Surely two distinct Creators must have been [at] work; their object, however, has been the same, & certainly the end in each case is complete'. Whilst thus thinking, I observed the conical pitfall of a Lion-Ant [predatory insects, not ants at all, usually now called antlions] – A fly fell in & immediately disappeared; then came a large but unwary Ant, his struggles to escape being very violent, the little jets of sand described by Kirby[107] were promptly directed against him. This fate however was better than that of the

> poor fly's – without a doubt this predacious Larva belongs to the same genus, but to a different species from the Europaean [*sic*] one – Now what would the Disbeliever say to this? Would any two workmen ever hit on so beautiful, so simple, & yet so artificial a contrivance? It cannot be thought so – the one hand has surely worked throughout the universe. A Geologist perhaps would suggest that the periods of Creation have been distinct & remote the one from the other; that the Creator rested in his labour.
>
> * NB: the pitfall was not above half the size of the one described by Kirby.[108]

They set sail again on 14 March. 'Farewell Australia!' said Darwin. 'I leave your shores without sorrow or regret.'[109] He was psychologically ready for home, but he had a long wait, sailing first to the Keeling Islands, to Mauritius, round the Cape of Good Hope and on to St Helena. Then, to Darwin's dismay, FitzRoy decided that they must return home via South America in order to complete his circle of chronological measurements of the world. 'This zig-zag manner of proceeding is very grievous,' he told Susan in August; 'it has put the finishing stroke to my feelings. I loathe, I abhor the sea, & all ships which sail on it.'[110]

He was no less seasick on the way home than he had been on the voyage out, and when eventually the *Beagle* approached the coast of Cornwall, on 2 October 1836, Darwin was numb. 'After a tolerably short passage, but with some heavy weather, we came to an anchor at Falmouth. To my surprise and shame I confess the first sight of the shores of England inspired me with no warmer feelings than if it had been a miserable Portuguese settlement. The same night (and a dreadfully stormy one it was) I started by the Mail for Shrewsbury.'[111]

7
The Ladder by Which You Mounted

TWO YEARS AND three months elapsed between Darwin's return to England and his marriage to his cousin Emma Wedgwood. In his *Autobiography* he described these months as 'the most active ones which I ever spent, though I was occasionally unwell'.[1]

The activity was imposed upon him by the sheer scale of his collections, made in the course of the previous five years – 1,529 species in spirits and 3,907 labelled skins, bones and other dried specimens.[2] Darwin was always a slow worker. Moreover his intense ambition as a scientist made him diffident, ever afraid of error. While the jars and packing cases which he had sent back from South America could have provided him with a lifetime's research, their very broad extent prompted him to seek out experts in every field to examine them. The spring of 1837, for example, found him approaching Thomas Bell (Professor of Zoology at King's College, London) to examine the reptiles and higher orders of *Crustacea*.[3] Leonard Jenyns, Henslow's naturalist brother-in-law, was asked to examine the fish. Botanists were consulted. There were a 'few dozen' drawers full of seashells. Many of the mammals and the creatures preserved in alcohol had been sent to the Zoological Society of London.

Describing himself as 'active' was certainly no exaggeration. The Darwin who returned to England in 1836 was a very different being from the seemingly rather lazy boy who had hesitated about accepting Captain FitzRoy's invitation to sail with the *Beagle* in December 1831. Because Darwin, after marriage, became a domestic recluse in Kent, cogitating his theories and conducting his experiments; and because he was, in person, of a retiring, almost apologetic manner, it would be possible to think of him still as the amateur naturalist who bumbled into world fame as a scientist almost by accident. Nothing could be further from the truth. We see him, from the moment he returned to England, actively promoting himself. That he had collected so very many specimens – of plants, bones, preserved mammals, birds, reptiles, fish, as well as a prodigious quantity of insects, shells, stones and fossils – was an indication of how wide the net of his ambition was cast. To Leonard Jenyns, an ichthyologist to whom Darwin sent fish specimens, he spelt out his ambition of writing 'the Zoology of the *Beagle*'s voyage on some uniform plan'. He said that 'The plan would resemble on a humbler scale Rüppel's Atlas or "Humboldt Zoologie" where Latreille, Cuvier &c &c wrote different parts.' (Eduard Rüppel was a German naturalist whose *Atlas* of 1826 chronicled his journeys in Africa.) Although he told Jenyns, 'I myself should have little to do with it,'[4] this was clearly not going to be the case. While Alexander von Humboldt did not write every word of his encyclopaedic scientific studies, his is the name on the spine. When Darwin set sail with FitzRoy he still had it in mind to be a naturalist-clergyman such as Leonard Jenyns. Those who have chronicled his life have sometimes written as if this was, in effect, what he did become, only without taking holy orders: a bearded eccentric with no particular desire to 'get on'. The Darwin of 1836–9, however, comes across in his letters as a man with towering ambitions.

We see this in the dedication with which he sat down at once to complete his *Journal of Researches into the Natural History and Geology of the Countries Visited during the Voyage of HMS Beagle Round the World*. In those days before photography, before *National Geographic* magazine, before natural history films, books of travels were the only way in which the ever-expanding reading public could look at the world. Louis-Antoine de Bougainville's *Voyage autour du monde* (1771) with its descriptions of the South Pacific was a book which made the late eighteenth-century world a larger place for all who read it. James Cook, a greater cartographer and navigator than Bougainville, sailed into some of the same territory when he led the expedition in HMS *Endeavour* to Tahiti to observe the passage of Venus in 1769. Cook was enraged, when he visited Cape Town in 1773 (during his second voyage of discovery in the *Resolution*), to discover that the journalist Dr John Hawkesworth had gone into print, using among other sources Cook's own journals, with his own *Voyages*. Cook was determined that this would not happen a second time. Apart from wanting the kudos of writing up his explorations in Antarctica before anyone else, Cook was looking forward to the money. Yet again, however, he was pipped to the post. Though the Admiralty, to protect Cook, put a gagging order on the official naturalist of the *Resolution*, Dr Reinhold Forster, preventing him from writing up his journals of the voyage, they could not prevent Forster's young son George from publishing the tearaway best-seller *A Voyage Round the World in his Britannic Majesty's Sloop, Resolution* in 1777.

The tension between the professionals of the Royal Navy and the naturalists, artists, journalists and writers they took on board would always be strong. FitzRoy, though aristocratic, was not especially rich and he wrote his own account of the voyage of the Beagle. As well as wanting some money therefrom, he was justly hopeful that his work as an expert cartographer would achieve public recognition. When Darwin sent to FitzRoy his finished draft of *Journal of Researches*, the Captain read it with considerable chagrin. What took FitzRoy's breath away, and that of the other officers to whom he showed Darwin's Preface, was the sense conveyed by the young man's words that he had been the hero of the entire journey, and that the prodigious accumulation of specimens had been all his own achievement, all his own work. Such was Darwin's genius for self-promotion that this is *still* how the *Voyage of the Beagle* is perceived, even by those scholars, for instance, who edited the second edition of the book (*The Voyage of the Beagle*) for Penguin in 1989. Captain FitzRoy drew the map of South America, patiently tacking in and out of inlets, braving storms and enduring the frustrations of calms, measuring, drawing, returning to the coast to redraw, while for much of the time Darwin was in his cabin being sick, or ashore, hunting, shooting and, of course, observing natural history. Yet although the voyage of the *Beagle* made such a difference to all subsequent geographical study of South America, it is not FitzRoy's voyage, but Darwin's *Voyage of the Beagle* which went down in the history books.

FitzRoy at first acknowledged the receipt of the proofs of Darwin's *Journal of Researches* in a curt note, written in the third person: 'Captain FitzRoy begs to say that he has spoken to Captain Beaufort – to Sir Edward Parry' (Rear Admiral and the explorer who made unsuccessful attempts to find the Northwest Passage in voyages between 1819 and 1824). FitzRoy declined to give any opinion 'at present' on Darwin's book, but it is clear that not only he, but Beaufort and Parry believed that Philos had overstepped himself in pushingness.

Next day, however, FitzRoy was unable to restrain himself and burst forth – on 16 November 1837. First, he reminded Darwin that the very idea of inviting the young naturalist on board the *Beagle* had been his own, FitzRoy's. Then he recalled how his officers 'gave you the preference upon all occasions – (especially Sulivan, Usborne, Bynoe and Stokes) and think – with me – that a plain acknowledgement – without a word of flattery – or fulsome praise – is a slight return due from you to those who held the ladder by which you mounted to a position where your industry – enterprise – and talent could be thoroughly demonstrated'.[5]

The letter perspicaciously notes Philos's ruthlessness in self-promotion and in particular his willingness to airbrush out of his story those who had 'held the ladder'. FitzRoy went on:

> I was also astonished at the total omission of any notice of the officers – either particular – or general. My memory is rather tenacious respecting a variety of transactions in which you are not aware that the ship which carried us safely was the first employed in exploring and surveying whose officers were not ordered to collect – and were therefore at liberty to keep the best of all – nay, all their specimens for themselves. To their honour – they gave you the preference.[6]

The shock FitzRoy felt is palpable ('Believe me Darwin, I esteem *you* far too highly to break off from you willingly – I shall always be glad to see you').[7] The diffident, tall Philos, whom FitzRoy thought he had come to know and love during five years at sea, was certainly there. He was the friend he would always be glad to see. There was, however, another Darwin, scrambling up the ladder and ignoring those who had held this useful device.

The accusation, which came, not from FitzRoy alone, but from the other officers concerned, was levied in 1837, the year in which we know that Darwin started his First Notebook on the Transmutation of Species. It is in these notebooks that Darwin began the observations and meditations which would lead eventually to his *Origin of*

Species. As has already been pointed out, Edward Blyth expounded the idea of transmutation in the January 1835 issue of the *Magazine of Natural History*. In Darwin's Second Notebook, never intended for publication, he referred to Blyth's question, 'May not, then, a large proportion of what are considered species have descended from a common parentage?'[8]

Blyth's three essays on evolution were published in the 1835, 1836 and 1837 issues of the *Magazine of Natural History*. Blyth believed that nature was operating a system of selection of species which was analogous to a human breeder selecting particular dogs or pigeons. He saw what could be called – what Loren Eiseley did call – a 'Malthusian struggle of population against natural resources'.[9]

> It would be easy to point out additional hindrances to the more extensive spread of species of fixed habit, by treating on the fraction which are allowed to attain maturity, even in their normal habitat, of the multitude of germs which are annually produced; and in what ratio the causes which prevent the numerical increase of a species in its indigenous locality would act where its adaptations are not in strict accordance will sufficiently appear, on considering the exquisite perfection of those of the races with which it would have to contend.[10]

Blyth, in these three essays, was not merely putting forward the theory of the origin of species which would one day be claimed as 'Darwinian'. He was also wrestling with the clear objections to the theory. One of these was contained in the second volume of Lyell's *Principles of Geology*: 'where a capacity is given to individuals to adapt themselves to new circumstances, it does not necessarily require a very long period for its development; if indeed, such were the case, it is not easy to see how the modification would answer the ends proposed, for all the individuals would die before new qualities, habits or instincts were conferred'.[11] This was one of the objections to his own idea which Blyth found insuperable.

While Blyth discussed these matters in a magazine which Darwin quotes in his notebooks, at the very period when Darwin began in earnest to answer 'the species question', we confront a problem. The first fifty pages of Darwin's First Notebook on the Transmutation of Species have been cut out. We know how carefully and deeply

Darwin cherished his notebooks. 'I remember', said his son Francis, 'when some alarm of fire had happened, his begging me to be especially careful, adding very earnestly, that the rest of his life would be miserable if his notes and books were to be destroyed.'[12]

On the first page of the notebook, Darwin wrote, 'all useful pages cut out, Dec 7/1856 (and again looked through April 21 1873)'. So far, searches in Cambridge University Library, Down House, the Royal College of Surgeons and the British Museum of Natural History have been in vain. The missing pages relate to the year 1837, the year when Edward Blyth was expounding his theory of the origin of species, the year when Captain FitzRoy denounced Darwin, who had just completed his journal of *The Voyage of the Beagle*, for not giving sufficient credit to those who 'held the ladder by which you mounted'. It has been questioned whether Loren Eiseley was right in accusing Darwin of plagiarism, but it seems inconceivable that Darwin, who, we know, read the *Magazine of Natural History*, had not read the articles. This is not to mention Darwin's very obvious debt – already mentioned, and to be explored later – to Lamarck.

One ladder-holder to whom Darwin always did acknowledge his debt, however, was Charles Lyell. His revised, second edition of the *Voyage* was dedicated to Lyell. 'The chief part of whatever scientific merit this journal and the other works of the author may possess has been derived from studying the well-known and admirable *Principles of Geology*.' The notes which Darwin sent home from the voyage for Henslow's safe-keeping had caused great excitement. Henslow had communicated their contents to Sedgwick at a meeting of the Geological Society on 18 November 1835, and when Lyell had absorbed them he wrote to Sedgwick in December, exclaiming, 'How I long for a return of Darwin! I hope you do not mean to monopolize him at Cambridge.'[13] In his Presidential Address to the Society in 1836 Lyell paid fulsome tribute:

> Few communications have excited more interest in the Society than the letters on South America addressed by Mr Charles Darwin to Professor Henslow. Scattered over the whole [area examined] and at various heights above the sea, from 1300 feet downwards, are recent shells of the littoral species of the neighbouring coast so that every part of the surface seems once to have been a shore and Mr Darwin

> supposes that an upheaval to the amount of 1300 feet has been owing to a succession of small elevations, like those experienced in modern times in Chili.[14]

FitzRoy's loyalty to his unfortunate friend Captain Seymour, and the testimony which got Seymour acquitted at his court martial – for the loss of HMS *Challenger* – provided some of the crucial evidence for Lyell, whose ideas had considerably developed since he published the first volume of *Principles*. 'Did you see', he wrote to Sedgwick in October 1835,[15] 'that Captain FitzRoy had borne witness in a court-martial that the late Chilian earthquake had altered the whole coast? That a north-westerly had become a southerly current: thus the island of Mocha was *upheaved* 10 feet? and this upon *oath*! and the Captain of the "Challenger" acquitted on this evidence!' Yet it was Darwin, rather than FitzRoy, whom Lyell cited in his Presidential Address to the Geological Society on 14 January 1837 as his principal witness for recent earth-movements, and it was Darwin's collection of the bones of extinct mammalia 'near the banks of the Rio Plata, in the Pampas of Buenos Ayres and in Patagonia'[16] which excited special mention. 'These fossils . . . establish the fact that the peculiar type of organization which is now characteristic of the South American mammalia has been developed on that continent for a long period.'[17]

Lyell was the most widely respected scientist in the eyes of the public, as was attested by Darwin's discovery of intelligent English merchants in South America discussing *The Principles of Geology*. Lyell's own views, however, were adapting all the time and he had by no means wholly convinced the scientific establishment. William Whewell, reviewing the second volume of *Principles*, had coined the terms 'uniformitarian' and 'catastrophist' to describe the opposing sects among geologists. Lyell, as the chief uniformitarian, was concerned to demonstrate that gradual changes in the environment, and the inherent relation between organisms and their environment, were enough to explain geological change, without recourse to imagined 'catastrophes' interrupting a natural process. (Buckland, Professor of Geology at Oxford, for example, remained a convinced catastrophist.)

That the young Darwin, who had collected so many specimens

and made so many innovative observations on his recent voyage, should be a keen supporter of Lyell's was of tremendous encouragement to the older man. (Lyell was twelve years Darwin's senior.)

Though Lyell's work was primarily concerned with geology *per se*, it clearly had a direct bearing upon other branches of science. The fossil evidence provided tantalizing proof of the repeated extinction of species: tantalizing since Lyell rejected Lamarckian notions of 'transmutation', and rejected catastrophist theories which could explain, for example, why prehistoric creatures ceased to be – not enough room for a *Megatherium* in Noah's Ark. Lyell could see that the origin of species must be attributable to just some such 'natural' process as explained the geological formation of the planet, but *how* species evolved eluded him.

> I have admitted that we have only data for *extinction*, and I have left it rather to be inferred instead of enunciating it even as my opinion, that the place of lost species is filled up (as it was of old) from time to time by new species. I certainly wish it to be inferred that . . . the extinction has been going on in the last 6,000 years, and that the substitution of species to supply the vacancies, which must always be occurring, has also been going on; though *how*, is a point we are as ignorant of as of the manner of God's creating the first man.[18]

In this letter to Professor Sedgwick, Lyell in effect wrote the blueprint for what would be central to Darwin's scientific journey for the next twenty years. Darwin set out to answer Lyell's questions. He was to do so using the arguments which had been spelt out by Edward Blyth, but he needed twenty years, and more than twenty, to provide what he considered convincing support for the theory.

The latest news, Lyell wrote to his sister in May 1837,

> is that two fossil monkeys have at last been found, one in India contemporary with extinct quadrupeds, but not very ancient – Pliocene perhaps – another in the south of France, Miocene and contemporary with Palaeotherium. So that, according to Lamarck's view, there may have been a great many thousand centuries for their tails to wear off, and the transformation to men to take place.[19]

While Charles Lyell lampooned Lamarckian evolutionism, he and Darwin were deep in geology. 'I am very full of Darwin's new theory of Coral Islands . . . the last efforts of drowning continents to lift their heads above water.'[20]

History labels Lyell as the pioneer of modern geology, Darwin as that of evolutionary biology. It is fascinating to see, during those very active years after Darwin's return to England and before his marriage, how much Lyell looked to Darwin for geological research, and how at this stage it appeared to be Lyell who was most concerned with the riddle of human origins. So, for example, in 1838 we find Lyell writing to Sedgwick on the knotty issue of whether nature 'has at length, stopped short in her operations' – that is, whether development or mutation has ceased since the creation of man. Lyell put it to Sedgwick that there were insuperable difficulties in this point of view. The Cambridge scientists Whewell and Sedgwick appeared to be saying that 'God rested on the seventh day. Thereafter nature ceased to develop or adapt.' 'To me', wrote Lyell, 'it appears that the line you are represented to have taken is to hazard a far bolder hypothesis than I should have dared to do, viz. that no new creatures have begun to exist for the last 6,000 years, or for such time as man has existed, although geology has now brought to light the proofs of an indefinite series of antecedent changes, such repeated failures of species of animals and plants, and their replacement by others.' He went on, 'The burden of proof rests on him who ventures to affirm that Nature has, at length, stopped short in her operations, and that while the causes of destruction are in full activity, even where man cannot interfere, she has suspended her powers of repair and renovation.'[21]

To Darwin, by contrast, Lyell turned, not for his views on such generalized or quasi-metaphysical notions, but for his thoughts on coral reefs and volcanoes. In his revision of *Principles of Geology*, Lyell was asking how far Darwin considered 'gradual risings and sinkings of the spaces occupied by corraline and volcanic islands in the Pacific as leaning in favour of the doctrine that many parallel lines of upheaval or depression are formed contemporaneously'.[22]

As we follow Darwin's footsteps during these very active years, then, nothing prepares us, quite, for the direction they were to take. Darwin, after a brief spell with the family in Shrewsbury and at Maer, took rooms at 36 Great Marlborough Street, just down the road from the house of his brother Erasmus who was, in the judgement of Thomas Carlyle, 'idle'. Darwin was the opposite. He had gone abroad 'idle', but he returned as a frenzied, driven person, a marked contrast from Erasmus who, beyond his friendship with bluestocking Harriet Martineau and the Carlyles, actually 'did' very little. Darwin allowed himself to be escorted about London by Ras, but only to a very limited degree was he a social animal at this stage. He liked Carlyle at first meeting, but told his cousin Emma Wedgwood, 'It is high treason, but I cannot think that Jenny [Mrs Carlyle] is either quite natural or lady-like.'[23] On another occasion he confessed himself unable to understand 'half the words she speaks, from her Scotch pronunciation'.[24] Where Erasmus, for all his eremitical tendencies and his self-protective fusspot bachelordom, had all the time in the world for these bluestocking friends, Darwin was seized with an insatiable appetite for work. For the first three months of 1837 he was in Cambridge – 'Read paper on "sand-tubes" at Cambridge Philosophical Society'[25] – writing up his researches on the voyage, and preparing volumes on the zoology and geology of those researches. At the same time he was making the Cambridge academics – Henslow, Sedgwick, Whewell – aware that he was now a professional scientist. Whewell responded by making Darwin one of the two Secretaries of the Geological Society. In London, while continuing to write up his researches and catalogue his prodigious collections, he was also very visibly setting up his brass plate as the brightest young scientist in Britain. At the Geological Society he heard Lyell's laudatory allusions to his work. He himself read notes on *Rhea americana* and *Rhea darwinii* at the Zoological Society of London.[26] He applied, successfully, to the Chancellor of the Exchequer to finance the publication, with engravings, of his *Beagle* researches. 'The whole of the Collection made by Mr Darwin either has already been, or will hereafter be distributed to the Public Museums, where they will be of acknowledged Service; and under the circumstances of the case, the Chancellor of the Exchequer feels justified in recommending to the Board [of the Treasury] to give their sanction to the Application of a Sum not exceeding in the whole £1000 from Civil Contingencies in aid of this Publication.'[27] At one bound, Darwin had liberated his scientific endeavours from dependency on his father's patronage, and established himself as a figure of national consequence.

More momentous, however, than any of the published work, or any of the papers and lectures which Darwin gave in this period, were the notebooks relating to the transmutation of species, the first of which he remembered as beginning 'about July 1837'. (This is the notebook on which he had written, in 1856, 'all useful pages cut out'.)[28] In the first surviving page, he went back to his grandfather's *Zoonomia*, and by the fifth surviving page he was searching for the 'final cause of life'.[29] It is clear that his thoughts had been moving along the lines outlined in Blyth's essays on evolution. In the notebooks we watch Darwin wanting to test the veracity of the theory with complete thoroughness. Some of the observations are charming – 'Gnu reaches Orange River & says so far will I go and no further.'[30] Others – 'Strong odour of negroes, a point of real repugnance'[31] – are less so. The notebooks of this period[32] demonstrate the fullest possible engagement with the evolutionary questions which would make Darwin's name. They are notebooks, not essays. They are fragmentary. They reveal a naturalist with a prodigiously wide knowledge – of botany, zoology and geology – and a wide reading in his subject, trying to put together the idea of 'mutation'. How could such a strange idea work in practice, however slow the 'mutation' was? 'We need not think that fish & penguins really pass into each other.' But did not Blyth's idea actually require just such a happening? If a 'large proportion of what are considered species have descended from a common parentage', would we not be obliged to believe that mammoths, woodpeckers, chimpanzees and water voles all descended, somehow or other, from a common ancestor? The notebooks find no answer to such a strange puzzle, but they are aware of its strangeness. Notebook B notes the differences in the Académie Royale des Sciences in Paris between Georges Cuvier, who insisted that there were great gaps in the supposed chain linking molluscs and vertebrates, and Etienne Geoffroy Saint-Hilaire, who believed in the fundamental 'unity' of species.[33] Most of the notebooks are made up of short pithy observations. 'Fish never become a man – does not require fresh creation! – If continent had sprung up round Galapagos in Pacific side, the Oolite order of things might have easily formed.' By now, whenever he refers to evolutionary theories of the origin of species he speaks of 'my theory'.[34] 'With belief of ~~change~~ transmutation & geographical grouping we are led to endeavour to discover *causes* of change.'[35] These words were probably written in 1838.[36] This was the year in which Darwin believed he had indeed lighted upon the *cause* of evolution: or at any rate, the dominating factor in the natural world. He found it not in his many observations of natural history, nor in the work of his fellow scientists, but in that contentious work of economics – the Revd Thomas Malthus's *Essay on the Principle of Population* first published in 1798 but subsequently revised. Malthus, a wealthy Surrey clergyman, died in 1834 of heart disease, aged sixty-three. He had very much belonged to the same intellectual and social milieu as Darwin himself. Ras Darwin's friend Harriet Martineau, one of the economist's most fervent disciples, wrote to Darwin's friend W. J. Fox (William Johnson Fox, preacher and author; a different figure from Darwin's cousin William Darwin Fox, Rector of Delamere.), 'I cd scarcely help laughd [*sic*] that day to think what the *Age* wd have said if it cd have seen us sitting after dinner; I, dancing Mrs Wedgwood's baby & Malthus patting its cheeks.' (This was Fanny Wedgwood, married to Hensleigh. Malthus's daughter had been a bridesmaid at their wedding.[37] Ras was at the dinner.)[38]

Darwin read Malthus fifteen months after his return to England, at the very moment when he was cramming the notebooks with miscellaneous observations about the species question. Malthus provided him with the key. The implications of Malthus's book, he would later recall, 'struck me at once'. That is, it was not just the strongest or the most robust who would get ahead in the 'struggle for existence'. Rather, it was those who possessed some particular attribute, or variation, which made them suited to living in a particular environment. Those possessing such attributes got ahead. Those lacking them went to the wall. Little by little by little, over many generations, these attributes would be refined, until 'The result would be the formation of a new species . . . Here at last I had a theory by which to work.'[39]

Absolutely central, then, to Darwin's story is this epiphany, this revelation, that Malthus explained the secrets of biology's puzzles. The puzzles would continue to be scientific ones. The answer, however, was found in a socio-economic theory. Another word for 'a theory by which to work' would be a myth. Darwin had found the central consolation myth which he would give to his Victorian contemporaries. What Malthus had believed applied only to the specific question of human population growth and hunger actually applied to everything.

Malthus is one of those distinguished intellectuals of whom Darwin would be one – Marx, Freud and John Maynard Keynes come to mind as others – whose surnames, morphed into epithets, became symbols for ways of looking at the world. For those who hated the 'Malthusian' economic idea – such as Charles Dickens, who lampooned it in the figure of Scrooge inveighing against the 'surplus population' – it would always encapsulate meanness and cruelty. William Cobbett denounced the 'barbarous and impious Malthus'.[40] For Malthusians, such as Harriet Martineau, his *Essay* simply explained the nature of the world. His *Essay on Population*, in short, was one of those books in which the nineteenth century delighted, which could be said to explain not merely a specific question – how can populations increase without an increase in the food supply? – but every question. Malthus wrote his *Essay* after extended conversations with his father, who had espoused just such a Utopian view of politics as had attracted old Josiah Wedgwood and Dr Erasmus Darwin. Malthus's father, a disciple of the Marquis de Condorcet and William Godwin, had believed that human wretchedness was caused by bad governance. Malthus himself, although he later joined the Church of England in order to get to Jesus College, Cambridge, had been sent by his father to the Dissenting academy at Warrington.[41] Malthus, talking to his father in the aftermath of the French Revolution, had seen that, however enlightened your system of government, the simple mathematics of population and food supply remained the fundamental fact of human existence. Populations increased on a 'geometric ratio' (2–4–8–16), doubling every twenty-five years. The food supply can increase, at its utmost, only in an 'arithmetic' ratio (1–2–3–4). Hence, every few years, the population finds that it does not have enough to eat. Here, Malthusian 'checks' take place to cull the 'surplus' population. These are famines, plagues or wars.

The paradox of Malthus's *Essay*, coming when and above all where it did – in Britain – is that while it had been partially true for most of European history, it was about to become untrue. While he wrote the first version of the *Essay*, Britain was suffering food shortages during the Napoleonic Wars. Shortages would continue after the wars. When the Industrial Revolution really began to take off in Britain, however, for the first time in economic history, the phenomenon became manifest of an increase of population actually increasing wealth. Victorian Britain would see a largely *urban* population growth, which created sufficient surplus wealth to feed the swelling populace, if necessary by importing food. By the time Malthus died, Britain was ceasing – or at any rate urban England was ceasing – to be Malthusian, though Scotland and, most disastrously, Ireland remained Malthusian economies, as do, to this day, the countries of sub-Saharan Africa.[42]

Darwin started to read Malthus's *Essay* on 28 September 1838. He can scarcely have been ignorant of the central Malthusian argument, which had been incorporated by Paley. William Whewell had discussed the Malthusian ideas in a paper delivered to the Cambridge Philosophical Society in 1829. In 1838, moreover, largely thanks to the tariffs imposed on imported wheat which artificially inflated the price of bread – keeping the great landowners rich and the poor hungry – there had been food riots in Britain. Darwin was living in a country where it appeared that Malthus's doctrines were demonstrable social facts. A clergyman Malthus may have been, but Darwin in his notebooks seized upon the essential godlessness of the idea. He writes, 'Epidemics – seem intimately related to famines [*sic*], yet very inexplicable,' and then quotes Malthus, adding his own underlinings and exclamation marks: 'It accords with the most *liberal*! spirit of philosophy to believe that no stone can fall, or plant rise, without the immediate agency of the deity. But we know from *experience*! that these operations of what we call nature, have been conducted *almost*! invariably according to fixed laws. And since the world began, the causes of population & depopulation have been probably as constant as any of the laws of nature with which we are acquainted.'[43] Darwin (who slightly misquotes Malthus, who wrote 'divine power' rather than 'deity') then commented, 'I would apply it not only to population & depopulation, but extermination & production of new forms – their number & correlations.'[44] Lyell's geological revelations had shown, in fossil evidence, the plenitude of extinct species. The argument against transmutation had been that these species had simply died out to be replaced by new ones. The system by which they were replaced could only be by a new creation. By embracing Malthusianism as a general principle of nature, Darwin could see, not only how less adequate species became extinct, but how a process could be at work *within species* which enabled them to adapt and survive. Thus, while some species had become extinct, others – whose fossils showed such affinity and comparability with later species – were in the process of transmuting themselves, adapting to Malthusian circumstances in order to make themselves capable of survival. Nature, the whole history of the natural process, was nothing less than a struggle for survival. The hind legs of a hare grow longer and longer to escape being eaten, and the giraffe's neck becomes longer to reach foliage. Those who don't eat die, or go hungry, just as weaker, feebler classes of people, unable adequately to feed their families, ground down by the draconian Poor Laws so pleasing to Miss Martineau, were confined to the workhouse, whereas the Darwins and the Martineaus and the Wedgwoods had large comfortable houses and incomes. As the *Quarterly Review* put it when considering the Malthusian question, 'It would be a real blessing if the working classes could be made acquainted with some of the fundamental principles of Political Economy: such as the laws of population; the causes of the inequality of mankind . . . They would then perceive that inequality does not originate in the encroachments of the rich or the enactments of the powerful, but has been necessarily coeval with society itself in all its stages.'[45]

Darwin would be able to offer to the nineteenth century something even more irrefutable than the principles of political economy which had apparently been immutable since history began. He could offer scientific proof that the clever and the strong and the adaptable triumphed over the weak by a law of nature. Harriet Martineau deemed herself to be 'radical' in politics, as did the Darwins and the Wedgwoods. Her father, a cloth manufacturer in Norwich whose business collapsed after the banking crisis of 1825, no doubt contributed to what seems, to a modern eye, a confusion in her mind – on the one hand thinking of herself as 'progressive', while at the same time, in her famous tales, written as 'Illustrations of Political Economy', advocating free trade and an absolute non-interventionist approach to industrial law. She stridently opposed any attempt by the legislature to impose safety regulations in factories, for example. Mill hands were among the happiest and healthiest of the working classes, 'while the woes of others are attributable to ignorance, bad dwelling, crowding – in short *town* not mill evils'.[46] She greatly opposed Lord Ashley's Ten Hours Bill, believing that the working man could take care of himself.

At the same time she believed society was moving ever onward and upward. The light which was present at Runnymede when Magna Carta was signed 'shone in the eyes of Cromwell after Naseby fight . . . and is now shining down into the dreariest recesses of the coal-mine, the prison, and the cellar' – and would eventually 'vivify' the whole world.[47] There was no need for Lord Ashley to pillage 'the rights of one class of people at the expense of the earnings of another'.[48] When the working classes adapted, improved themselves, they would be as cheerful as the prosperous mill operatives she imagined living in the North. The Victorian rentier class of which Miss Martineau was a shabby-genteel representative, Darwin a well-heeled one, had to persuade themselves that there was something inexorable, natural, about their superiority to the working class on whom their wealth in point of fact depended.

8

Lost in the Vicinity of Bloomsbury

On 27 November 1838, the naturalist wrote in his notebook, 'Sexual desire makes saliva to flow "yes, *certainly*" – curious association: I have seen Nina licking her chops . . . ones tendency to kiss, & almost bite, that which one sexually loves, is probably connected with flow of saliva, & hence with action of *mouth* & jaws . . . The association of saliva, is probably due to our distant ancestors having been like *dogs* to bitches. How comes such an association in man – it is bare fact, on my theory intelligible.'[1]

The notebook contains reflections upon matters which would interest Darwin for the rest of his life, some of which would resurface in published form, in *The Origin of Species* (1859), in *The Descent of Man and Selection in Relation to Sex* (1871) and in *The Expression of the Emotions in Man and Animals* (1872). It contains in its early pages material which derived from conversations with his father about insanity. Later, Dr Darwin was displaced as confidant by Emma Wedgwood, and she herself made an annotation on page 113 of the book: 'A child born on the 1st March was frightened on the 24th May at Cresselly by the boys making faces at it, so much so that the nurse had to carry it out of the room, nearly 3 months old.'

Notebook N, as it is now called, is an object which contains between its rust-coloured leather covers and in its mere ninety-two leaves so much of Darwin. It contains the range of his reading in the late 1830s: Johann Caspar Lavater on physiognomy, John James Audubon on the birds of America, William Gardiner's *The Music of Nature*, James Mackintosh's *Dissertation on the Progress of Ethical Philosophy*; Lamarck, Hume (*Natural History of Religion*), Oliver Goldsmith's *Essays*, as well as old Sir Thomas Browne. There is, in other words, a wide range, and the need at all times to collect and annotate and question every detail of experience – the spectacle of a restless dog, evidently dreaming, a panther at the zoo uncovering its teeth to bite, 'the *senseless* grin of passion'.[2] Then, how typical, there are the everlasting questions: 'Has an oyster necessary notion of space?';[3] 'Does a negress blush? I am almost sure Fuegia Basket did.'[4] Then, we find the tendency which would become ever more marked as Darwin grew older, that is the yearning to generalize from the particular, not least to find in the particular confirmation of his general evolutionary ideas. 'Our distant ancestors having been like dogs to bitches.'[5] And then again, Notebook N contains the handwriting of the Wedgwood cousin who would become Darwin's wife. Meanwhile he, observing her increase of saliva flow, and his own, as their kisses became more sexual, did so with the same degree of detachment as Henslow had seen a chimpanzee pout and whine when a man went out of the room,[6] or as the Keeper at London Zoo had noted that wolves wag their tails 'a little when attending to anything or excited'.[7] Darwin's and Emma Wedgwood's degree of excitement in one another's company could be noted with clinical accuracy: 'Blushing is intimately concerned with thinking of one's appearance, does the thought drive blood to surface exposed, face of man, face, neck – "upper" bosom in woman: like erection.'[8] Emma's own addition to the notebook – about the three-month-old baby crying when boys made faces at it – locates all these thoughts, buzzing in the brain of Darwin, within the family circle: in this case in the Allens' house, Cresselly, in Pembrokeshire. The baby in question was one of the many members of the family called John: Emma's nephew (son of her brother Henry Allen Wedgwood – Harry).

When Darwin's mind turned to marriage, there was never much doubt that for him, as for so many members of his family, this would mean marriage to a cousin.

From the spring of that eventful year 1838 he had begun to think earnestly about marriage. Much has been made, in our story, of the importance of money. Although Dr Darwin and Josiah II had plenty of it, Charles Darwin did not feel especially rich, and the prime objection which he put down on paper when mulling the question over was: 'If marry – means limited. Feel duty to work for money. London life, nothing but Society, no country, no tours, no large

Zoolog. Collect, no books.' It was certainly enormously to Baron Alexander von Humboldt's advantage that his homosexuality obviated the need to shackle himself with domesticity and dinner-parties. Darwin was differently made. It has often been noticed how often sex comes not only into his thoughts, but into the titles of his books. Yet the need to get on was, if anything, as strong; as was the purer thirst for knowledge which now consumed all his waking hours. The sheer time-wasting involved in courtship was an agonizing prospect – yet another reason for marrying a cousin whom he knew already and who would require no initiation into the family ethos. The notes on this question of marriage were infinitely bound up with his future career. One possibility would be to put in for a Cambridge professorship, though interestingly enough he considered that there he would be a 'fish out of water'. This was 'better'[9] than poverty, of course.

He visited Shrewsbury in July and it seems clear that he discussed the whole matter with the 'Governor', who 'says soon for otherwise bad if one has children – one's character is more flexible – one's feelings more lively & if one does not marry soon, one misses so much good pure happiness'.[10] The pros and cons of matrimony were duly written out in note form. Cons included 'forced to visit relatives and to bend in every trifle' and – underlined twice – '*Loss of time*'. Advantages, rather engagingly, included 'Charms of music & female chit-chat. These things are good for one's health – *but terrible loss of time.*'[11]

In 1837 Darwin's sister Caroline had married her cousin Josiah Wedgwood III of Maer Hall. It was to Maer that Darwin returned in the summer of 1838, and there he began to woo Josiah's sister Emma. Bessy Wedgwood, Emma's mother, had teased Dr Darwin by pretending that she hoped Charles would marry Harriet Martineau,[12] so it was a hearty relief to realize that his son had seen sense and had kept matters in the family.

'Emma having accepted Charles gives me as great happiness as Jos having married Caroline, and I cannot say more,' the Doctor wrote to Josiah II, signing himself 'your affectionate brother'.[13] Announcing his betrothal to Lyell, Darwin said, 'The lady is my cousin Miss Emma Wedgwood, the sister of Hensleigh Wedgwood,

& of the elder brother, who married my sister, so we are connected by manifold ties.'[14] The old aristocracy, the landed classes, whose stranglehold on political power in Britain had been somewhat loosened by the Reform Act of 1832, could preserve its power in part by rigidly selective matrimonial processes, Lady Catherine de Bourgh doing all in her power to limit the number of Miss Bennets who could capture the fancy of a Mr Darcy. The new 'aristocracy', the Victorian upper-middle class and intelligentsia, would have to guard its breeding stock no less jealously – if anything rather more.

The letters which Emma wrote from Maer to her cousin and future husband during November and December are full of humour and affection, but they make no pretence to passion: 'Write to me soon like a dear old soul.'[15] One letter is signed, 'your affectionate Grandmother Emma W'.[16] She writes to 'my poor old Charley',[17] 'My own dear old geologist',[18] and 'my own dear Nigger'.[19] Emma was a year older than Darwin. The high, flat brow, hooded eyes and thin straight lips which curled in a humorous smile marked her instantly as a Wedgwood. Slightly untidy ringlets fell on either side of the intelligent face. She wore a faintly crooked central parting. The brown eyes and nose are almost comically identical to those of her brother Jos.

It would quickly turn out to be a stable, harmonious partnership, though it demanded of Emma two painful sacrifices. One was that she was electing to share her life with a man who lacked her belief in Christ. She had clearly winkled this out of him in their conversations before marriage. Quite what Darwin did or did not believe at the beginning of 1839 we can only conjecture, but it is probably safe to define his position as 'theism, in which God is the first cause and creator of a universe that operates entirely according to laws'.[20]

Emma who, together with her sisters ran the Sunday school for local children at Maer, was a believer in Jesus Christ, in the redemption of the world through his blood and in the Resurrection. His father had counselled Darwin to keep his doubts a secret from Emma, but there was too much mutual respect in the relationship for that to be a possibility. She implored him to read 'our Saviour's farewell discourse to his disciples which begins at the end of the 13th chap of St John. It is so full of love to them & devotion & every beautiful feeling. It is the part of the New Testament I love best.'[21]

Equally, however, she wanted Darwin to continue being honest with her. She knew that 'you are acting conscientiously & sincerely wishing, & trying to learn the truth . . . May not the habit in scientific pursuits of believing nothing till it is proved, influence your mind too much in other things which cannot be proved in the same way, & which if true are likely to be beyond our comprehension[?]' She feared the danger of 'giving up revelation . . . I am rather afraid my own dear Nigger will think I have forgotten my promise not to bother him, but I am sure he loves me & I cannot tell him how happy he makes me . . .' These words were written a month after marriage. Darwin wrote at the end of it, 'When I am dead, know that many times, I have kissed & cryed over this.'[22] The knowledge that they were a grief to his wife added poignancy to his researches, but neither of them lost their affection for the other nor their own integrity.

The second sacrifice made by Emma was, from a practical point of view, more demanding. The courtship letters saw him as a patient who needed looking after. Often, as when she signed herself as his 'Grandmother', she viewed the situation as a comedy. 'It is very well I am coming to look after you my poor old man for it is quite evident that you are on the verge of insanity & we should have had to advertize you, "Lost in the vicinity of Bloomsbury a tall thin gentleman &c &c quite harmless whoever will bring him back will be handsomely rewarded."'[23] Darwin, for his part, had clearly not spared Emma the details of his mysterious condition. After a railway journey back from Staffordshire, in which he had changed trains at Birmingham and not had time for dinner, he added, 'I have no very particular news to tell you, as you will guess by my having written so full an account of my stomachic disasters.'[24]

Throughout the autumn which culminated in his betrothal, Darwin had been suffering from intense headaches, cardiac palpitations and repeated gastric upsets. Emma, a humorous and healthy young person, was unable at first to take these symptoms entirely seriously. Her satirical self-casting, before she married him, as Darwin's mental nurse, granny and helpmeet would prove to be prophetic. Darwin had come to believe that 'a test of hardness of thought' was felt as 'weakness of my stomach'. It was the period when, while still unpacking the collection, he was filling the notebooks with thoughts and with extracts from his reading, as well as with his melancholy regurgitations of Malthus. Which of these mental exertions – the contemplation of volcanic eruptions in South America, or the idea of Malthusian struggle being the explanation for all forms of animal life – led to his outbursts of nausea and flatulence he does not appear to have known. His medical advisers urged him 'strongly to knock off all work'. It is advice which the young Darwin, the pre-*Beagle* Darwin, would have accepted with gusto. The prematurely aged, bald, bewhiskered scientist, 'lost in the vicinity of Bloomsbury', would have been quite incapable of emulating the young loafer of The Mount or Christ's College and simply taking his gun and rods for a few happy days of country sport. He had become a man possessed. Some have even wondered, from the evidence of the notebooks, whether Darwin, even at this early stage of his thought, feared exposure to criticism, even persecution, if he were able beyond question to verify Blyth's theory of species origin.[25] 'Mention persecution of early Astronomers – then add chief good of individual scientific men is to push their science a few years only in advance of their age (differently from literary men), must remember that if they *believe* & do not openly avow that belief they do as much to retard as those whose opinion they believe have [*sic*] endeavoured to advance cause of truth.'[26]

The dynamic of their relationship seems to have been entangled with Darwin's psychological need, on some level, to be a semi-invalid. Somehow, the testing of the theory – which had become 'my Theory' – needed an inordinate amount of time and seclusion. A normal life, which included travel and social life, would have eaten into his research time. The body rescued him and with a set of debilitating psychosomatic conditions it confined him to base. This would have been no use, indeed would have led to loss of time, had his transmutation from world traveller to reclusive invalid not been accompanied by finding a willing nursemaid.

Darwin decided that the marital home should be in central London. His initial ambition was for 'small house near Regent's Park – keep horse'.[27] Fossil-collecting and botanizing were more in his line than house-hunting, however, and having 'been all round' Regent's Park he found nothing suitable. 'After many long walks, Erasmus & myself are driven to the conviction, that our only resource will be in the streets or squares near Russell Square.'[28]

Over a thousand new houses had been built on the Duke of Bedford's estate in Bloomsbury between 1792 and 1828, the year when the University of London opened in Gower Street. Moreover it was during the 1820s, 1830s and 1840s that Robert Smirke's majestic British Museum was being built on Great Russell Street. University College Hospital (originally the North London Hospital), a centre of pioneering work in medicine and anatomy, was near by. By choosing to live in Bloomsbury, Darwin, as well as placing himself conveniently close to Euston Station (handy for trains to Staffordshire), was in the very centre of the march of mind. The house which he eventually lighted upon, 12 Upper Gower Street, had been the residence of the first Warden of University College, Leonard Horner.[29] (Horner, Edinburgh born and bred, had known Darwin in Scotland. His appointment as Warden was unwise, and having quarrelled with most of the professors and the University council, Horner had resigned in July 1831.[30]) The advantage of becoming the tenant of this Bedford-owned property was that they took it fully furnished. 'We shall not have much to buy – even the crockery & glasses are very perfect.'[31] The solicitor acting for them 'examined all the tables & chairs & said they are made of excellent wood & must have cost a great deal of money. In fact I am convinced we have been most fort[unat]e & I am in great triump [*sic*] at having come to so good an end.'[32] He was cutting things fine, for arrangements were not finalized until after Christmas 1838, only weeks before the wedding. On New Year's Day 1839 large vans pulled up outside 12 Upper Gower Street and began to unload. 'I was astounded, & so was Erasmus, at the bulk of my luggage & the Porters were even more so at the weight of those containing my Geological Specimens.'[33] Because of the yellow curtains and bright garishness of the new furnishings, No. 12 was nicknamed Macaw Cottage.[34]

In preparation for the master's new way of life, Syms Covington, who was becoming slightly deaf,[35] left Darwin's employ at the end of 1838. His last act of service had been to make a fair copy of

Darwin's paper on the geology of Glen Roy – nearly ninety pages, which were submitted to the Royal Society of London. It is not clear why Covington decided to leave. Emma, who variously spelt his name Cavington and Cuvington, told Darwin in a letter of 30 November 'I am afraid poor Cavington will hate the sight of me.'[36] Half-hearted efforts by the young people to get Covington a new situation were unavailing, and the man emigrated to Australia in 1839. Before they were married, Emma had engaged a housemaid and was looking round for a new manservant for Darwin. Covington was given £1 as a leaving present. A butler was engaged who turned out to be a thief; he was replaced by Joseph Parslow, who remained with the Darwins until their old age.

Darwin and Emma Wedgwood were married at Maer on 29 January 1839. The vicar, John Allen Wedgwood, took what was a subdued service – indeed an 'awful ceremony' in Darwin's eyes.[37] Darwin's sister Caroline was preoccupied by the sickness of a baby, a 'poor puny little delicate thing'. Mrs Wedgwood, Emma's mother, was confined to bed with flu. Emma's elder sister Elizabeth was in a gloom at losing her. Having considered the idea of a honeymoon journey, Charles and his bride decided they did not have the 'steam up' for it.[38] Even the modest scheme of a day-trip to Warwick Castle was abandoned, and they set off almost directly for London sustained by nothing more festive than sandwiches washed down with a bottle of water.

They soon settled into a routine of comforting dullness in Macaw Cottage. 'I fear poor Emma must find her life rather monotonous,' Darwin confessed to his sister. He rose at seven promptly ('following Sir W[alter] Scott's rule, for, as he says, once turn on your side, & all is over'), leaving Emma 'dreadful sleepy & comfortable'.[39] He then worked for three hours. At ten, they breakfasted together. Then they sat 'in our arm chairs' until half past eleven. Emma then sat 'quiet as a mouse' in his room while he worked until luncheon at two. In the afternoons they sometimes sauntered out together. They dined at six, then sat 'in an apoplectic state' until half past seven.[40] Darwin was learning German, to which he devoted his evening while sipping tea. Occasionally they gave a dinner-party, with the butler in his best livery. They sat for their portraits by George Richmond (who did drawings of them, not paintings). Jos II gave them a piano, which Emma played each day. They went to church at King's College in the Strand, rather than to any of the local churches in Bloomsbury, presumably in the hope of getting an intellectual sermon. It was after church one morning in June that, rather to Darwin's horror, Captain FitzRoy surfaced – 'more especially to make my bow to Mrs Darwin'. This letter was left at their door, 'for fear you shd be absent from home'.[41] It was clear that Darwin had successfully avoided the Captain, and thereby was not in danger of confronting FitzRoy's view that his account of the voyage paid insufficient tribute to his debts to the Captain and other officers of the *Beagle* in helping him amass his collection of specimens. Nearly a year later, Darwin was writing to FitzRoy and attributing his failure to meet to illness – 'for I do not go to the west end of town more than once a week, and I believe I have only seen Hyde Park once during the last two months . . . My stomach as usual has been my enemy.'[42] It was joyous news when FitzRoy (himself now a married man, since 1836) decided to move out of town, though Darwin thought 'how very inconvenient you must find it, living at so great a distance as 20 miles from your weekly journey's end . . . for my own part . . . I do not think I shall ever venture out even as far as a suburban cottage'.[43]

Within only a short time in Bloomsbury, Darwin, whose youth had been entirely rural, felt rooted in the protection of urban solitude. 'You enquire about my taste for the country,' he told his sister Caroline, '– my last visit I consider a very fortunate one – it has cured me of much sentiment & silliness – in fact I hate the thought of the country – the ennui & rain of Maer has effected a thorough cure. I shudder when I think of a damp, dull green view: London is so cheerful; thank goodness we shall not leave it for 6 months.'[44]

That year, 1839, as well as being the year of his marriage, and the year when his first son was born – William Erasmus Darwin, on 27 December, just eleven months after the wedding – was also when Darwin came before the public as a man of science. The scientific world had been aware of his work ever since he had begun, from the *Beagle*, to send back reports to Henslow, Fox, Lyell and others, to dispatch palaeontological and geological specimens and to contribute, as one of Lyell's most loyal supporters, to the current state of geological knowledge. It was in May–June 1839 that he came before the reading public as the author of his *Journal of Researches*. He did not come forward alone. Captain King, who had commanded HMS *Adventure*, surveying the coasts of South America from 1826 onwards, and who had been in overall command of the first expedition, before FitzRoy took command of the *Beagle*, and Captain FitzRoy himself, contributed two volumes, and Darwin one volume. The navy were not risking a repetition of the publishing fiasco in which both Captain Cook's two completed voyages were written up by other hands before he had found his way into print. As the *Quarterly* reviewer of the three-volume set remarked, 'Self-immolation is a term which we have more than once heard applied to the course pursued by those officers of the British navy who have given themselves up to nautical surveying and discovery.'[45] While FitzRoy had been appalled by reading Philos's first version of his *Journal*, the reviewer noted that 'Mr Darwin . . . speaks in the most grateful terms of the treatment which he received throughout from Captain FitzRoy, who may well be satisfied with the results.'[46] The *Athenaeum* complained that there was something unsatisfying about reading three separate accounts. 'They exhibit occasionally a want of unity and continuous interest.' Nevertheless, it recognized that Darwin's, of the three journals of the set, was the most interesting. It respected his 'ardour and the readiness with which he converses with nature in every variety of situation'. It praised Darwin's 'vigour of mind'. At the same time, it detected two Darwins. On the one hand was the punctilious observer, the accumulator of impressions, the unforgettable word-painter, the great naturalist. On the other, there was Darwin the theorist. It took issue with his geological speculations.

> It is true that he intends to disclose his facts as well as his theoretical views, completely, in a series of works, one of which is now in the course of publication; but in the mean time, the journal which is the subject of our comments labours under this disadvantage, that, stripped of details, it exhibits a predominating spirit of bold generalization, of which the world, not without justice, is exceedingly distrustful.[47]

This referred, not to Darwin's thoughts on the origin of species, which he did not reveal in the *Voyage* journal, but to his theories about the origin of volcanoes. From the President of the Geological

Society, William Henry Fitton, now a venerable figure in the field, Darwin had high praise. 'Of much of its natural history, I cannot judge: – but your Geology seems to me to be excellent – & a great part of it new . . . What I like best – however – is the tone of kind & generous feeling that is visible in every part: so that one sees that it is the work of a plain English gentleman – travelling for information, and not for Effect.'[48]

Though recognition from the British scientific establishment was of paramount importance in Darwin's life, there was one figure in the world who mattered to him even more, and that was Alexander von Humboldt. Not only was the old man the most famous man of science in the world, he was a direct role model. His early work, like Darwin's, had been in the sphere of geology, investigating the volcanic origin of basalt in the Rhine region – a journey he had made with George Forster, one of the naturalists who sailed in HMS *Resolution* with Captain Cook. Humboldt had then undertaken stupendous journeys in South America and published accounts of its natural history. His journey laid the foundations of modern geography and of meteorology. When he was later lured back to serve the Prussian government in what he considered to be the provincialism of Berlin, he had begun to work out the giant scientific fresco which he entitled *Kosmos*, a great vision of *everything*. It was not yet in print when Darwin's *Voyage* journal was published, but versions of it had been given as lectures for the previous two decades. Its scope was intended to demonstrate the 'unity amid the complexity of nature'.[49]

To be the new Humboldt! That was Darwin's ambition. That was what, potentially, was contained in the notebooks in which he was slowly and patiently testing Blyth's transmutation theory. If the theory was true, then it would indeed explain the complexity and simplicity of nature. It was therefore with the greatest trepidation that Darwin sent a presentation copy of his book to Potsdam. Despite the expenditure of evening oil (or gas) learning German, he had not advanced sufficiently in the language to address the Master in his own tongue. In September, however, his wildest dreams came true and he received a very long letter, over 1,500 words, in French. Like the President of the Geological Society in London, Humboldt recognized Darwin's originality, in botany, geology, meteorology. He reminded Darwin that he had been sailing and treading in the wake of 'l'immortel Cook'. What progress in the sciences had been made since then! This was demonstrated, for Humboldt, by the vast superiority of Darwin's journal to that of Reinhold Forster, the ship's naturalist on HMS *Resolution* in 1772–5. Humboldt paid Darwin the compliment of disagreeing with some of his geological speculations: 'Il me reste aussi bien des doutes sur le transport des bloc de nos plains baltiques sure les rideaux de glace!'[50] ('I still have doubts about blocks of our Baltic tides being transported on rafts of ice!') In general, however, Humboldt's letter was an enthusiastic endorsement, not just of a book but of a life. He said that he had come to the end of his career, rejoicing without any regret that he had given his life to science. He only wished that circumstances permitted him to converse in person with 'Monsieur Charles Darwin'. 'Vous êtes placé bien haut dans mon esprit . . . Vous avez une belle carrière a parcourrir [*sic*].'[51] Even allowing for the fact that Humboldt had become an old (seventy-year-old) courtier who was used to larding his correspondence with compliments, and even recognizing that Humboldt was a gay man who found it thrilling to endorse a young man's book, this was a stellar accolade.

'Few things in my life have gratified me more than hearing of his approbation, although I have swallowed the dose quite readily as if it had been a little less strong: even a young author cannot gorge such a mouthful of flattery.'[52]

The birth of his first son on 27 December 1839 coincided with, or provoked, more illness in Darwin. He had a headache every day for the first week of 1840.[53] He told Lyell in February that he had been unable to work for nine weeks.[54] He consulted his second cousin Dr Henry Holland three times during February and March. The symptoms, described by Maria Edgeworth in a letter to her half-sister, were totally debilitating: 'His stomach rejects food continually, and the least agitation or excitation brings on the sickness directly so that he must be kept as quiet as it is possible and cannot see anybody.'[55] He consulted another doctor in April – his own father in Shrewsbury. Robert Darwin, in passing, sent word to his daughter-in-law, who stayed behind in London, that she should continue breastfeeding William. The Doctor always weighed any member of the family visiting The Mount, and he feared that Darwin had lost ten pounds in the last year, eighteen pounds since his return from the voyage. He now weighed in at 148 pounds (10 stone 8 pounds). It would seem that Dr Darwin had limited sympathy with his son's condition. On a later visit to The Mount, 'I told him of my dreadful numbness in my finger ends, & all the sympathy I could get, was "yes, yes, exactly – tut-tut – neuralgic, exactly, yes, yes".'[56] Briskness, however, seemed to have its efficacy. Darwin reported to his wife in London, 'I enjoy my visit & have been surprisingly well & have not been sick once.'[57] As soon as he became reunited with Emma, he was able to provide his baby son with constant rivalry for her attention by trembling, shaking, feeling 'numb', vomiting or simply feeling 'languid'. Emma noted that there is nothing that 'marries one so completely as sickness'.[58] It is impossible to tell whether she was simply unselfish or whether she actually derived some emotional satisfaction from being his everlasting nursemaid. 'It is a great happiness to me when Charles is most unwell that he continues just as sociable as ever and is not like the rest of the Darwins who will not say how they really are, but he always tells me just how he feels and never wants to be alone but continues just as warmly affectionate as ever, so that I feel I am a comfort to him.'[59]

The year 1840 was dominated by illness: so much so that they spent the whole of the summer at Maer, only returning to London in November. The Swiss geologist Louis Agassiz came to the Geological Society in that month. Two years older than Darwin, Agassiz had followed a parallel path, in so far as he was pre-eminent in the fields of geology, palaeobiology and biology. His groundbreaking work, appearing in parts from 1833 to 1843, was *Recherches sur les poissons fossiles*. He raised the number of named fossil fish to 1,700. His studies enormously enriched human knowledge of extinct life: they led inexorably, from 1836 onwards, to geological reflections; and inevitably, given his position as Professor of Natural History at the University of Neuchâtel (a chair he took when he was just twenty-five), his mind turned to the Swiss Alps and the movements and effects of glaciers. 'Great sheets of ice, resembling those now existing in Greenland, once covered all the countries in which unstratified gravel (boulder drift) is found.'[60]

English scientists were thrown into turmoil by Agassiz's address to the Geological Society. Greenough, De la Beche, Whewell, Sedgwick, Hopkins and Murchison[61] all rejected the glacier idea, though it was confirmed later in the century by the Scandinavian geologists Otto Torell and Gerard De Geer and by the Germans Albrecht Penck and Eduard Brückner. 'By the 1880s decisive evidence was available for not just a single glaciation but repeated advances and retreats of the ice sheets, their borders mapped in detail by the terminal moraines of debris they left behind.'[62] Old Buckland was an early, indeed almost an immediate, convert to Agassiz's glacial theory. Lyell himself would come around to it. It now finds general acceptance among geologists.

Darwin's attitude to Agassiz was equivocal. He sent the Switzer his own paper on Glen Roy ('I have lately enjoyed the pleasure of reading your work on glaciers, which has filled me with admiration').[63] To Lyell he admitted that Agassiz's book on glaciers was 'capital', yet he stubbornly persisted in thinking the boulders on Glen Roy and Jura had been carried on floating ice by sea. As late as September 1843 he was telling Fox that he would like to revisit Scotland to find evidence of 'roads' carried by prehistoric seas.

'My marine theory for these roads' – originally Lyell's theory, but Darwin now saw it as his own – 'was for a time knocked on the head by Agassiz ice-work – but it is now reviving again . . . even Lyell for a time became a catastrophist . . .'[64]

His other area of interest – the transmutation of species – continued as an almost secret preoccupation transcribed into notebooks. Whatever his motives were in cutting out the opening pages of his Notebook B, his rival in the field of transmutational theory, Edward Blyth, was conveniently removed from the scene. It is sometimes said that Blyth, having devised his theory of the origin of species in the *Magazine of Natural History*, did not really see its implications. His essays had specifically addressed natural selection, sexual suggestion, the role of hereditary variation as ingredients in artificial breeding, but he believed that macro-mutations would 'very rarely, if ever, be perpetuated in a state of nature'.[65] It is not true to say, then, that Blyth did not see the implications of his theory: he merely thought it was a theory which explained certain phenomena in nature, rather than explaining everything. Luckily for Darwin, Blyth's small chemist's shop in Tooting failed, and Blyth was obliged to take what employment he could find for his talents. In the previous few years he had submitted a number of papers to the Zoological Society which were printed in their *Proceedings* – on the osteology of the Great Auk, and on fifteen species of sheep, including the newly discovered *Ovis poli* from Pamir. He had also shown members of the Society drawings and specimens of the yak, the Kashmir stag and other Indian ruminants, such as the Himalayan ibex. All these papers, demonstrating his interest in the zoology of the subcontinent, made him the ideal curator of the Museum of the Royal Asiatic Society of Bengal in Calcutta, a post he took up in 1841.[66]

Notebook D, the third in Darwin's series of notebooks on transmutation, makes it clear that Darwin had met Blyth at the Zoological Society before he sailed to India. 'Mr Blyth remarked the greater difference in the 4 Struthionidae, than in many large orders of birds. The Emu & Cassowary closest. Ostrich & Rhea closer (& two Rheas still closer). Mr Blyth asked whether structure of pelvis was not adaptive structure, like little wings of Auks which does not make that bird a penguin.'[67] This shows that micro-evolution (the Galápagos finches identified by Gould, Blyth's little wings of auks) was a phenomenon of which Blyth was speaking to scientific colleagues before he set sail. Darwin also noted that Blyth had seen that 'only near species or varieties produce heterogeneous offsprings'.[68]

It is of interest that Darwin was recording here, not merely the observations of Blyth, to whom he owed so much, but the answers to those observations made by Richard Owen. Although Darwin would later accuse Owen of being at one with those diehards who insisted upon the fixity of species, we find in this entry in Notebook D – the date is 1838, remember – 'Owen says relation of Osteology of birds to Reptiles shown in osteology of young Ostrich.'[69] In his paper 'On the Anatomy of the Southern Apteryx (*Apteryx Australis*, Shaw)' in the *Transactions of the Zoological Society*, Owen would publish his presentation of April 1838: 'The close resemblance of the Bird to the Reptile in this skeleton is well-exemplified in the young

Ostrich, in which even when half-grown the costal appendages of the cervical region of the vertebral column continue separate and moveable, as in the *Crocodile*.'[70]

Was it mere accident that, when Blyth had been safely dispatched to Calcutta, Darwin began work in earnest on the origin of species? The little man from Tooting, abandoning his pharmacy to tell the men of science about the gradual evolution of birds' wings and beaks, was in danger, in his artless, generous way, of making his insights part of the accepted currency of zoology. Darwin's grandfather Josiah Wedgwood had been so obsessed by the dangers of industrial espionage that, for example, during the production of his famed jasper ware, he had separated the workers involved in the process into different worksheds, so none but he knew the production secret in its entirety. This was an understandable precaution since his considerable fortune depended upon others not being privy to his scientific and technological discoveries. Blyth, like many an innocent before and since, did not see scientific ideas of discoveries with quite such jealousy. He never berated Darwin for appropriating 'his' ideas because he probably did not think in such possessive terms. Darwin's notebooks, however, made it clear that he now regarded the ideas which Blyth had so artlessly thrown out as being 'my theory'. Like Alberich stealing the ring of power from the Rhinemaidens, Darwin wanted sole possession of the dangerous idea. Blyth had not had the time, the money nor, in fairness, the breadth of mind to give a comprehensive answer to his world-changing question – *May not, then, a large proportion of what are considered species have descended from a common parentage?* When the botanist J. D. Hooker was planning a visit to India later in the 1840s, Darwin suggested that he should look up Blyth in Calcutta. Patronisingly, he wrote, 'he is a very clever, odd, wild fellow, who will never do, what he could do, from not sticking to any one subject'.[71] Thus the heir to great riches dismissed the Tooting pharmacist. Not sticking to one subject – if that was a fault, Darwin was determined not to be guilty of it. Agassiz had (quite unintentionally, for he had never read Darwin on the boulders of Glen Roy when he presented his glacier theory in London) put a stop to Darwin's advance as a geologist. The way forward was through the transmutation theory.

If Blyth at the Zoological Society – founded in 1826 – was a minor figure, Richard Owen, who befriended Darwin during his phase of living in London, was a major one. In his *Autobiography* Darwin wrote, 'I often saw Owen, whilst living in London, and admired him greatly, but was never able to understand his character and never became intimate with him. After the publication of the *Origin of Species* he became my bitter enemy, not owing to any quarrel between us, but as far as I could judge out of jealousy at its success.'[72]

History is written by the victors, and for much of the twentieth century, after the revival of Darwin's reputation, Owen was forcibly diminished by the Darwinians – in some quarters airbrushed altogether from the story. This imbalance of perspective was largely corrected by Nicolaas Rupke's magnificent biography of 1994, which reminded us, not only of Owen's actual views on evolution, but also of his many achievements in other spheres, as a museum director and as a scientist. In his long life he was idolized as 'Britain's answer to France's Georges Cuvier and Germany's Alexander von Humboldt'.[73] Anyone who has visited the Natural History Museum in London is looking upon Owen's achievement. It opened in 1885, when Owen was full of years, but he had presided over this collection when it was still housed at the British Museum, and he had campaigned tirelessly for a separate museum of natural history. Before that, he had set in order the great museum of anatomy, the Hunterian collection, at the Royal College of Surgeons. The cataloguing of Hunter's huge collection of skeletons, foetuses and freaks in formaldehyde was a prodigious undertaking and could have been undertaken only by a great scientist who combined anatomical knowledge with enormous organizational skill. As well as being a key player in the glory age of the foundation of museums, he was an anatomist/palaeobiologist of historic and international standing.

Tall, spindly, dandified, Owen was entering his prime when Darwin got to know him. (He would start the work of cataloguing the Hunterian collection in 1842.) Nothing that Darwin wrote at the time suggested hostility to Owen, and indeed friendship with the slightly older man put Darwin on a good footing not only with the London scientific establishment, but also with the largely Oxford-based circle of William Buckland's Christ Church pupils. Hitherto, Darwin's academic contacts had been predominantly Cambridge, but the Oxford dimension was useful to him.

Owen had not himself been to Oxford. His father had been a West India merchant who died when Owen was only five years old. Owen briefly studied anatomy at Edinburgh University and subsequently at St Bartholomew's Hospital in London. The rigours of hospital life had been his hard school. He went on to study anatomy under Cuvier in Paris, and returned to London as a lecturer and a foremost authority on fossils. When Darwin's stupendous collection, formed during the *Beagle* voyage, began to arrive in London piecemeal before the return of the young naturalist himself, it had been Owen who was among the first to examine them. On Darwin's return, Owen, whose formal knowledge of fossils far outsoared Darwin's, helped him to classify the collection.

Though not an Oxford man, Owen, through his museum work, met those who were. The trustees of the Hunterian Museum – Philip de Malpas Grey Egerton, a keen amateur palaeontologist, Viscount Cole and others – had been pupils of William Buckland, the redoubtable canon of Christ Church who was made Dean of Westminster by another pupil, Robert Peel, when he became Prime Minister. (He succeeded another clergyman-scientist, Samuel Wilberforce.) Owen introduced Darwin to Buckland, who enjoyed discussing the Galápagos land iguanas and marine iguanas.[74] We mistake the nineteenth-century scientist-clergy if we suppose they were all bigots with closed minds. While at the beginning of his life as a scientist Buckland reconciled the biblical accounts of creation with his geological research, by taking the Deluge as historical – thereby allowing time for all the extinct species to leave behind fossil traces of themselves – he discarded these views when he had read Agassiz and became convinced by the glacier theory. He had also long ago recognized by reading Lyell that the earth was vastly older than the mythological creation narrative of Genesis, as interpreted by fundamentalists, would suggest. He later praised an early essay of Darwin's (9 March 1838) on the role played by earthworms in soil-formation, while rejecting Darwin's clearly false suggestion that chalkland had been formed in a similar way.[75]

Poor Buckland, who died of tuberculosis in 1856, had for some time been insane. Owen heard, as early as 1850, that the Dean of Westminster had committed 'much greater excesses than ever in respect to his own person – beating his head and scratching himself so as to produce alarm'. His younger children supposed that 'Papa must be acting.'[76]

Long before mental illness set in, Buckland had been gloriously eccentric. He always wore an academic gown when out on field trips collecting palaeontological specimens. His houses were filled with bones and fossils. One of his passions was zoophagy. He 'used to say that he had eaten his way straight through the whole animal creation, and that the worst thing was a mole, that was utterly horrible'. He had eaten bluebottle flies, panthers, crocodiles and mice. At Nuneham, when presented with the heart of Louis XIV in a silver casket he exclaimed, 'I have eaten many strange things, but have never eaten the heart of a king before' – and down Buckland's gullet it went.[77]

The crucial thing about Owen, though, in the story of Darwin is to be found in Darwin's First Notebook on the Transmutation of Species from 1837. 'Mr Owen suggested to me that the production of monsters (which Hunter says owe their origin to very early stage) & which follow certain laws according to species, present an analogy to production of species.'[78] What this makes clear is that as early as 1837 Owen believed in some form of evolution.

Epigenesis is the process by which plants, animals and fungi develop from a seed, spore or egg through a series or sequence of steps in which cells differentiate and organs form. Saint-Hilaire accepted Lamarck's notion that an evolutionary process was at work in nature, but he disputed his theory about its mechanism. Rather than Lamarck's idea of acquired characteristics, Saint-Hilaire (1772–1844), Professor of Vertebrate Zoology at the Museé d'Histoire Naturelle – later one of the scientists who accompanied Napoleon to Egypt – believed that living organisms are adapted to their environment. Interference at some stage in the early development of the organism – the embryo or foetus – could lead to the advancement or holding back. A new species could thereby arise as a 'monster'. Owen, whose early training had been as an anatomist, had, as stated, begun the catalogue of the

Hunterian Museum in London (now housed in the Royal College of Surgeons) where dozens of monsters, hybrids and natural oddities are on display. It was here that his preoccupation with some of the central issues of evolutionary science began, and his engagement with the French forerunners in the field.

It is extraordinary, given this fact, and given the numbers of scientists who had been wrestling with evolutionary ideas since his grandfather's day, that Darwin apparently believed that he had made the subject his and his alone. With the rigour of an industrialist or retailer branding a particular product, he was determined that the concept of evolution should be stamped with his surname, though it was not a word he used until the final edition of *The Origin*, preferring the more accurate 'modification'.

In January 1842 his hero Humboldt came to London. Darwin met him at Sir Roderick Murchison's house. Flatteringly, the request to meet Darwin came from Humboldt himself. It would be tempting to think that Darwin, who had now advanced to this belief that species mutate by means of natural selection – would have tried out his idea on the famous German. So costive was he, however, that he said nothing about it. 'I was a little disappointed with the great man,' he said later. 'I can remember nothing distinctly about our interview, except that Humboldt was very cheerful and talked much.'[79]

By now, Darwin was tiring of London life and pined for anti-social rural seclusion. Both he and Emma wanted a garden such as they had known in their own childhoods in Shrewsbury and at Maer. A second child, Annie, was born on 2 March 1841. Even if Darwin could endure the social round of breakfasts, dinners and 'calls' for much longer, Macaw Cottage, though a substantial residence by the standards of most Londoners, felt crowded now that it had to accommodate a full nursery staff.

House-hunting began again. During Annie Darwin's first months of life, her mother started to note properties and their acreages in her diary – forty acres at Langley, six near Reading, twenty-five in Harrow Weald. They were not looking to return to Shropshire or Staffordshire. What was needed was a substantial house, rurally secluded but within easy reach of the capital. Then they found the ideal place, in Woking, Surrey. Hesitation made them lose it to another tenant.

Before they leave London with the children, we should mention one last acquaintance in the capital. When house-hunting in his bachelor days, Darwin had specifically wanted to find a place near Regent's Park, and it is clear from the notebooks that he was a frequent visitor to the Zoological Society of London. Mr Hunt (first name lost) and Alexander Miller became friends. So did Jenny the orang-outang and her male companion Tommy. (Did Darwin smile at the thought of the orang-outang sharing a first name with Mrs Carlyle? There was no doubting which of the two females he preferred.)

'Mr Yarrell [a London bookseller friend of Darwin's, keen amateur naturalist] has seen Jenny, when Keeper was away, take her chair & bang against the door to force it open, when she could not succeed of herself. It was very curious to see her take bread from a visitor, & before eating *every time* look up to Keeper and see whether this was permitted & eat it [*sic*].'[80] He noted her fondness for listening with 'great attention' to the harmonica, '& readily put it, when guided to her own mouth'. He also liked her fondness for the scent of verbena. She enjoyed the peppermints he gave her. 'Will take & give food to Tommy.' Jenny was the attentive, musical type of female whose company he enjoyed among his own family members. Indeed, as he watched her, and then returned to Upper Gower Street to see William and Annie, the thought was inescapable that Jenny *was*, in a sense, a family member.

Then they found the house. The Kent village of Down (from the 1850s onwards it was spelt with an 'e', though the Darwins retained the old spelling for their house) was a hamlet, sixteen miles from London, 'about 40 houses, with old walnut trees in middle where stands an old flint Church . . . Inhabitants very respectable – infant school – grown-up people great musicians – all touch their hats as in Wales, & sit at their open doors in evening. No high road leads through the village. There are butcher & baker & post office.'[81]

Down House was in Darwin's view 'ugly, looks neither old nor new'. Nevertheless, 'three stories, plenty of bedrooms' and 'in good repair'. They decided to take it, assuming that it would be possible to rent it for a year to see how they liked it. The owner, however, insisted upon a sale and Edward Cresy, a local architect and civil engineer who was acting on the Darwins' behalf, urged Darwin to buy at the bargain price of £2,500. 'I have therefore, after many groans, offered £2150 or at most 2200£.'[82] All the other houses they had viewed were over £3,000. Darwin's father advanced him £3,000 against his future inheritance, charging him 4 per cent interest.

Emma, heavily pregnant, moved into the house on 14 September, and Darwin followed a few days later, having bought a horse and carriage for £100.

On 23 September Emma went into labour, and gave birth to a sickly girl, Mary Eleanor. No sooner was her labour over than she heard that her father Josiah, who had suffered some kind of stroke and had been ill for a year, was close to death. Her sister Elizabeth wrote that he was 'very weak', though she continued to feed him.[83]

Old Dr Darwin squeezed himself into a phaeton and drove over to minister to his old friend. When he saw the condition of Jos, he broke down and wept.[84] In fact, Jos lingered for another year. It was the baby – Mary Eleanor – who died, aged three weeks. Emma supposed that the child had a look of Bessy, their mother. They buried her in Down churchyard. She said, 'It will be long indeed before we either of us forget that poor little face.'[85]

No wonder Erasmus, when he visited, called the place Down-in-the-Mouth. It was an inauspicious beginning. They remained devoted to one another, however, and the children. They had William, and they had Annie, who had already become the love of Charles Darwin's life.

9

Half-Embedded in the Flesh of their Wives

JESSY BRODIE, A tall Scotswoman with carroty hair, was forty-nine years old when she became the Darwins' children's nurse. Her pale face was smallpox-scarred. Her bright-blue eyes saw and had seen much. She had been the nurse to the children of William Makepeace Thackeray. She had seen one of Thackeray's daughters, Jane, die of a chest infection. She had assisted at the birth of his daughter Minnie and she had witnessed the effects of postnatal depression on his wife – who tried to drown their three-year-old, Anny, in the sea at Margate. When Charlotte Brontë, who had never met Thackeray, artlessly dedicated to him her novel *Jane Eyre* it was assumed that 'Currer Bell', her nom de plume, had been a governess in the famous novelist's house and that the mad wife in the attic, Mrs Rochester, was based on poor Mrs Thackeray, who, on a packet-steamer to Ireland, tried to throw herself overboard. Thackeray was overwhelmed by Brodie's unselfishness while this drama ensued. 'She was sick every quarter of an hour, but up again immediately staggering after the little ones, feeding one and fondling another. Indeed, a woman's heart is the most beautiful thing that God has created and I can't tell you what respect for her I have.' In the following weeks, while the writer had to deal with his wife's mental breakdown, there was 'only poor Brodie of whom I can make a friend; and indeed her steadfastness and affection for the little ones deserves the best feelings I can give her. The poor thing has been very unwell, but never flinched for a minute, and without her, I don't know what would have become of us all.'[1] Brodie, during all this tragedy, had been planning to get married, but her intended deserted her and went to Australia. Thackeray could not afford her wages on top of his many other expenses, and a new position had to be found. She came to Down House in the year the Darwins moved in. The staff consisted of a butler (Parslow), a footman, two gardeners, a cook, a laundry maid, a housemaid and one or two nurserymaids. Bessy Harding, Willy's nursemaid, was 'pert' to Brodie, who relied on Emma's friendship and support to ease her into the household. She continued to miss the Thackeray children acutely. Thackeray knew that 'she longs to come back to them'.

Brodie was always known as Brodie because she was the nurse – or nanny as it would be termed in modern English parlance. The governess (when the children were old enough to have one) would be Miss So-and-So. The cook, regardless of marital status, would be Mrs. The maids were known by their first names. Thus was the hierarchy of things maintained. Darwin was a polite, kindly employer, always prefacing his requests to them with the words, 'Would you be so good . . .' Nevertheless, the practically minded Brodie, whose father from the north-east Scottish coast had been a ship's master – a prisoner of war during Napoleonic times – found the quietness of Down stultifying. She considered it a pity that Mr Darwin, unlike Mr Thackeray, had not something to *do*.[2] One of the gardeners, Henry Lettington, a deacon of the local chapel, made a comparable observation. 'Oh, my poor master has been very sadly. I often wish he had something to do. He moons about in the garden, and I have seen him stand doing nothing before a flower for ten minutes at a time. If only he had something to do I really believe he would be better.'[3]

In a confessional letter to J. D. Hooker, Darwin in January 1844 described himself as 'engaged' in a very presumptuous work.

> At last gleams of light have come, & I am almost convinced (quite contrary to opinion I started with) that species are not (it is like confessing a murder) immutable. Heaven forfend me from Lamarck nonsense of a 'tendency to progression' 'adaptations from the slow willing of animals' &c, but the conclusions I am led to are not widely different from his – though the means of change are wholly so. I think I have found out (here's presumption!) the simple way by which species become exquisitely adapted to various ends.[4]

In Darwin's long, silent cogitations at Down, we watch him turn into a contemplative. On the one hand, he became an intensely introverted and concentrated version of what he had been since boyhood – the detailed observer of natural phenomena in all their minute detail, in all their prodigious variety. On the other hand he was a man who had seen an all-containing vision of things. It was like confessing a murder because the theory could see off, not just the Revd William Paley, but also the 'Watchmaker' whose intricate design Paley had described with such eloquent reverence. The German scholars of the New Testament in Tübingen had shaken the nineteenth century's belief in the divinity of Christ. Lyell's *Geology* had made it impossible to maintain a literal belief in the Book of Genesis as the work of science, or history, which the seventeenth century had tried to make it. 'The simple way by which species became exquisitely adapted to various ends' thus 'murdered' not just traditional faith but the Creator himself. Or at least the 'murder' removed the logical necessity of belief in such a figure. What appeared to be intricate design was a system which made the exquisite adaptations by itself. Others – including Goethe, Erasmus Darwin, Lamarck, Blyth – might have mused upon the possibility of evolution, upon the mutability of species, upon the emergence of new species by processes of adaptation. Ever since 1789, the British Establishment, which included the universities and the Church of England by Law Established, had associated new ideas with progressivism, science with revolution, speculations upon mutability with a deliberate sans-culottish desire to overthrow the hierarchies and values of the old century. A generation had passed, however, and power, once in the preserve of the landed class alone, was now possessed by new capital. Science and scientific knowledge had advanced – partly in the universities, quite largely outside them. It was by no means certain that the comfortable upper-middle class, which was the backbone, intellectually, politically, socially, of Victorian England, any longer needed the Bible, or its God.

In his 1844 essay on 'The Variation of Organic Beings', having established how humanity improves plants and animals by means of selective breeding, Darwin posited just such selective process at work in nature as being the work of the Creator.

> Let us now suppose a Being with penetration sufficient to perceive differences in the outer and innermost organization quite imperceptible to man, and with forethought extending over future centuries to watch with unerring care and select for any object the offspring of an organism produced under the foregoing circumstances. I can see no conceivable reason why he could not form a new race (or several were he to separate the stock of the original organism and work on several islands) adapted to new ends.[5]

But although Darwin here pays lip-service to the idea of a Creator, and although Christians who believe in the theory fall back on this, they do so by overlooking what is shocking about Darwinism, what is destructive to faith. In the very next paragraph after he had supposed there was a 'Being' who had devised natural selection as the 'creative' process, Darwin was referring to 'this imaginary Being'[6] and by the time *Origin of Species* was first published, incorporating much of this material, the imaginary Being has been replaced with the word 'Nature'.[7] Philosophically speaking, at least since Spinoza had coined his simple and destructive phrase *Deus sive Natura*, there had been a usage of the word 'God' which was a form of scientific laziness. It meant 'There appears to be such and such a complex organism or astronomical phenomenon – how do we account for it? why, God *made it so*.' That having been said, nature does follow 'laws' whatever theology we embrace or discard. The species theory developed by Darwin has no direct bearing on physics, nor on the extraordinary complexity and neatness of the solar system. Within the sphere of biology, however, it does appear to remove any necessity for religious explanations. In a sense once it has been espoused, his evolutionary system was essentially destructive to theology, since to personalize the 'imaginary Being' who allows or moves or permeates, or who set in motion the process, is merely to be indulging in tautology. In old time it was said You Cannot Serve God and Mammon. The Victorians in general, Darwin's rentier class in particular, had no doubts, given that not very difficult choice, which of the two they served. Having decided to worship Mammon they had little enough room for its demanding Alternative. The theory of natural selection neatly smothered this troubling Alternative (who in the old book had called for the feeding of the poor, the clothing of the naked, the establishment of justice on earth). Darwin would conclude his 1844 essay, as he would his 1859 book, with the claim that 'there is a simple grandeur' in the theory. Darwin, however, by 1844 no more really believed in the creative power of God than a frock-coated Victorian gentleman, climbing into his carriage to be taken to a well-rendered Morning Prayer on Sunday morning, took seriously the ragamuffin Gospel injunction to sell all and give to the poor. The silences of Darwin, which his servants so fascinatedly observed, were present also in his prose, even in his notebooks, even when in rhetorical flourishes he was trying to reassure himself that there was 'grandeur' in his vision. However we interpret him, it will always be hard to know which caused him the greater anxiety: the fear that his theory might be true – thereby dismissing the God of the Bible, perhaps any God – or the fear that it might be false – thereby diminishing him from the status of greatest scientific mind of the nineteenth century to a mere mortal, one who had tried out an idea of great ingenuity, but, like the majority of scientists in history, one who had not proved his case.

Meanwhile, Emma Darwin, who was establishing the family at Down and managing the household, kept Sunday as a firm duty. She had family prayers at Down House, attended by the servants, the children and her husband. Twice that day for Morning and Evening Prayer, they attended the parish church, sitting in a large pew lined with green baize near the parson's desk. A small trace of Emma's nonconformist ancestry was noticed in the phenomenon that whereas the choir and anyone seated sideways to the altar turned east, as to the Heavenly Jerusalem, to say or sing the words of the Creed, the Darwins did not move, so found themselves staring into the eyes of the other churchgoers. Francis Darwin, recollecting this eccentricity of his mother's, commented, 'We certainly were not brought up in Low Church or anti-papistical views, and it remains a mystery why we continued to do anything so unnecessary and uncomfortable.' Parslow, the Darwins' butler, sang in the choir.[8]

Francis – who became a botanist in grown-up life – was born in 1848. The children of Charles and Emma Darwin were William (born 1839), Annie (1841), Mary Eleanor (1842), Etty (1843), George (1845), Betty/Bessy (1847), Francis (1848), Leonard (1850), Horace (1851) and Charles (1856). The 1840s were thus dominated, for Emma, by reproduction. Nonetheless, she had time to oversee the reordering of the house and garden, and the day-to-day running of the household. In their time at Down they added a bow window and veranda at the back of the house, a new hall, and the study where Darwin worked for the rest of his life. To stop the intrusion of people walking down the lane which adjoined the property, they lowered the road by two feet and built a flint wall. In the first three years at Down, they also greatly extended the servants' quarters. A schoolroom, and accommodation for a governess, were a necessity. The somewhat tired sixteen acres of land which surrounded the house became a beautiful garden which was also a botanist's laboratory and exercise-ground. Along its north side, beyond a high hedge and shrubbery, was a kitchen garden, surrounded by a high flint-and-brick wall, its beds further protected by box hedges. The south side of this long wall housed the greenhouse. Darwin was challenged by his cousin William Fox to see who could grow the biggest peas from the latest varieties. Each year the crop was carefully measured. On the other side of the long wall was the orchard. Beside the orchard was a red-brick potting shed where Darwin tested his breeding programmes and hybridizations. In addition Darwin leased a strip of land a quarter of a mile from the house from his neighbours the Lubbocks. Sir John Lubbock, squire of the parish, was a Whig banker, mathematician and astronomer. His house was called High Elms. Darwin planted his rented strip with a mixture of hardwoods and had a path made round its perimeter, ever afterwards known as the Sandwalk. Darwin's children, when they grew up, would remember their father on his 'thinking path'. Every day at noon, he walked five times round the Sandwalk, swinging a walking stick heavily shod with iron, with which he would strike the ground. The rhythmic click of the stick spoke of his approaching presence.

As well as the five-times circumambulation of the Sandwalk, Darwin would sometimes walk in the woods at dawn. Later in the day, often accompanied by the children, he would walk down the hill to the Big Woods. 'He seemed to know nearly all the beetles and was immensely interested when any of the rarer sort were found.' The children grew up close to nature. Francis found 'something impressive and almost sacred' in the changing seasons, and they all recalled a father who was 'the most delightful play-fellow'. When they hid in the shrubbery, he commented, 'This is analogous to young pigs hiding themselves, and [is the] hereditary remains of savages' state.'[9] Like many family men he had begun to address his wife as if she were his mother – 'Mammy'.

So the routines – ceaseless work and cordial family relationships – attached themselves like ivy to Down House. The village of Down had never had a parsonage. Down House had been the personal property of a previous parson. Darwin, with his love of the garden, his burgeoning family, his religious wife and regular churchgoing, was, to all outward signs, scarcely distinguishable from the naturalist-clergyman he had intended to become before the *Beagle* voyage changed his destiny. There was more than purely intellectual curiosity in the face of nature, there was an intensity of delight.

> The clover fields are now of a most beautiful pink and from the number of Hive Bees frequenting them, the humming noise is quite extraordinary. Their humming is rather deeper than the humming overhead that has been continuous and loud during the last hot days, over almost every field. The labourers here say it is made by 'air-bees' and one man seeing a wild bee in a flower, different from the kind, remarked that 'no doubt it is an air-bee'. This noise is considered as a sign of settled fair weather . . . There were large tracts of woodland that were cut about every ten years, some of which were very ancient. Larks abounded, and their songs were most agreeable, nightingales were common.[10]

These observations could have been made by the Revd Gilbert White at Selbourne, though they were recorded by Charles Darwin at Down in June 1844. Yet the same mind – aged only thirty-five – could, in the very next month, jealously and aggressively guard what had become the 'Darwin' brand. Here he is writing from Down on 5 July – note the repetition in the first sentence of that possessive pronoun:

> My Dear Emma. I have just finished my sketch of my species theory. If, as I believe that my theory is true and if it be accepted even by one competent judge, it will be a considerable step in science. I therefore write this, in case of my sudden death, as my most solemn & last request, which I am sure you will consider the same as if legally entered in my will, that you will devote 400£ to its publication & further with yourself, or through Hensleigh [her brother] take trouble in promoting it.[11]

And so on, at great length, including the request to get Lyell to edit it. It is a remarkable letter. Presumably his wretched state of health made him feel likelier than most mortals, despite his youth, to die prematurely. It prompts the question, however, why, if he cared so passionately about 'his' theory, he did not publish his essay 'The Variation of Organic Beings'. It is a perfectly coherent work of 164 printed pages (in Sir Gavin de Beer's learned edition published in 1958).

It is, in essence, a shorter version of *The Origin of Species* which, fifteen years in the future, he *would* publish. One reason for not publishing, unquestionably, was the state of his health. To a Swiss correspondent, Adolph von Morlot, stratigrapher and archaeologist, who had written to him that summer on the old question of boulders and glaciers, Darwin had said, 'My health during the last three years has been exceedingly weak, so that I am able to work only two or three hours in the 24.'[12]

One can imagine a scientist of a temper less cautious than Darwin's, however, publishing the essay in an almost interrogating tone: what if this theory of the transmutation of species occurred in this way? What do my colleagues think of it? What are the objections to the theory, beyond non-scientific blind prejudice? How does it match what we know of fossil evidence?

This, though, was not Darwin's way. He wanted to establish it as *his* theory, and although he told his wife that he believed it was true, he knew he could not yet prove it beyond reasonable doubt. So he hesitated.

There was also, naturally enough, an element of prudent fear holding him back. The theory had not yet led Darwin himself to a complete rejection of Christianity, though he could feel belief ebbing away. He was aware that figures whom he esteemed in the scientific world, such as Professor Sedgwick at Cambridge or Dean Buckland, would deplore the materialism of the theory. Did Darwin, reclusive, shy, neurotic, prone to 'stomachic catastrophe', have the temperament to weather the controversy which publication of such a theory would inevitably ignite? So Hamlet paced the Sandwalk of Elsinore/Down, and the essay though completed by the summer of 1844 was unpublished.

Thought of his own death, before 'his' theory was shown to the world as Darwin's Theory of Evolution, had been enough to make him pen the nervous letter to Emma. What he could not have foreseen was that in the autumn of that very year, with the noise of trumpets, and the razzmatazz of a publishing genius, a book was about to shock Victorian England to the core, with an energetic survey of the current state of scientific knowledge and a promotion of the evolutionary idea.

In October 1844, a total of 150 persons of influence were sent a book, free of charge, at the request of its anonymous author. Its title was *Vestiges of the Natural History of Creation*. The only name to appear on its title page was that of the publisher, John Churchill of Princes Street, Soho. Although Churchill was a religious man, attending chapel twice every Sunday, his list of publications included various titles considered at the time to be outré, including a translation of Friedrich Tiedemann's *A Systematic Treatise on Comparative Physiology* (1834), a work which advocated serial transformation, a gradual evolutionary process at work in nature. Churchill had received brickbats for publishing (very successfully) a popular work of anatomy by William Carpenter – *Principles of Human Physiology* – which had openly declared a belief in a machine-run universe with no need of a Creator. Churchill had the nonconformist's belief in speaking one's mind and a clever publisher's eye to the main chance. He was also good at keeping secrets, which was why the author approached him in the first instance. Even so, Churchill was kept in the dark about the author's true identity. All correspondence about the publication of *Vestiges* had to be conducted via Alexander Ireland in Manchester, a journalist who was a champion of new fads and advanced ideas such as phrenology and mesmerism.

Ireland had the good journalist's ability to create a story. It was presumably he who, having started the storm of excitement on the book's publication, began the rumour that the mystery author might be none other than Prince Albert.[13] Even though this particular idea might have been fanciful hindsight on Ireland's part, there was still a highly satisfactory buzz of excitement, first in London, then throughout literate England, about the book, its adventurous contents and its unknown author. When Sedgwick was eventually persuaded to read and ultimately to review *Vestiges*, he told Lyell, 'I cannot but think the work is from a woman's pen, it is so well dressed . . . I do not think the "beast man" could have done this part so well.' The author's reading was extensive but 'very shallow' – another sign of feminine origin. In his notice in the *Edinburgh Review* he developed the thought, even likening the impulsive female author to Eve in the Garden of Paradise. She 'leaps to a conclusion as if the toilsome way up the hill of Truth were to be passed with the light skip of an opera-dancer. This mistake was woman's from the first. She longed for the fruit of the tree of knowledge, and she must pluck it right or wrong.'[14]

Vestiges of the Natural History of Creation was, and is, an immediately readable and superbly comprehensive work of natural science. Though Sedgwick would dismiss it, in conversation with another Fellow of Trinity, as 'rank materialism',[15] its very title pays lip-service, at least, to the traditional explanation for the origin of all things. 'Science leaves us, but only to conclude, from other grounds, that there is a First Cause to which all others are secondary and ministrative, a primitive almighty will, of which these laws are merely the mandates. That great Being, who shall say where is his dwelling place, or what his history! Man pauses breathless at the contemplation of a subject so far above his finite faculties, and only can wonder and adore!'[16]

Vestiges in fact was to say that theology and metaphysics on the one hand had their sphere, in which the author chose, respectfully, not to dabble, while science and the march of mind had *theirs*. This book is a sort of popular Humboldt's *Kosmos*, a compendium for the ever-expanding book-reading and book-buying public, of the current state of scientific knowledge. It is a book which a brisk reader could easily digest in two or three evenings, or – going more slowly – in a week. By the time you had read it, you would have been conducted through the marvels of astronomy, as revealed by Herschel's telescopes. You would have been given a potted history of geology, largely culled from Lyell, and you would recognize the great antiquity of the earth – of far greater antiquity than the conventional method of reading the Bible would imply. From the fossil records you would start to see the multitude of species no longer to be found on this planet. You would then be asked to speculate on that most momentous of questions, the origin of present species, noting an advance, in both plants and animals, 'along the line leading to the higher forms of organization'.[17] As in geology, so in biology, we should look not for direct divine intervention to explain the development of things, but rather for 'natural laws which are expressions of his will'. The author saw the evolution of life, the gradual transmutation of one species into another, almost like an embodiment of Ovid's metamorphoses myths.[18] Alluding to old eighteenth-century Lord Monboddo's 'much ridiculed' theory that mankind evolved from monkeys, the author pointed out that fossil evidence now suggests just such a possibility.[19] 'The whole train of animated beings, from the simplest and oldest up to the highest and most recent, are then, to be regarded as a series of *advances of the principle of development*.'[20] The last hundred pages of the book, drawing heavily upon Lamarck, while mocking Lamarck's conclusions, but also on the work of contemporary medics, anatomists and biologists, see biology and history as an everlasting progress or improvement. Though the early Victorian reader might be dismayed to be shaken in the old Bible certainties, might indeed even feel horrified by the sheer size and pitilessness of the universe depicted here, the investor in *Vestiges* could end on a note of uplift. The story had begun violently, with the clash of meteoric substances in the furthest reaches of space. Earth had crashed with volcanic eruptions. Lumbering prehistoric monsters then vied for mastery. This story would end in brightness.

John Phillips, a young Cornish mining engineer and geologist and Professor of Metallurgy at the College for Civil Engineers at Putney, is quoted as saying, 'There is no break in the vast chain of organic development till we reached the existing order of things.'[21] In the end, after the long march of monsters and apes and primitive men, came 'civilization'. In the end, came the Victorians. War did not cease, but 'shrinks into a comparatively narrow compass'.[22] The sex passion no doubt 'leads to great evils'. But 'the civilized man is more able to give it due control'.[23] There could be no evidence that Victorians had evil 'sex passion' under more control than, say, men and women in the time of Oliver Cromwell, but in the optimistic mood of its closing pages *Vestiges* saw everything as an improvement. The author does not mention the Reform Act of 1832. Somehow there is no need to emphasize the idea. What had begun as a work of popular science had ended like Samuel Smiles's *Self-Help*, as a work of supreme collective self-satisfaction.

It is not clear why Darwin was not in a position to buy a copy of *Vestiges*. True, the first print run of 750, of which 150 had been given away, was sold out quickly, but he could surely have laid hands upon a copy to purchase. Instead, he went to study it in the Reading Room of the British Museum at the close of the year 1844. It was a very great shock to his system. Since Blyth had been safely dispatched to India in 1841, Darwin, shrouded for much of the time in domesticity or the even tighter womb-wall of invalidism, had worked steadily, sharing his views on transmutation with very few intimate scientific colleagues: Jenyns, Lyell, Hooker . . . In all the speculation about the possible authorship of *Vestiges*, his name had not been mooted because he had not yet come out as a transmutationist, and his fame, such as it was, rested on his *Voyage* and (chiefly) his geology. He had quietly allowed himself to make his particular version of transmutation his own. It was 'my theory'.

The anonymity of the author of *Vestiges* added to the book's appeal. On the one hand, the anonymity suggested that the contents of the book were too dangerous to be owned. On the other hand, was there not a sense in which anonymity conveyed something larger than the individual, as if the author spoke for the Age? Sedgwick in his intemperate review (Whewell said of it, 'the material appears excellent, but the workmanship bad, and I doubt if it will do its work')[24] was also anonymous. He spoke, likewise, not only for himself, but for the old world when he told readers of the *Edinburgh Review*: 'If the book be true, then the labours of sober induction are in vain; religion is a lie; human law is a mass of folly, and a base injustice; morality is moonshine; our labours for the black people of Africa were works of madmen; and man and woman are only better beasts.' Who, having read such a review, could resist purchasing the book? Yet Darwin, reading it in the library, knew that these words came from his old friend and tutor Sedgwick, who had taken him on a geological tour of Wales. Could timid, albeit passive-aggressive Darwin endure such a reproof from a Cambridge professor? Could he subject Emma to the knowledge that his views reduced morality to moonshine? When the author of *Vestiges*, a father of eleven children, was asked by a knowing friend why he did not acknowledge authorship, he said there were eleven reasons.[25]

The author was a clever Edinburgh journalist, publisher and bookseller named Robert Chambers (1802–71). His father had been a cotton manufacturer who went bankrupt, but who had instilled in Robert and his brother William a love of books and learning. Robert had read the whole of the *Encyclopaedia Britannica* as a boy and would go on, in the 1860s, to oversee the publication of *Chambers's Encyclopaedia, A Dictionary of Universal Knowledge*. His early years in Edinburgh as a hack writer had been derided by Sir Walter Scott and by his son-in-law John Gibson Lockhart – editor of the *Quarterly Review* (the Tory rival to the Whig *Edinburgh*). Chambers, however, who wrote an unauthorized life of Scott, had learnt a thing or two from Sir Walter, and one of these was the appeal of anonymous publication. Scott's first novels were all published anonymously, and curiosity about the author's identity quickened the public appetite for the books themselves. Chambers's *Edinburgh Journal*, a low-priced educational weekly, was a publishing sensation when it began in the year of Scott's death, 1832. By April of that year it was selling 32,000 a week.

Darwin, as he sat reading *Vestiges of Creation* in the British Museum, was brought face to face with the uncomfortable truth that its general underlying ideas were not original. Chambers was not a scientist, but a clever man who had read Lyell, Lamarck and the majority of published scientists. Like any layman (like most scientists, come to that) he made some mistakes. Broadly, though, he scooped Darwin. What Darwin must have seen, as he anxiously read *Vestiges*, was the basic unoriginality of his own mind. Chambers, anonymously, but with tremendous pace and brio, spelt out a natural origin for species in a framework of purely material causation. This was an idea which might find resistance among older scientists and theologians, but for the majority of minds in the nineteenth century it was irresistible. The only big remaining question was: if Lamarck was wrong about the method by which species evolved, what alternative theory could be found?

The actual method by which nature went to work, allowing some species to become extinct while others mysteriously refined themselves and strengthened themselves, was one which still eluded the scientific mind. Any reader of *Vestiges*, however, could see that this idea, of mutation, development, evolution, was central to the Victorian mind-set. A year after *Vestiges* was published, the spellbinding Fellow of Oriel and Vicar of St Mary's Oxford, John Henry Newman, wrote a book on *The Development of Christian Doctrine*, attempting to explain how the modern Church, so apparently different in ethos and practice from early Christianity, had reached its current position. Newman in old age must have read with mixed emotions a letter from his feline ex-disciple Mark Pattison – by then the agnostic, cynical Rector of Lincoln College, Oxford – who wrote to his old master, 'Is it not a remarkable thing that you should have first started the idea – and the word – Development, as the key to the history of church doctrine, and since then it has gradually become the dominant idea of all history, biology, physics and in short has metamorphosed our view of every science, and of all knowledge?'[26] The very fact that Pattison got it all slightly wrong – Hegel, surely, if anyone, and not Newman was the greatest proponent of development philosophy – shows that the development idea was in the air. For once the over-used word *Zeitgeist* is useful. Darwin, who lived in a bubble, did not quite realize, until he read *Vestiges*, how much of a symptom of the *Zeitgeist* he was. Darwin's friend Henslow wrote breezily in December 1844, 'I have been delighted with *Vestiges*, from the multiplicity of facts he brings together, though I do [not] agree with his conclusions at all, he must be a funny fellow.'[27] Rather heavily, after Christmas, Darwin answered, 'I have also read the *Vestiges*, but have been somewhat less amused at it, than you appear to have been.'[28] His driving ambition, to become the English Humboldt, was frustrated. With many a groan, he would rush to the privy which had been rigged up behind a curtain in his study at Down. When gastric upset permitted, he resumed his patient and exhaustive study of the barnacle.

The anonymous author of *Vestiges* having pipped him to the post as the popularizer of evolution, Darwin decided to pare his *Journal of Researches* from its two naval companions – the accounts of Captains FitzRoy and King – and to issue his account of his voyage of HMS *Beagle* as a separate volume with a commercial publisher. Lyell introduced him to John Murray, his own publisher. It was the house which published the *Quarterly Review*. In an earlier generation, John

Murray II had published Jane Austen and Lord Byron. This John Murray – the III – had published Lyell. Educated at Edinburgh University, he was well versed in science. He would have noted that in this revised, second edition of the *Beagle* journal, Darwin had introduced evolutionary reflections which were absent from the first published version. On the Galápagos archipelago, he had added:

> Considering the small size of the islands, we feel the more astonished at the number of their aboriginal beings, and at their confined range. Seeing every height crowned with its crater, and the boundaries of most of the lava-stream still distinct, we are led to believe that within a period, geologically recent, the unbroken ocean was here spread out. Hence, both in space and time, we seem to be brought somewhat near to that great fact – that mystery of mysteries – the first appearance of new beings on this earth.[29]

He incorporated into the book matter from his published work on the zoology and geology of the voyage, while reducing the number of words in the book by cutting descriptive passages. Darwin sold the copyright to Murray for £150 and the book was published with the title *Journal of Researches into the Natural History and Geology of the Countries Visited during the Voyage of HMS Beagle Round the World*. Between 1 January 1847 and January 1848 Murray sold 236 copies: the total number sold, including the early edition, was 4,100.[30] Not a great sale, but it was satisfying to see the *Beagle* reflections detached from those of Captain FitzRoy. The real work, however, in the late 1840s, was on the barnacles. As far as his relationship with the firm of Murray was concerned, the early stages of working with them were unfortunate. In addition to his *Beagle* journal, Murray was the publisher of the Admiralty's *Manual of Scientific Enquiry* to which Darwin had contributed the chapter on geology. By a printer's error, two pages of Darwin's chapter were transposed. It was highly characteristic of Darwin, both that he was tortured by being associated with error and that he felt extreme personal sympathy for the printer responsible. 'Do you chance to know', he asked Murray, 'whether the unfortunate Reader or head Compositor will be fined heavily, for I cannot bear the thought of this, & if you happen to *know* that such is the case & would out of charity direct one of the clerks to inform me I would write to Mr Clowes [the head of the printing firm] & make the poor workman some present.'[31]

As for the barnacle studies, these had begun on board the *Beagle*. Of the 1,529 species bottled in wine spirits and sent back from the voyage to London, there was a single bottle containing over a dozen very rare, tiny South American barnacles.[32] What made them rare? Up to 1830 barnacles had been defined by the shape of their shell houses and not by their softer bodies. The Chilean barnacle Darwin collected had no house of its own, no cone-house or stalk. By the 1840s many naturalists, thanks to the dramatic improvement in the quality of microscopes, had begun to work on the smaller creatures of the sea. T. H. Huxley, the assistant ship's surgeon on HMS *Rattlesnake* bound for Sydney via the Cape, was at work on molluscs. Edward Forbes, lecturer in botany at King's College, had written a book on starfish, and was about to write *A Monograph of the British Naked-Eyed Medusae* – jellyfish.

The fascination of the barnacles, for Darwin, was that they were small enough, collectable enough and perhaps in a sufficient state of transition to serve as potentially useful demonstrations of species mutation. If the species theory were true, then the human race, and all other animal life-forms on the planet, had developed from some marine invertebrate such as the barnacle. Darwin's interest in marine zoology went right back to his Edinburgh days as a student with Drs Grant and Coldstream, but the barnacle research began in earnest in 1846 and continued for a good eight years until, in 1854, he completed his monograph on pedunculated cirripedes. The work demonstrated, among other things, Darwin's prodigious skill in the art of anatomical dissection, for he was able to demonstrate features of barnacle life which had never previously been guessed at. For example, he discovered the existence of an ur- or proto-eye with a pair of tiny ophthalmic ganglia in adult Lepads. Organs which could later develop as potential means of hearing and smelling were also detectable. Rightly did he receive the Royal Medal of the Royal Society of London in 1853 for his work on the barnacle. The work had been exhaustive and exhausting. It put him in touch with a whole network of scientific researchers throughout the world, for there was almost literally not a stone unturned in his analysis of 'Mr

Arthrobalanus': a creature no bigger than a pin-head which could be dissected only under the microscope.

It was Owen who had suggested that Darwin make a comparison between Arthrobalanus and other species. As well as the work putting him in touch with scientists in Europe and Australia, he enlisted the help of his old servant Syms Covington, now settled, with a family in Sydney, from whom he had not heard in years. 'I am now employed on a large volume, describing the anatomy and all the species of barnacles from all over the world. I do not know whether you live near the sea, but if so I should be very glad if you would collect me any that adhere (small and large) to the coast rocks or to shells and corals thrown up by gales.'[33] Luckily for Darwin, Covington came up trumps. Darwin acknowledged that his old servant had taken great trouble to search for specimens and was able to tell him he had found 'a new species of genus of which only one specimen is known to exist in the world, and it is in the British Museum'. (The 'curious' specimen was probably *Catophragmus polymerus*.)[34]

In acknowledging this valuable specimen, Darwin confided in Covington,

> You have an immense, incalculable advantage in living in a country in which your children are sure to get on if industrious. I assure you that, though I am a rich man, when I think of the future I very often ardently wish I was settled in one of our Colonies, for I have now four sons (seven children in all and more coming) and what on earth to bring them up to I do not know. A young man may here slave for years in any profession and not make a penny.[35]

Darwin's really innovative discovery about the barnacle was its hermaphrodite sexuality. In the genus *Ibla* he found a species in which small rudimentary males were parasitic on the female, and in both *Ibla* and *Scalpellum* he found 'complemental' males attached not to a female but to a hermaphrodite. This discovery was unique in the animal kingdom, and it was clearly of the first relevance in determining the pattern of evolution: as he put it in his finished monograph, *Living Cirripedia*, 'how gradually nature changes from one condition to another, in this case from bisexuality to unisexuality'.[36] It was as if – the comparison is mine, not Darwin's – the myth of human sexuality expounded in Plato's *Symposium* was being played out beneath the lens of his microscope. By April 1848 he could tell Henslow:

> all the Cirripedia are bisexual, except one genus, & in this the female has the ordinary appearance, whereas the male has no one part of its body like the female & is microscopically minute; but here comes the odd fact, the male, or sometimes two males, at the instance they cease being locomotive larvae become parasitic within the sack of the female, & thus fixed & half embedded in the flesh of their wives they pass their whole lives & can never move again.[37]

So he mused to Henslow, in 1848, as he himself became increasingly incapacitated by his mystery illness and ever more dependent upon Emma, whom he now openly called 'Mammy'.

Nothing so forcefully reminds a grown man of his unfulfilled childhood gaps than the death of a parent. Having been motherless since 1817, on 13 November 1848 Darwin lost his father. Robert Darwin, immense in height and weight, bronchitic and well past his eightieth year, had been failing for months. For most of the summer he had been bed-bound, or confined to a wheelchair in his Shrewsbury hothouse. Darwin had visited his father in May 1848, clearly expecting him to die soon. Emma, heavily pregnant with her son Francis (who was born on 16 August), was unable to accompany him, not least because her 'baby' George (less than three) was ill and keeping her awake at nights.[38] Darwin himself, once installed in The Mount, threw himself on his sisters' nursing skills but, as he wrote to 'my dearest dear old Mammy . . . today am languid and stomach bad . . . Without you, when sick I feel most desolate.'[39] 'My poor Father had a wretched night last . . . Oh Mammy I do long to be with you, & under your protection.'[40] Darwin returned to Down and remained there for the rest of the year until his sister Emily Catherine wrote of their father, 'he is weaker and weaker . . . there is no sign of rallying'.[41] When the old man died, on the Monday of that week, the 13th, the sisters arranged the funeral to give their brothers Erasmus and Charles time to reach Shrewsbury by Friday.

Both brothers were seized by psychosomatic symptoms upon the receipt of this news. Although Catherine had suggested Bobby start his journey on Thursday, Friday found him only having got as far as London from Down. From Ras's house in Park Street, Darwin wrote to his 'ever dear Mammy': 'I often fear I must wear you with my unwellnesses & complaints. Your poor old Husband.' So wrote a man who had not yet reached the age of forty.

Ras, who had also been ill, at least managed to reach Shrewsbury the day before the funeral and to follow the procession to the church, with the four Parker boys, children of Darwin's eldest sister Marianne, acting as honorary pallbearers. After them came Caroline and Jos Wedgwood, Dr Parker, with Catherine, Ras and Susan Darwin. By the time Charles arrived in Shrewsbury, the service had already begun. He did not go to the church but stayed behind at The Mount with his sister Marianne, who was too distraught to attend. Ras wrote later to Fanny and Hensleigh that everyone at The Mount had been ill but, as Dr Parker observed, from nervous feelings rather than from any known disease.[42]

By now, 'nervous feelings' and their debilitating physical consequences took up the greater part of Darwin's waking life. In 1849 he wrote to his old servant Syms Covington, 'I have not been able to walk a mile for some years.' Illness, as for so many of the rentier class in the nineteenth century, had become a full-time occupation. Hence, it was a great era of quack cures. Sir Walter Scott lampooned the vogue for spas in *St Ronan's Well* (1824). Spa towns all over Europe sprang up offering relief to valetudinarians with limitless time at their disposal. One of the fads which gripped the Austrian mountain clinics in the early 1840s was the water cure, pioneered by Vincenz Priessnitz in Gräfenberg. Dr James Wilson, a fashionable general practitioner in the heart of London – Sackville Street, Piccadilly – returned from Austria in 1842 bowled over by Priessnitz's methods and, presumably, by his commercial success. He shared his obsession with a medical neighbour in Sackville Street, Dr James Manby Gully. Both young doctors (Wilson was born in 1807, Gully was born in 1808, the son of a coffee-planter in Kingston, Jamaica) were liberal in politics, forward-thinking in their medical ideas. There is no evidence that Gully or Wilson were quacks, if by that is denoted medics who are deliberately fraudulent. Gully had suffered grave financial hardship after the abolition of slavery when his Irish father went bust, unable to make a profit from coffee if he paid his former slaves. He was not averse, therefore, to making money, even though he probably genuinely believed in the healing properties of hydrotherapy. He and Dr Wilson decided to find a spa town which would be 'an appropriate locality for the practice of hydro-therapy'. They settled upon Malvern, a spa village, nestling upon the stretch of majestic hills from which, in the fourteenth century, Piers the Ploughman had seen his visions of the 'fair field full of folk'. Its medieval priory has some of the most magnificent fifteenth-century glass in England. Though its healing wells refreshed visitors in the eighteenth century, it was still a tiny place when Wilson and Gully descended upon it. Its population in 1817 was numbered in the hundreds, and when Wilson arrived in 1842 there were 477 houses and 2,768 souls.[43] Wilson, for his clinic, purchased a bankrupt hotel. Gully bought two large houses in Wells Road. Dr Wilson was undoubtedly the cleverer of the two medics. He spoke seven languages and his vast library contained over 700 volumes on the water cure alone.[44] His patients were more than a little in awe of him as he rode about the hills on his thoroughbred bay mare, an autocrat who insisted with great fierceness on his regimen being followed to the letter. It was inevitable that he and Gully should quarrel since, although Gully always credited his colleague with having pioneered the treatment, it was he who made the water cure into a real craze, he who put Malvern on the map. While Dr Wilson, with his tall bearing, high brow and side-whiskers, seemed like an authoritative university professor, Dr Gully, rather short and smooth, looked like a chancer, a conman. It seems entirely fitting that his career ended shadily – when his mistress, Mrs Charles Bravo, was accused of murdering her husband in 1876. This was long in the future, however, and by then he had attracted a string of famous men and women to Malvern, many of whom had taken his cure – Carlyle, Tennyson, Florence Nightingale among them.

Darwin became attracted to him when he read Dr Gully's book, *The Water Cure in Chronic Disease: An Exposition of the Causes, Progress & Terminations of Various Chronic Diseases of the Digestive Organs, Lungs, Nerves, Limbs & Skin; and of their Treatment by Water and Other Hygienic Means*. It was published in 1846 by the canny man who had bought out *Vestiges* – John Churchill, of Princes Street, Soho. It is a relatively long book, 692 pages, and like almost all medical books written more than a few decades ago, it reads like mumbo-jumbo. Gully cunningly directed his sights, however, on sufferers from *chronic* disease – that is, those who would live long enough to make their illness a hobby, and return again and again for expensive treatment. He noted that disease had 'a greater tendency to become chronic in some than in other persons', realizing that those who devoted any time to reading his book would almost inevitably belong to the 'some' rather than to the 'other' category. In particular he was able to observe that these individuals tended to have the same symptoms, 'chronic irritation of the stomach and bowels'. Also, 'they are nervous and fidgety, excessively anxious about all they are or are not concerned in; are for the most part bad sleepers, and wake with a sense of sinking'.[45] Darwin was hooked.

Expanding on his theme of nervous headaches and nervous dyspepsia, such as had plagued Darwin ever since his return from South America, Dr Gully honed in on symptoms which plainly matched the great naturalist's own. Mucous indigestion was a sort of asthma of the bowel, a build-up of mucus in the lower bowel. 'You may give fictitious, temporary appetite, by bitters &c,' Gully told his eager readers: 'you may send blood to the surface for a period with various stimulants, but you can neither *maintain* appetite until you have got rid of mucous inflammation, nor *keep* blood on the surface until you have made it, and directed it thither, and these two ends can only be fulfilled by the hygienic means of the water treatment.'[46] In this section of his book, Dr Gully could have been describing Darwin. To his generalized diagnoses of the conditions he claimed to cure he appended case histories, such as the woman who had hardly known a day without sickness since adolescence and was unable to stir from her sofa until she came to Malvern. She was wrapped in wet sheets and 'The effect on the nervous system was immediate; she declared she had not known such calmness for thirty years.' After three weeks, a woman who had been vomiting for thirty years was cured.[47]

The death of Dr Robert Darwin had left all his children far better off than they had anticipated. It transpired that he had converted all his canal stocks into railway stocks, so that he was able to leave each child with an annual income of over £8,000. (Thanks to his advice, his Wedgwood relations had also shifted round their investments and became richer, though Emma for some reason had retained some of her canal shares, and their poor performance caused Darwin anxiety.)[48]

Money could not eliminate mortality, however, and a month after Robert died, Darwin sent his three eldest children up to London to have their likenesses immortalized in Daguerreotype. Emma's uncle, Tom Wedgwood, had been a pioneer of photography and together with his friend Humphry Davy had made experiments with 'silver pictures'. He got as far as reproducing images on paper which had been moistened by silver nitrate, but he did not know how to fix them. He died of drugs and drink aged thirty-three in 1805.

It was a French inland revenue official, Louis-Jacques-Mandé Daguerre, who discovered that exposing an iodized silver plate in a camera would result in a lasting image, if the latent image on the plate was developed by exposure to fumes of mercury and then fixed with a solution of salt. This was ten years before Dr Robert died, and Daguerreotypes were now common. The three children, all notably plain with round, all but double chins, disdainful cupid's-bow lips and snub noses, are frozen in boredom by the long exposure. William, the eldest, clutching a book as a stage prop, was destined to become a banker. Henrietta, Etty, in a check frock fringed with a lace collar, has no stage prop – her hands rest on her lap. Short lank hair fell strictly combed from a central parting. She would marry Richard Buckley Litchfield and live until 1927. The photo which haunts us is that of Annie, holding a little basket of flowers. Her dress is identical to her sister's. Her hair, plaited and bowed, hangs like catkins on either side of her face. Because we know her fate – death aged ten – she appears to be staring reproachfully at Death itself, asking why she is to be given so little of life.

It was of his own life, however, that her father was starting to despair. Strange as it may appear to the reader of the twenty-first century that a great scientist could fall for Dr Gully's mumbo-jumbo, Darwin was not alone in hoping to become well through the water cure. For the children, the thought of their father leaving Down other than for a hurried visit to town was novel. The idea of the whole family moving was revolutionary. Etty could still remember, sixty years later, 'the exact place in the road, coming up from the village, by the pond and the tall Lombardy poplars, where I was told'.[49]

They were all to move, Jessy Brodie, the children, Miss Thorley the governess, the whole boiling. Emma and Darwin left Down on 8 March and travelled via London, taking the London & North Western Railway to Birmingham, changing to the British & Birmingham Railway to Worcester and travelling the eight miles or so from Worcester to Malvern by four-horse coach. (Great Malvern station was not opened until 1860; the present beautiful railway buildings, in which English Gothic meets Austrian Tyrol, by the architect E. W. Elmslie, were completed in 1862. Malvern was originally on the Worcester & Hereford Railway, later absorbed by the Great Western Railway.) For the first few days, the Darwins stayed at the grandest hotel in the town, the Foley Arms (its full name being the Royal Kent & Foley Arms because Queen Adelaide had once stayed there), before moving into the Lodge, a white stucco villa on the Worcester Road, set in its own grounds, on a wooded slope. (Everything in Malvern is on a slope.) Darwin, writing to his cousin Fox, and offering him a berth in the spare bedroom, described it as 'a very comfortable house, with a little field & wood opening on to the mountain, capital for the children to play in'.[50]

From the first, Darwin was captivated by Dr Gully, and for his sake was even prepared to reduce his snuff intake to six pinches daily.[51] Those who frequent spas, adopt cranky diets or pursue supposedly curative regimens care only partially for the efficacy of the treatment. The commercial purveyor of such programmes knows that a large part of the attraction is the time it allows the patient to concentrate uninterruptedly upon themselves and their symptoms. Meanwhile all those employed at the clinic, from the doctors to the nurses to the meanest orderly, are being paid to concentrate on their patients, and their patients alone, so that the valetudinarian, however devoted their partners, carers or family at home, has the satisfaction of knowing that every flickering attention must be devoted not merely to *them*, but to those tormentors with whom they are partially in love, their symptoms. Clever little Annie noticed this. Emma wrote to Fox: 'Annie was telling Miss Thorley all her Papa had to do about the water cure and how he liked it. "And it makes Papa so angry."

Miss T must have thought it a very odd effect. He said it did make him feel cross.'[52]

Darwin described the exacting regime to his sister Susan.

> ¼ before 7 get up, & am scrubbed with rough towel in cold water for 2 or 3 minutes, which after the first few days made & makes me very like a lobster – I have a washerman, a very nice person, & he scrubs behind, whilst I scrub in front – drink a tumbler of water & get my clothes on as quick as possible & walk for 20 minutes – I cd walk further, but I find it tires me afterwards. I like all this very much. At the same time I put on a compress, which is a broad wet folded linen covered by mackintosh & which is 'refreshed' – ie dipt in cold water every 2 hours & I wear it all day, except for about 2 hours after midday dinner; I don't perceive much effect from this of any kind. After my walk, shave & wash & get my breakfast, which was to have been exclusively toast with meat or egg, but he has allowed me a little milk to sop the *stale* toast in. At no time must I take any sugar, butter, spices, tea, bacon or anything good – At 12 o'clock I put my feet for 10 minutes in cold water with a little mustard & they are violently rubbed by my man; the coldness makes my feet ache much, but upon the whole my feet are certainly less cold than formerly.

This excerpt from a letter which takes the reader through every moment of the day – the patient wrapped in a wet sheet, the patient wrapped in a blanket, the patient offered a hot-water bottle, fed homeopathic medicines ('which I take obediently without an atom of faith')[53] – gives the flavour of the self-obsession. Only those who were abnormally self-preoccupied could endure not merely the discomfort of such routines, but also their tedium. Charles Dickens tried the cure, but, unlike the other Charles, he was able to see its absurdity. 'Oh Heavens, to meet the Cold Waterers (as I did this morning when I went for a shower bath) dashing down the hills, with severe expressions on their countenances, like men doing marches and not exactly winning.'[54]

While at Malvern, Darwin continued to do small amounts of work and to continue learned correspondence – on 9 April he thanked Hooker for 'two interesting gossipaceous & geological letters'. Hooker was in the Himalayas, and wrote detailed accounts of glacial formations and 'Malarious valleys'.[55] Sadly, the long hours of the water treatment did not allow Darwin the time, even had he possessed the energy, to make explorations of the Malverns, whose complex geology is of interest – the central core containing some of the oldest (Pre-Cambrian) surface rocks in Britain, so old that they contain hardly a trace of fossilized living creatures. Some of the Pre-Cambrian rock is fused with later molten material – green-tinged hornblende, pink gabbro, black diorite and other granites – from which the garden walls and villas of mid- to late Victorian Malvern were so characteristically constructed.

By June, however, he was telling astronomer-polymath Herschel that the water cure had:

> an astonishing renovating action on my health; before coming here I was almost quite broken down, head swimming, hands tremulous & never a week without violent vomiting, all this is gone, & I can now walk between two & three miles. Physiologically it is most curious how the violent excitement of the skin, produced by simple water, has acted on all my internal organs. I mention all this out of gratitude to a process which I thought quackery a year since, but which now I most deeply lament I had not heard of some few years ago.[56]

Most of the time in Malvern could be summarized by his phrase 'perfectly idle. Health greatly improved'.[57] On 30 June 1849 they returned to Down and he resumed his work on barnacles and geology.

As soon as he reached home, he determined to continue with the water treatment, and he engaged a local builder named John Lewis to build a wooden hut near the well in the garden where he could have 'showers'. Lewis's fifteen-year-old son was taken on as Darwin's page. Every morning he would go to the hut and pump gallons of water into a little steeple attached to the roof. Darwin would undress in the hut and young Lewis would pull a string, releasing water on to his master's back 'with great force'. Etty remembered how she and Annie would hear their father groaning inside the hut. Then he would come out, 'half running and half frozen', to walk with the two little girls in the Sandwalk.[58] From July 1849 to June 1851 he took some form of water treatment every day: a daily douche if at home, or, if he were staying with his sister Caroline or with the Wedgwoods, he would merely be wrapped in dripping sheets.[59] He also dosed himself with hydropathic remedies prescribed by Dr Gully, and at Dr Gully's suggestion he consulted a clairvoyante whose powers, she alleged, enabled her to 'see the insides of people & discover the real nature of their ailments'. Darwin showed her a sealed envelope and said, 'I have heard a great deal of your powers of the reading concealed writings and I should like to have evidence myself; here is this bank note & if you will read the number I shall be happy to present it to you.' She replied that she had a maidservant at home who could do that. She informed him that 'the mischief' was in his stomach and his lungs and described to him 'a most appalling picture of the horrors which she saw in his inside'.[60] Darwin concluded she had been tipped off by Dr Gully: what is of interest is not that Darwin retained some scepticism about her clairvoyant powers, but that he agreed to consult her in the first instance. Although he tried to convince himself that the water cure had been effective, it was not long before his old symptoms returned, and to judge from the careful note he took of his flatulence, this seems to have troubled him on a daily basis, the 'fits' of it occurring anything up to seven times daily, to a severity which ranged from 'slight' to 'excessive'.[61] This being the case it would not have required very powerful clairvoyant powers to discern that his insides were 'full' – and at intervals less full – of 'horrors'.

Coincident in time with the growth of flatulence and the varying successes and failures of Dr Gully's treatments, Darwin's vestiges of glimmering Christian belief ebbed away. He wrote later that in the 1840s 'disbelief crept over me at a slow rate but was at last complete. The rate was so slow that I felt no distress, and have never since doubted even for a single second that my conclusion was correct.'[62] On other occasions, however, later in life, he would admit 'I am in thick mud . . . yet I cannot keep out of the question.'[63] Although acknowledging that 'the whole subject is too profound for the human intellect. A dog might as well speculate on the mind of Newton,' Darwin was unable completely to leave the subject alone.[64] Whereas the naturalist Darwin, slow, patient, contemplative, could not cut corners and lived for the intellectual pleasure of research, the other Darwin, the man with one big simple idea, started first with the theory, and was doing his best to make the evidence prove the theory.

The trouble with this, as Darwin himself recognized, was not just that 'a dog' was speculating about the mind of Newton. It was more that, as Wittgenstein was to put it, if a lion could speak we could not understand him.[65] The dog and Newton did not have a common language. Science could neither prove nor disprove theology, however much the observable nature of things, and mercilessness of things, might make religious faith and practice impossible. The sheer indifference of the material universe was about to hit the Darwin family in the most painful way possible.

It now seems overwhelmingly probable[66] that Annie Darwin contracted tuberculosis some time in 1850, when she was aged nine. Tuberculosis is caused by a slow-working bacillus, *Mycobacterium tuberculosis*, which can be picked up in infected milk or passed through the air when a carrier coughs. Could she have picked up the disease when she was shown round the Wedgwoods' Etruria Works in 1848, and seen the children working there? Out of 387 workers, 103 were between the ages of ten and seventeen, and thirteen were under ten. Jos II told an inquirer that he had no information about the health of these children, though many employed in delicate painting-work with lead glazes were said to have 'that sort of delicacy which is universal in sedentary employments',[67] and many were feared to be consumptive – the word 'consumption' arising out of the disease's wasting effect. The existence of the tubercle bacillus was not discovered until 1882, by the German bacteriologist Dr Robert Koch.

Dr Thomas Yeoman, a London physician, wrote, 'Consumption, Decline or Phthisis, is the plague-spot of our climate; amongst diseases it is the most frequent and the most fatal; it is the destroying angel who claims a fourth of all who die.' Sir James Clark, one of Queen Victoria's doctors and a leading authority, believed that a third of all deaths in England arose from tuberculous diseases. These took many forms. It was not simply a lung disease. It could enter the abdomen causing tuberculous peritonitis: it could infect the blood, leading to meningoencephalitis, which would induce vomiting and coma.

When, exactly, Annie was touched by the 'destroying angel' cannot be known. Emma, who gave birth to the eighth child, Leonard, in January 1850, had many other things to notice. In May, Miss Thorley took Annie to London to see Obaysch at the Zoo, the first hippopotamus in Britain (or so palaeontologists observed) for half a million years. Queen Victoria visited the creature five times and, with her gift for making unlikely favourites, found his eyes 'very intelligent'. Richard Owen noted that it 'now and then uttered a soft complacent grunt, and lazily opening its smooth eyelids leered at its keeper with a singular protruding movement of the eyeball'. The hippo had been obtained from Abbas Pasha, Viceroy of Egypt, by the British Consul General in Cairo, and tickets to see him were in great demand. Darwin obtained them as a Fellow of the Zoological Society.

With June, there was evidence of the tribal tendencies of the Wedgwoods to meet in great packs. Five Wedgwood cousins came to Down – Ernie and Effie travelled from London. Cecily, Amy and Clement from Barlaston. In addition Darwin's aunt Sarah Wedgwood, his mother's last surviving sibling, had taken a house called Perleys in the village. She was tall, upright, skinny and unbending. 'It is my misfortune', wrote this child of the great manufacturer, 'to be not of an affectionate disposition, though affection is almost the only thing in the world I value.' Annie and Etty enjoyed the 'mysterious charm' of Aunt Sarah's rather neglected garden.[68]

It was while the Wedgwood cousins were staying, during a hot dry summer broken by thunder and lightning, that Emma began to notice that Annie was not well. 'Annie first failed about this time,'[69] she added as an annotation to her diary. The child began to find her lessons a strain, and often wept after going to bed. The physician Richard Cotton, describing tubercular symptoms in children, would note, 'the child is peevish, irritable, and indisposed to exertion'.[70] It would seem that Annie was in fact an exceptionally sweet-natured child which made her exhaustion and frequent cascades into distress the more noticeable. Late summer brought outings which plainly tired her. In August, Emma and Miss Thorley escorted the elder children on a nine-mile ride in the phaeton to Knole. A little later, Darwin and his wife took Willie, Annie, Etty and baby Leonard to stay with Uncle Jos (Josiah III) and his wife Caroline (Darwin's sister) at Leith Hill Place, a lovely house on the Surrey Heights. Expeditions searching for bilberries in the low-growing shrubs of that sandy upland should have been just the kind of day these children most loved, with Willy, now ten, showing himself a chip off the old block by carrying his entomological box with him on each outing and displaying a lepidopteral mania. Annie, though, was 'overfatigued'.[71]

On the last Sunday in August, the Archbishop of Canterbury, John Bird Sumner, came over to Down from Addington, his summer palace near Croydon, to conduct a confirmation service for the elder Lubbock children in the parish church. He was a moderate evangelical, and he spoke in a 'very plain and easy' manner to the children (Lady Lubbock noted), meditating on the lines from Bishop Ken's hymn –

> Teach me to live that I may dread
> The grave as little as my bed.

At the beginning of October Miss Thorley took Annie and Etty to the then popular coastal resort of Ramsgate. In those days you could board ferries from Ramsgate to France and Belgium, which partly added to its attraction. Queen Victoria had stayed there in her teens. Late as it was in the year, the Darwin children bathed in the sea, handing in their names in the bathing-room on the promenade. The names were then entered on a slate. Swimmers took it in turns to enter the bathing-machines, strange devices like gypsy caravans which enabled the holidaymakers both to preserve their modesty and to enter into deep water without the long walk through shallows. A horse would pull their machine to waist-high waves, a door at the back of the machine would open and an umbrella of canvas would then unfold to conceal the bathers from the eyes of anyone watching from the beach.

When the children had been in Ramsgate a fortnight, their parents arrived. Emma was pregnant yet again. Darwin had promised to give himself a break from work, but could not resist examining the barnacles on Ramsgate pier. Two days after their arrival – Emma would afterwards remember their sick child's 'bright face on meeting us at the station'[72] – Annie became feverish, headachy, and it was clear that she was seriously unwell.

In November, they took her to see Dr Henry Holland, who had attended at her birth. Annie's 'nights became worse about this time', noted Emma.[73] At home, the sickly child was allowed to have her tea in her father's study, drinking from blue teacups on the mahogany

Pembroke table, rather than being with the other children in the nursery. She and Etty made arrangements of seashells gathered at Ramsgate, and Darwin gave them some of the shells he had brought back from his *Beagle* voyage. In December they took Annie for a second visit to Dr Holland, who appeared powerless to help her. A little while later, Emma ominously wrote in her diary 'Annie began bark.'[74]

It was at this juncture that the Darwins turned to Dr Gully. Darwin consulted the Malvern doctor by letter; Gully seems to have suggested that they apply a version of the water treatment at home.[75]

Christmas came and went. At first Darwin had wrapped Annie in a wet sheet and rubbed vigorously for five minutes. Next he tried the 'spinal wash', rubbing a wet towel up and down her spine. Then he tried 'packing' her in damp towels and sheets for as long as an hour and a half. Next, 'shallow baths' were tried in which hands and feet were soaked and scrubbed. 'The feet and hands, the soles and palms especially,' wrote Gully, 'contain an accumulation of animal nerves and of blood vessels . . . in order to bind them by the closest sympathies with the great centres of thought and volition, so that their applications and movements may be accurately directed by the mind.'[76] These gruesome routines were followed, with no curative effect, for the weeks of January and February 1851.

Darwin went up to the London Library and borrowed books to read to the child – *Geneviève* by Lamartine, and *The Book of the Seasons* by William Howitt, a natural history book for children. Darwin found that his brother Erasmus and Hensleigh and Fanny Wedgwood were all talking about a book called *Phases of Faith* in which Francis Newman, brother of John Henry, recounted the collapse of his Christian belief. Newman gave classes on geometry at the Ladies College (later Bedford College, London) which were attended by, among others, Mary Ann Evans – who would become famous writing novels as George Eliot. Evans lodged (and slept) with John Chapman, the proprietor of the forward-thinking *Westminster Review* of which she was the editor, and enlisted Newman, Herbert Spencer, John Stuart Mill and James and Harriet Martineau – anonymously – to propagate 'enlightened radicalism'.[77] They believed, in her words, 'in the Theism that looks on manhood as a type of godhead and on Jesus as the Ideal Man',[78] and she recruited George Lewes – the man who would eventually become her life-companion – to write on the evolutionary ideas of Lamarck.

Darwin found Francis Newman's *Phases of Faith* 'excellent'.[79] Reading the Bible, Newman wrote, 'The further I inquired, the more errors crowded upon me, in History, in Chronology, in Geography, in Physiology, in Geology. Did it *then* at last become a duty to close my eyes to the painful light?'[80] Newman concluded, 'The law of God's moral universe, as known to us, is that of progress.'[81] It is clear to the set who wrote for the *Westminster*, most of them known to Erasmus and the Wedgwoods, that 'God' was now a word used in a sense which deprived it of its old meaning and was to be enshrouded, either actually or metaphorically, in inverted commas. For Mary Ann Evans, Newman was 'our blessed Saint Francis'.[82]

Meanwhile, Annie became ever weaker, and more tearful. March was cold, and by Annie's birthday, on the 2nd, it was clear that the water treatment was a torture. Her hacking cough turned to influenza as rain battered the windows of Down House and winds howled. Emma and Charles Darwin were entering the phase of despondency when parents clutch at straws. Seven months pregnant, Emma was not in a position to travel. It was decided that Darwin and Brodie should go with Annie and Etty to Malvern.

They took lodgings in the Worcester Road at Montreal House, whose landlady was Eliza Partington. It was on the opposite side of the Worcester Road from their previous Malvern house, so that, rather than backing on to the hills, its gardens looked towards the Vale of Evesham. On 31 March Darwin left Brodie and the children here, and returned to London, quite unaware of the very grave nature of Annie's illness. Miss Thorley was sent down to Malvern to give the girls their lessons between the intervals of Annie's treatment at the hands of Dr Gully. She found a child who had rapidly deteriorated in two weeks. Annie started to vomit and to run high fevers. Gully wrote to Darwin warning him that his daughter's life was in danger. While Darwin journeyed to Malvern, Miss Thorley and Dr Gully sat by the child's bed, feeding her spoonfuls of white wine and watching her pass in and out of delirium.

By the time Darwin arrived, Gully had returned to his clinic and was busy with other patients. There was a swift and efficient postal service between Malvern and Down so that Darwin was able to send daily bulletins to Emma between 17 April – when he wrote, 'She looks very ill: her face lighted up & she certainly knew me'[83] – until her death on 24 April. The flickerings, sometimes surges, of hope which characterize both correspondents are perhaps the most pitiable feature of these excruciatingly painful letters. By 20 April, he could write, 'I do not know, but think it is best for you to know how every hour passes. It is a relief to me to tell you: for whilst writing to you, I can cry.'[84] It was natural to him to observe, so that the whole painful death, in which the child made rallies and relapses, vomited and was fed wine or brandy, slept, wept, woke and yet again was sick, is punctiliously noted. On 23 April he wrote:

> She went to her final sleep most tranquilly, most sweetly at 12 o'clock today. Our poor dear dear child has had a very short life but I trust happy, & God only knows what miseries might have been in store for her. She expired without a sigh. How desolate it makes one to think of her frank cordial manners. I am so thankful for the daguerreotype. I cannot remember ever seeing the dear child naughty. God bless her. We must be more & more to each other my dear wife.[85]

Fanny Wedgwood suggested that she should take Darwin's place at the funeral, and he accepted, with misgivings, but also with the knowledge that he and Emma needed one another. He left Malvern the day after Annie's death, and was able to write to Fanny from Down on Friday the 25th, 'It is some sort of consolation to weep bitterly together.'[86]

The funeral was arranged hurriedly, but no expense was spared. Darwin paid £57 12s 6d to Cox and Co. for a full-blown ceremonial, with a hearse, a coach for family mourners, black horses with ostrich-feather plumes, and two 'mutes', paid mourners, swathed in black gowns, kid gloves, silk hairbands and strands of crape.[87] Fanny arrived in Malvern with her lady's maid, who was sent home to Leith Hill Place immediately, taking Etty to be with her cousins. So it was that on the day of the funeral in Malvern, Darwin and Emma were weeping at Down, and Etty was on her way to a resumption of healthy child-life. The following Wednesday, Caroline

Wedgwood wrote of the children, 'They are all gone cowslip-gathering in the fields . . . Etty seems quite content and excellent friends with all the cousins.'[88]

Those left in Malvern on Friday 25 April were Fanny Wedgwood, Hensleigh, Miss Thorley and Brodie. The hearse drew up in the drive of Montreal House, and the coffin was stowed. They then made their slow progress for the funeral service at the Priory conducted by Mr Rashdall, the vicar. Annie was buried in the churchyard near the Abbey Gate. Darwin eschewed pieties or quotations from the Bible when he chose the wording on her stone: 'ANNE ELIZABETH DARWIN, BORN MARCH 2, 1841. DIED APRIL 23, 1851. A DEAR AND GOOD CHILD'.

Both Brodie and Miss Thorley were very much discomposed by the funeral. Miss Thorley went for a drive, alone, among the hills, and said when she returned that she was better. Brodie appeared inconsolable. Fanny urged her to go back to Down as soon as possible, hoping that once she was settled into routines with the other children 'she will be able to put restraint on herself'.[89] Brodie's grief could not, however, be restrained. Etty remembered that she 'quite lost her self-control, and indeed, almost, her reason, and insisted on leaving'.[90]

Three weeks after Annie had died, Emma gave birth to a son. They named him Horace.

10

An Essay by Mr Wallace

THE 1850S WAS the patient, slow decade in which Darwin cogitated upon, and tested, his theory of the origin of species. Before the decade was out, he would be bounced by events into declaring the theory before the world. Illness, natural reclusiveness and love of family all combined to keep him much at Down. For months on end, he was by way of being a hermit. He was not, however, intellectually isolated – not entirely. Two men who played a vital role in the story were Edward Blyth, the Tooting druggist who was now the curator of the museum at Calcutta, and Thomas Henry Huxley (1825–95).

Huxley was the son of that mathematics master at the grammar school in Ealing who had taught the Newman brothers, John Henry and Francis. When Huxley was eight, the headmaster of this school died, and George Huxley the mathematician, disliking his successor, moved the family to his native Coventry. Strangely enough, Thomas Huxley's formal education seems to have been rather neglected by his father. He was his parents' seventh child, the youngest to survive infancy, and perhaps by then tiredness and grief had sapped them. If he had scant formal schooling, however, Huxley grew up in a clever, well-read household. He read Sir William Hamilton's *Logic* at an early age and he mastered German. When he was fourteen, two of his sisters married doctors, and this determined the boy's future. The Coventry brother-in-law, Dr Cooke, excited young Thomas's interest in anatomy. The other medic, Dr Scott, was London based, and in 1841 Huxley went to the capital to become his apprentice. He became an accomplished anatomist, and obtained a Free Scholarship at the Charing Cross Hospital and Medical School in 1842. It took four years in those days to qualify as a doctor, and by 1846 Huxley was able to join the Royal Navy and sail with HMS *Rattlesnake*, which left England for the southern hemisphere in that year.

Fine boned, dark haired, bright eyed, impulsive and charming, Huxley was to be one of Darwin's doughtiest allies: so doughty and so loyal that it is sometimes easy to overlook the fact that he was never in the strictest sense an orthodox Darwinist. For the first seven years of his acquaintance with Darwin he was a non-believer in the transmutation of species, and in the years after the publication of *The Origin of Species* he was really closer to the views expressed in Chambers's *Vestiges* than he was to Darwin's view of natural selection. His importance in the story, though, cannot be exaggerated because he was all the things which reclusive, nervous, slow Darwin was not. Huxley loved debate, he was unafraid of controversy and he was an exuberant man, as extravert as Darwin was introvert. Moreover, he was far more widely read than Darwin, had the command of several languages, had mastered philosophy and, unlike Darwin, completed his medical training and was a qualified academic scientist. As the ship's surgeon on HMS *Rattlesnake* to Australia, Huxley had plenty of time for research, and concentrated on hydrozoa. The papers which he sent home to the Royal Society on the *Medusae* led to his being greeted, when he returned to London, as a scientific anatomist of the first rank. He came home in 1850, but it was five years before he could afford to marry Henrietta Anne Heathorn, the girl he had met, and fallen in love with, in Sydney. In that time, he had published over thirty learned scientific articles. For most of his life he would remain under the shadow of Owen, who was the leading anatomist in Europe, and one reason for this was the diffuseness of Huxley's intelligence: he liked to write, and to lecture, on all manner of subjects other than anatomy, venturing into ethics and philosophy. Since, however, his medal from the Royal Society was awarded principally for his work on marine biology, it was inevitable that the man who had spent four years investigating sea urchins, speculating on the embryology of marine life and peering at barnacles should have wanted to meet Darwin, whose work on barnacles proved him a kindred spirit. Darwin finally completed his work on these arthropods in 1854 after eight years devoted to the subject. He nervously sent a copy of his book to Jermyn Street where Huxley, a poor man who needed to work for his living, was a lecturer at the School of Mines.

In fact, when the book arrived, Huxley was in South Wales investigating marine invertebrates on the beach at Tenby – the charmingly pretty Pembrokeshire resort where George Eliot first began to write fiction. Darwin told Huxley, 'If ever so practised a hand as you sets to work on the Cirripedia I have *no doubt whatever* you will discover many errors on my part.'[1] While tentatively endorsing Darwin's view on the connection between the cement glands and the ovaria of the *Cirripedia*, however, Huxley was unconvinced by Darwin's assertion that the gut-formed glands *were* the ovaria. (Cement glands excrete a proteinaceous adhesive, or cement, which enables barnacles to stick to rocks, boats, etc.) He was, nevertheless, impressed by Darwin's skill as an anatomist. Since Huxley's return from Australia, Darwin had liked the man and had written references for him. 'You are excellently qualified for a Professorship in Natural History.'[2] It does not appear, though, that the two men were sufficiently close for Darwin to win him over, during the 1850s, to a belief in evolution, still less in the theory of natural selection. Huxley had been irritated by the slapdash methods, the 'unscientific habit of mind' of *Vestiges*; 'it set me against Evolution'.[3] He was not converted until he read *The Origin of Species*.

If Huxley was the new friend, and if the old friends and patrons of Darwin's genius – Lyell, Hooker, Henslow – remained unconvinced or only partially convinced by his views, there was one naturalist in the world with whom Darwin could trust that he had intellectual kinship – that is, the curator of the Museum of the Royal Asiatic Society in Bengal, Edward Blyth. When Hooker had gone to India in 1848, Darwin had recommended him to look up Blyth in Calcutta.

Blyth had been far from idle. In 1849, he had published a *Catalogue of the Birds in the Museum Asiatic Society in Calcutta*, with, the following year, a *Supplement*. Until this book was written, containing 1,816 species of bird with addenda, very little work had been done on the taxonomy of Indian birds, and as Blyth put it, 'the nomenclature of Indian birds' was 'so recently in a state of chaos'.[4]

This book would have been a prodigious achievement in itself, finished as it was against bouts of illness in author and in his wife, and in the absence of other ornithologists with whom to discuss his work, time for observation in the field and adequate reference books. It was not all he did, however, as the fifth volume of the Cambridge University Press *Correspondence of Charles Darwin* makes plain.

Blyth was a compulsive cataloguer, a joyful naturalist who was interested in absolutely every species. Darwin could not have hoped for a more encyclopaedically voracious observer, or a more willing sharer of his observations. Fire a question from Down, about otters in South India, about rabbits and whether they were indigenous to the subcontinent, about mules or about Malayan cats or pigeons, and one could be certain of a detailed enthusiastic response in thousands of words. The letters of Blyth to Darwin during the 1850s amount to notes, not only for *The Origin of Species* but also for *The Variation of Animals and Plants under Domestication* and *The Descent of Man*. Blyth's letters were heavily annotated in pencil by Darwin:

> Cormorant
> Otters do they breed those that are
> trained Canary birds bred
> Prolifickness of Rabbits
> ——— of Races
> What a memory you have
> Will acknowledge everything . . .

Conclusions to which Darwin moved with reluctant slow paces, such as the kinship between man and the apes, were cheerfully assumed by Blyth:

> I think we have run mankind *home* to the tropical regions of the old world, probably (so I think) to both Asia & Africa (the regions of the Orang & Chimpanzee respectively, which are the two most nearly affined genera); and it seems to me that specifical distinctions are more likely to be [truly] detected among the quasi-primitive races referred to, than among the infinitely commingled & variously modified races which are more or less civilised (i.e. domesticated); while among these latter, the next grand point of interest is to trace the stages & phases of development, upon emerging from the primitive forest life. – How is it that most of the Papuans, with the Bushmen & Earthmen of S. Africa & others, are so very diminutive? I am far from being satisfied that insufficiency of nutritional food is the cause of this.[5]

Whereas Samuel Thomas Sömmerring and the anatomists of the liberal Enlightenment in the eighteenth century had established the kinship of African slaves with European humanity and had inspired old Josiah to his question, 'Am I not a Man and a Brother?', mid-nineteenth-century colonial Europeans wanted to create an hierarchical taxonomy, with civilized, 'domesticated' human beings at the top, wearing silk hats and stiff collars, stays and crinolines, whereas the subject 'races' of humanity, squat, naked and brown, were plainly closer to our common ancestors the orang-outangs. That Blyth threw out this observation casually, not as the central platform of some crackpot racialist manifesto, is what makes it so revealing of its time. Besides, with his omnivorous collector-mentality, Blyth was no longer primarily interested in the ideas about the origin of species which he had so fructiferously suggested in articles written nearly twenty years in the past. He would probably have agreed with Hooker, who wrote to Darwin in July 1855, 'The more I study, the more vague my conception of a species grows, & I have given up caring whether they are all pups of one generic type or not . . .'[6] Blyth was more than happy, by this stage, for Darwin to puzzle out the implications of all the abundant data which he was able – 'What a memory you have!' – to supply. In September 1855 he re-read the second volume of Lyell's *Geology*, finding that he in fact remembered most of it, though it had been eighteen years since he had last turned its pages. What had altered was the extent of his own knowledge, so he wrote out – in all humility acknowledging the author's greatness – a list of Lyell's mistakes. He did not want his name to be associated with the emendations, and in expressing this modest thought, he set down what was evidently his credo: 'truth is what we seek, & the establishment of it is the more important in proportion to the high scientific rank of the authority we presume to call in question'.[7]

The central tenet of Darwin's theory of the origin of species, the tenet which he was so patiently attempting to establish, was not merely that species evolve, or mutate, but *how* they do so. And the clue lay in those early essays of Blyth's – that, through the process of natural selection, species were adapted to their environmental needs: a finch that needed to obtain nourishment from tree-bark developed more woodpeckerish tendencies than one which could eat insects from leaves. A mammal that ate foliage from trees needed the ever-longer neck of the giraffe. This was the principle. Twenty-first-century scientists can test the process by which certain characteristics are strengthened or perfected by computer-generated, speeded-up evolution. Was it possible for Darwin, in the 1850s, to learn anything from artificial breeders of racehorses, dogs or pigeons?

Blyth freely, and to us perhaps ominously, used the word 'race' as a synonym of species, tending to use the noun 'breed' for hybrids. 'It occurs to me to add, that no "varieties" have ever sprung up in America analogous to the *humped Ox*, the *fat-rumped Sheep*, or new races of *fowls* or *Pigeons*!!! I distinguish *races* from *breeds* artificially produced by the intermixture of the latter; which latter, like hybrids generally, have little tendency to become permanent.'[8]

Blyth's breezy, chatty sentences enunciated some facts which challenged Darwin's *idée fixe*: it seemed not so much that he had forgotten his analogy between artificial hybrids and the processes of nature as that he no longer considered it. What interested Blyth was an accumulation of fact. And here was a fact which, if invariable, would knock the theory of natural selection off balance, if not destroy it altogether. If hybrid adaptations, artificially produced, were impermanent in their results, this surely provided a poor model for natural selection which was slowly handing on to the future single-hoofed as opposed to three-toed horses, sharp-beaked tree-finches and so on. An adaptation which did not last beyond a generation would be fatal. 'Mr Blyth makes a great distinction between "Breeds" artificially made & "Races". why I know not.'[9] The problem created by Mr Blyth's 'great distinction' was very much more than an irritating detail, for it cut to the very heart of the theory. If adaptations *within* species could not be explained by the analogy of artificial hybrid-breeding, how much less could actual *transmutation* in which one species evolved into a new one? This, after all, is what makes the Darwinian theory of evolution distinctive – the process of natural selection.

There was – is – literally no limit to the search which would be required to demonstrate the hypothesis, even in a plausible *majority* of species. Single-handed as he was, Darwin looked out for evidence wherever he could find it. Blyth's fascination with everything matched his own. Hooker provided patient, thorough and detailed answers to

Darwin's botanical inquiries, seed by seed, plant by plant. 'I most earnestly hope', Darwin could write, 'that at Vienna you will make particular enquiries about the *pure* Laburnum, which one year bore the hybrid flowers & on one sprig the C. purpurens – Dr Reissik (?) is name of man I think. [George] Bentham [Honorary Secretary of the Botanical Society, later President of the Linnean Society] will not believe that it was a *pure* Laburnum, & it does seem quite incredible . . .'[10]

Lyell, for his part, had *geological* doubts about the theory which he passed on sometimes in letters to Hooker, sometimes directly. Lyell wondered, if all species are in the state of flux which Darwin suggested, how natural history could be possible: it appeared to destroy the taxonomy of Linnaeus and throw everything back into a melting-pot. 'I fear much that if Darwin argues that species are phantoms, he will also have to admit that single centres of dispersion are phantoms also . . .'[11] Much of Lyell's geology was posited on the notion that groups of species spread out from a single geographical source while remaining constant in form. Fossils of identical species demonstrated that they were of the same age. Hooker, for his botanical part, as well as supplying Darwin with factual evidence, raised repeated questions. 'You say most truly about multiple creations & my notions,' Darwin wrote on 13 July 1856, 'if any case could be proved, I shd be smashed: but as I am writing my Book, I try to take as much pains as possible to give the strongest cases opposed to me, & offer such conjectures as occur to me . . .'[12] One of these objections, as Darwin candidly acknowledged the month following, was that 'the vegetable world does not appear in the confusion I should expect it to be in, were transmutation the law'.[13] This was to say nothing of another aspect of the theory, which would come markedly to the fore in its printed version, about nature being in a state of *struggle*: as he gazed at the tranquil water lilies in Kew Gardens, Hooker could be forgiven for not imagining them to be at war.

In this they differed from academic scientists.

Darwin dreaded fisticuffs, while contemplating the publication of a theory which could not fail to excite heated controversy. Towards the end of 1856, at a meeting of the Geological Society in November, Richard Owen read a paper on a much-debated fossil mammal named by him *Stereognathus ooliticus*. His case was that a single fossil tooth could legitimately lead to the determination of affinities and organization of an entire skeleton.

Huxley had written a paper taking a very different view, even though he had not acknowledged to himself how close he was to accepting a 'mutational' position with regard to species. Hearing Owen commit 'a cutting telling & flaying alive assault' was a bitter experience.[14] Huxley felt himself treated as an 'implacable foe' by Owen.[15] The battle lines were being drawn for the war that would divide the scientific academy when *The Origin of Species* was eventually published. 'The best natures insensibly deteriorate under such trials,' Darwin remarked sadly.[16]

Darwin meanwhile gave particular attention to his experiments with artificial hybrids. 'I have been astonished at differences in skeletons of domestic rabbits,' he told Hooker.[17] His young nephew Godfrey Wedgwood, who would one day take over the management of the Etruria Works,[18] quizzed the gamekeepers at Sandon Park,[19] the seat of Lord Harrowby near Stone, Staffordshire, on the domesticated rabbits which had 'gone native' and, having escaped their hutches, mated with rabbits in the wild, noticeable for their different colouring. (This would be of more use in *The Variation of Animals and Plants under Domestication* than it would be for *The Origin of Species*.)

A more obvious species to observe, partly because they were to hand, partly because artificial breeding played so large a part in the culture of their 'fanciers', were doves and pigeons. Darwin grew up in a house with a dovecote, and not long after settling at Down he began to keep some ornamental pigeons, as well as some specialist chickens which would also be of potential use to him. His twentieth-century biographer Janet Browne most attractively suggested that 'he could not help but anthropomorphize natural selection into a mating ceremony deftly engineered by a wise, all-seeing and sensible English gentleman'.[20] At one point in his enthusiasm he had as many as ninety pigeons at Down, of sixteen different kinds and eight or nine different types of fowl. By studying what the breeders and fanciers were trying to effect, 'Darwin could almost watch the process of artificial selection taking place.'[21]

Almost as soon as he began to acquaint himself with pigeon-fanciers, Darwin realized that they believed that each breed was a distinct one, deriving from a variety of wild stocks, whereas Darwin (in common with nearly all academic taxonomists) believed they all derived from a common ancestor, the rock pigeon. Here, then, was a possible template for nature itself, the whole of it: the apparently different species possessing in fact a common ancestry. Lyell could still not get his mind around it. Hooker, with the friendliest will in the world, had his doubts. The theory, however, with the stupendous attraction of its immense simplicity, could at least be tried. The wider Darwin cast his net and the more random his inquiries the better, so that whether he was contemplating orang-outangs or barnacles, rabbits or lilies, he was asking of them the same basic cluster of questions.

Pigeons, moreover, provided the chance of human diversion. Darwin was for much of his life a recluse, and it is hard to know whether it helps us to place this fact in a hedge of causative sentences. Was he ill because his intellectual concentration was so potentially revolutionary, so explosive? Was he reclusive because his work demanded so much concentration? Reclusive he may have been but exclusive, compared with Emma, he was not. Emma had two friends before she was married, at Maer. Beyond that, she never had any friendships outside the admittedly large circle of the Wedgwood and Darwin families. Her aunts, her sisters, her sisters-in-law, her nephews and nieces, these were her only intimates. Darwin had, at least, the intellectual companionship of his scientific colleagues, and in these years, up to and beyond the publication of his great work on *The Origin of Species*, this companionship was increasingly necessary, from both a cerebral and an emotional point of view. He could also take pleasure, as he developed his interest in pigeons, in the very different social life.

To William, his son, a boy at Rugby, he could write delightedly of his acquaintanceship with Bernard Brent, 'a very queer little fish' who was a well-known pigeon-fancier. Darwin paid John Lewis, the Down carpenter, to muck out his own collection of pigeons,[22] and there were occasional mishaps. Etty always resented the fact that her beloved cat had been killed at her parents' behest, and without her knowledge, for mauling one of the pigeons.[23] 'I am getting on splendidly with my pigeons: and the other day', Darwin proudly told William, 'had a present of Trumpeters, Nuns & Turbits; & when last in London, I visited a jolly old Brewer, who keeps 300 or 400 most beautiful pigeons & he gave me a pair of pale brown, quite small German Pouters. I am building a new house for my tumblers, so as to fly them in the summer.'[24] The brewer was Matthew Wicking, of the brewers Jenner, Wicking and Jenner in Southwark Bridge Road.

A most valued new friend was William Bernhard Tegetmeier, a journalist and naturalist, who would admit Darwin into the Philoperisteron Society, a club of gentlemen pigeon-fanciers that met in the Freemasons' Tavern in London. Darwin attended these meetings when health permitted. Tegetmeier, as well as supplying Darwin with company rather different from his family, was also the procurer of birds and animals. He sent laughing pigeons in a cage to Down by carrier on 4 June 1856, having previously dispatched an Angora rabbit – all grist to an evolutionist's mill – and Darwin remained what he had always been, a boy-naturalist, never happier than with animals. Only Emma's last confinement, giving birth to their son Charles Waring Darwin, prevented Darwin attending the Freemasons' Tavern to hear Tegetmeier read a paper on the development of the skull of Polish fowls.[25]

Charles Darwin Junior was born on 6 December 1856. In the year which saw the birth of this, the last of their children, they had also experienced – a mere month earlier – the death of their aunt Sarah. Born in 1778, she was the last surviving child of Josiah Wedgwood I, and had come to live at Petley, Down, to be near her niece and nephew. Her childish eyes had looked on Dr Erasmus Darwin as he sped up to Etruria in his phaeton, heard her father extol the glories of the French Revolution, and been taken to hear Mr Priestley, the great chemist, preach at the Unitarian Meeting House in Newcastle under Lyme. The father who had leant over her cradle had been born into poverty on the damp village hillside of Burslem. Enriched by his ingenuity, the family who followed her coffin were gentlefolk. There was a sense of it being a farewell, not just to one old lady, but to a whole generation: a great assembly of Wedgwoods – Jos, Frank, Hensleigh – joined the Darwins, donned black cloaks and affixed black crape to their silk hats.[26] Darwin considered that Mr Innes, the perpetual curate, 'did not read this very impressive service well'.[27]

Most of the old lady's servants, with their broad Staffordshire voices, were sent back to Barlaston to work in Frank's household. Darwin noted nervously that 'she has left a great deal of money', adding with perhaps a sense of let-down, 'to very many Charities'. Like Emma, Aunt Sarah had no social dealings outside the family. 'She had no gift for intercourse with her neighbours, rich or poor,' Etty Darwin remembered, 'and I do not believe ever visited in the village.'[28] Her failure to visit the poor was a matter of choice. Class surely played its part in her failure to visit, or to be invited by, the rich. Whereas in the towns, nouveaux riches moved relatively easily from one house to another, not pausing too much to ask the age of the money which bought the house in the country, there were still invisible but palpable hierarchies. A married couple might venture together upon their ascent, but an old spinster was safer at home among her own kind.

Emma's reasons for not mixing were more to do with family preoccupations than with the shyness of rank-category. The baby Charles was backward in learning to walk and talk. He was a solemn little person. He cried less than any of their babies, but he had no high spirits. It was clear that something was wrong with him from the beginning.

The sheltered childhood of Darwin's children continued its inward course. Darwin realized he had made a mistake sending his eldest boy William to Rugby – which he had done on the recommendation of Hensleigh, who had sent his eldest there. In spite of claiming to be 'glad to hear of your sixth form power',[29] Darwin was so absorbed in his own work that he seldom wrote to the boy and had no sympathy with the public-school ethos or with its all but exclusively classical syllabus. The headmaster was a dull young man called Goulburn ('a very mild fellow', as one Rugby master described him) under whose headship the school shrank in numbers. Much of the syllabus consisted in learning Latin verse and passages of the Bible by heart.[30] When he surfaced from his own work to consider the reality of this, Darwin said he could not endure 'to think of sending my boys to waste 7 or 8 years in making miserable Latin verses'.[31] The remaining boys were sent to Clapham Grammar School, a scientifically minded place whose head, Charles Pritchard, later became Professor of Astronomy at Oxford.[32]

The girls were kept at home at Down House, as was normal for the daughters of gentry. 'My mother took very little trouble about our education, and I was frankly bored with schoolroom lessons and schoolroom life,' Etty remembered. 'My aim was to escape as soon as possible either to look after some of my many pet animals, or to get a quiet corner to read.'[33] Playmates, when not her direct siblings, were visiting cousins. The girls were taught no science or mathematics. Etty believed her sister Elizabeth (Bessy or Lizzie) to have had some handicap such as cerebral palsy, though there survives no evidence to prove this. A succession of more or less inadequate governesses, as dithery and hopeless as those in an Ivy Compton-Burnett novel, came and went: after Miss Thorley her sister, Miss Emily Thorley (a 'magnet for unstable women', Etty later decided, which was presumably a phrase to cover lesbian inclination); then Miss Pugh, who wept at meals and was carted off to an asylum; Madame Grut, who left after only a few weeks, having quarrelled with Darwin; the beautiful German Miss Ludwig, with whom the twelve-year-old Horace fell in love, and who translated Darwin's correspondence into her native tongue. One of the governesses complained that the children used 'very bad language, so bad that she hardly liked to repeat it'. When told to convey the offending phrase on paper, she wrote, 'By George'.[34]

While the family pursued its life, Emma effected changes to the house. The year 1857 saw the construction of a new dining-room on the side of the house – later it became a drawing-room.[35]

Darwin's correspondence took up more and more time. Apart from his experiments with hybrid pigeons and other specimens, and his long slow cogitations on the Sandwalk, letter-writing – and letter-reading – was his means of retaining academic contacts. The regular visit of the postman enabled what was in effect a decades-long seminar with Blyth, Hooker, Huxley, Lyell, Henslow as the principal participants, Hewlett Cottrell Watson (botanist and phytogeographer), Herbert Spencer (journalist and thinker), John Lubbock (Darwin's neighbour, a politician and keen natural scientist), Thomas Eyton (a Shropshire naturalist), George Bentham (botanist) and others as keen occasional contributors. A key figure who entered the postal seminar in 1856 was Asa Gray, the Fischer Professor of Natural History at Harvard.

Gray was preparing a paper – 'Statistics of the Flora of the Northern United States' – and he wrote to acknowledge his indebtedness to Darwin's work. Darwin replied, in May 1856, at some length with suggestions and questions: 'With respect to naturalised plants; are any *social* with you, which are not so in their parent country?'[36] Gray was, like almost all Darwin's correspondents, subject to the lines of inquiry which all related to the theme of themes: adaptation and mutation of species, how and in what circumstance this phenomenon occurred. Gray replied in punctilious detail.

'It is extremely kind of you to say that my letters have not bored you very much,' Darwin wrote, '& it is almost *incredible* to me, for I am quite conscious that my speculations are quite beyond the bounds of true science.' The correspondence with a fellow truth-seeker whom he had never met created intellectual intimacy, so much so that in July 1857 Darwin opened his mind to the American botanist:

> Nineteen years (!) ago it occurred to me that whilst otherwise employed on Nat. Hist, I might perhaps do good if I noted any sort of facts bearing on the question of the origin of species . . . I think it can be shown to be probable that man gets his most distinct varieties by preserving such as arise best worth keeping and destroying the others. I *assume* that species arise like our domestic varieties with *much* extinction; & then test this hypothesis by comparison with as many general & pretty well established propositions as I can find made out, – in geograph. distribution, geological history – affinities &c. &c. &c. And it seems to me, that *supposing* that such hypothesis were to explain general propositions, we ought, in accordance with common way of following all sciences, to admit it, till some better hypothesis be found out.[37]

In concluding a long letter, he apologized for being 'horribly egotistical'.[38] He followed it, in September, with a four-page abstract of his idea and the motto which 'most satisfactorily' answers most of the objections: *Natura non facit saltum*.[39] Nature does not jump, it proceeds gradually. Gray's reply is lost, but Darwin's response to it suggests an initial degree of scepticism on the Harvard professor's part: 'What you hint at generally is very very true, that my work will be grievously hypothetical & large parts by no means worthy of being called inductive; my commonest error being probably induction from too few facts.'[40]

Darwin wanted the response to his theory from scientists whom he respected. He was ready to listen to Gray even if the American botanist disagreed with him. Helpful criticism could only be welcome, since, despite having worked on his theory for nearly twenty years, he did not want to make it public until its validity had been tested as thoroughly as it could be. A particular dread was that a 'popular scientist' such as the author of *Vestiges* would catch wind of the theory and publish a clumsy or distorted version of it. For this reason, he implored Gray to keep the ideas of natural selection to himself: 'If anyone like the Author of Vestiges, were to hear of them, he might easily work them in, & I then shd. have to quote from a work perhaps despised by naturalists & this would greatly injure any chance of my views being received by those alone whose opinion I value.'[41]

Only three weeks later, on 27 September 1857, Darwin received a letter from the Malayan Archipelago from an amateur naturalist named Alfred Russel Wallace. Only a tiny fragment of this letter survives. It refers to a letter which Darwin had written to Wallace in May 1855 praising his paper 'On the Law Which Has Regulated the Introduction of New Species'. This was an evolutionary tract which posited that 'every species has come into existence coincident both in time and space with a pre-existing closely allied species'. Wallace was therefore very close indeed to Darwin's own views of evolution. What he had not yet developed was a theory of *how* species evolved. The fragment of Wallace's September 1857 letter has been heavily annotated by Darwin, who did not reply to it until 22 December.

> You say that you have been somewhat surprised at no notice having been taken of your paper in the Annals. I cannot say that I am; for so few naturalists care for anything beyond the mere description of species. But you must not suppose that your paper has not been attended to: two very good men, Sir C. Lyell & Mr E. Blyth at Calcutta specially called my attention to it. Though agreeing with you on your conclusion[s] in that paper, I believe I go much further than you; but it is too long a subject to enter on my speculative notions.[42]

Darwin was not giving anything away. Whereas, in strictest confidence, he was prepared to divulge his theory of natural selection to a Harvard professor, he was understandably cautious about what he wrote to a stranger.

Wallace (1823–1913), fourteen years younger than Darwin, had not enjoyed the elder man's privileges. He was born in Usk, Monmouthshire, where his father was so careless as not to notice that the child's middle name, Russell, had been spelt Russel in the baptismal register. Wallace retained the misspelling throughout his life. Newman Noggs, the lawyer's clerk in *Nicholas Nickleby*, another decayed gentleman, declared, 'I have forgotten all my old ways. My spelling may have gone with them'[43] ('Mr Noggs kept his horses and hounds once').[44] The Wallaces had been swept into the socially downward capitalist drift. When the family moved to Hertford, Alfred Russel was sent to the grammar school, but worked as a pupil-teacher aged thirteen in lieu of fees. When he was fourteen he left school and went with his brother William to London where he learnt to be a surveyor and mastered the rudiments of geology. The next year, he spent as the apprentice to a watchmaker in Leighton Buzzard. Then he followed his brother into Herefordshire, worked as a surveyor and studied astronomy and botany in his spare time. His father died when Alfred was eighteen. His brother's surveying work was not bringing in enough to feed two, so Alfred took a job as a master at the collegiate school in Leicester.

It was in Leicester that Wallace made friends with another amateur naturalist called Henry Walter Bates (1825–92), who had started work as an apprentice in a hosiery warehouse and gained his scientific education at the Mechanics Institute. These Institutes were a post-war innovation of the 1820s and were a vital ingredient in the Victorian technological success story. By the 1850s, there were over 700 of them and they were to be found in most of the towns and cities of Britain. Charitable endowments, they had their own buildings which served as technical schools – teaching engineering and other skills – libraries and museums. They were the forerunners of the polytechnics.

One of the high points of Henry Bates's teenage years had been the exhibition at the New Hall in Leicester, showing 'preserved specimens of Foreign and English quadrupeds, birds and insects'. The sixpence entry fee opened his eyes to the possibility of becoming a full-time natural historian, and in the leisure hours of his later teens, when he was not being an aspirant hosier, he was collecting bugs and butterflies in Charnwood Forest. His father, who was successful in business and was rising in the world, encouraged his son's enthusiasms, and these continued after Bates had completed his apprenticeship, abandoned hosiery and taken a job at Allsopp's Brewery in Burton upon Trent.[45]

A little before his move to the brewery, Bates met Wallace, and the two men began a friendship based on a shared love of natural history. Bates introduced Wallace to entomology. It was also about this time that Wallace read Malthus's *Essay on Population*; the experience would eventually lead him to an idea of evolution by the process of natural selection which was all but identical to Darwin's. Wallace and Bates, despite their modest means, formed the bold ambition to explore the Amazon, defraying their expenses by the sale of specimens to scientists and museums back home. They set sail in April 1848 and by March 1850 they had quarrelled irreconcilably and parted company. Wallace was joined on his Amazonian adventures by his brother Herbert, who died of yellow fever in 1851. This was not the only calamity to befall Wallace. After four years in Brazil, he turned for home. The ship which bore him, the 235-ton brig *Helen*, contained all his notes, all his specimens and a cargo of 120 tons of rubber as well as cocoa, red annatto dye, piaçaba fibres and aromatic balsam. It was this last which began to smoulder in the heat and to turn into a bubbling cauldron after three weeks at sea. The Captain made the mistake of opening the hatches to quench the smoke, but in doing so he gave air to the fire, and the smouldering balsam burst into flames. The ship was 700 miles from Bermuda and managed to sail on, despite the fire, for another 500 miles before being rescued by a cargo ship, the *Jordeson*, bound from Cuba to London. Wallace lost, not only all his notes on the natural history of the Amazon, but also the specimens which would have financed the last lap of his journey. Some of his specimens had been sent to London in advance, so that in 1851, by the time he was back in England, Wallace was no longer a figure entirely unknown to the scientific establishment. He realized that the only way of continuing his life as a scientist-collector was to press on. So in 1854 he set off on an altogether different venture: to the Malayan Archipelago, where he would spend the next eight years. As he told his brother-in-law, he was determined to remain there until he had solved the 'whole problem' of the Archipelago, above all the puzzle of the zoological distinction between creatures who lived on either side of the strait between Bali and Lombok, later known as Wallace's Line.

It was a slow journey, but Wallace was advancing into a fully committed evolutionist. The question for him, as for any of his predecessors in the field, was *how*? Some of the dogged resistance to the idea of mutation in species came, in the scientific academy, from blind prejudice. Much of it, however, derived from the simple fact that no one except Lamarck had ventured to account for the process of mutation. Lamarck's theory had serious flaws. Then – how? In 1858, Wallace was struck down by an attack of fever at Ternate, in the Moluccas. It was while the fever was upon him that he recalled his reading of Malthus, and the actual method of natural selection flashed upon Wallace's mind. It took him only a few hours of concentrated thought. Over the next two days, he wrote down his idea. There was only one man in the world with whom he wished to share it, and he posted his hastily composed essay to Down House.

For Darwin himself, Wallace's letter from the Moluccas came as potential calamity, at least as far as his claim to be the originator of the theory was concerned.

Both men, Darwin and Wallace, behaved with honour: both realized that they had arrived at the theory independently. Wallace saw that Darwin had, in fact, thought of it before him. Darwin, while wanting this to be recognized, did not wish to diminish Wallace's achievement. Darwin might imagine Wallace saying, 'you did not intend publishing an abstract of your views till you received my communication, is it fair to take advantage of my having freely, though unasked, communicated to you my ideas, & thus prevent me forestalling you?'[46]

Darwin turned, as to a wise old Nestor, to Lyell, who came up with a solution of whose sagacity Jeeves himself would have been proud. Together with Hooker, he wrote to the Linnean Society setting out the story. Nothing could have been more appropriate than to make known the theory to that learned society, adjacent to the Royal Academy in Burlington House, Piccadilly, devoted to the memory of the great Swedish taxonomist of the eighteenth century. It was

Linnaeus, after all, who had been the first scientist in modern times to devise a system of classifying flora and fauna by species, thereby laying the groundwork for later scientists to investigate their origin. The Linnean Society, even more than the Royal Society, contained the great eminences of British biology. To them, Lyell and Hooker sent their joint letter, presenting 'the investigations of two indefatigable naturalists, Mr Charles Darwin and Mr Alfred Wallace'. They sent extracts from Darwin's essays on species, 'a very brief abstract of my theory' in 1842, some thirty-five pages, and the longer one, some 230 pages in 1844 – the last, in some ways, being the purest and most condensed version of the theory Darwin ever composed.[47] They included an abstract of Darwin's letter to Asa Gray written in September 1857 which not only repeated the hypothesis but made it clear that Darwin was engaged upon writing it up in book form. They sent these together with 'An Essay by Mr Wallace entitled "On the Tendency of Varieties to depart indefinitely from the Original Type". This was written at Ternate in February 1858 for the perusal of his friend and correspondent Mr Darwin, and sent to him with the expressed wish that it should be forwarded to Sir Charles Lyell, if Mr Darwin thought it sufficiently novel and interesting.'[48]

11

A Poker and a Rabbit

So Lyell and J. D. Hooker would present the papers of Wallace and of Darwin to the Linnean Society as a joint notion. 'These gentlemen having independently and unknown to one another, conceived the very same ingenious theory to account for the appearance and perpetuation of varieties and specific forms on our planet, may both fairly claim the merit of being original thinkers in this important line of inquiry . . .'[1]

Darwin had not sought – yet – to launch his theory. The letter from Wallace had bounced him into it. He was like an actor being pushed on to the stage before he had fully mastered his role. This would have strained nerves more robust than Darwin's. It is no surprise that the year which followed the meeting at the Linnean Society was marked by tension and anxiety, and all the physical symptoms which made Darwin's life so difficult. It was also, as it happened, a year marked by family illnesses and deaths. Darwin's most famous book came to birth prematurely, and in an atmosphere of misery.

Darwin was not a public performer. Speeches and lectures before audiences were never easy for him. As it happened, the summer of 1858 was a time when it would have been all but impossible for him to leave Down in order to make a momentous public utterance in London.

Building work had been in progress at Down House since September 1857. It was finished in June 1858, when, as Darwin wrote to William, his son at Rugby, 'we entered two days ago into the new Dining Room, & it is charming'.[2] The boy had sent him a cutting from *The Times* in which the Bishop of Oxford, Samuel Wilberforce, son of the great Abolitionist, had denounced the Evil Trade. The hour of the great confrontation between Wilberforce and Darwin over matters of science lay two years in the future. Darwin on this occasion pronounced the bishop's words to be 'capital'.[3]

Building works and bishops were less distracting than illness. The beginning of June found Darwin confined to a sofa with a boil.[4] Scarlet fever was raging in the village – three children died of it. On 18 June, Etty, now fifteen years old, was struck down with a high fever and a violent sore throat – diagnosed by Emma as a 'quinsy'. They feared it was diphtheria, a disease which appears to have come from France in 1858 and was sweeping England in an epidemic wave. 'No actual choking but immense discharge & much pain & inability to speak & very weak and rapid pulse, with a fearful tongue'.[5] The doctor who 'damped us yesterday much', eventually pronounced the attack to be mild, and Etty would recover. Her baby brother Charles Waring Darwin, however, not yet two years old, the youngest child, developed scarlet fever.

As Darwin sat beside his little son, he was agonizing, in correspondence with Lyell, about the presentation to the Linnean Society. It was to have been on 1 June, but that meeting was cancelled as a mark of respect to a former President, Robert Brown, who died on 10 June. The momentous meeting was therefore rescheduled for 1 July, and in the days which led up to it Darwin was sitting in his house of sickness. His professional destiny was in the hands of Lyell. He could not doubt that Lyell's solution, of presenting the outline of the theory to the Linnean Society at the same time as disclosing Wallace's identical conclusions, was the right one. This, however, was not a time which was conducive to thought.

Darwin's namesake, his little son Charles Waring, was the last child in the line. Emma was nearly fifty. On 29 June, Darwin wrote to Hooker, 'You will, & so will Mrs Hooker, be most sorry for us when you hear that poor Baby died yesterday evening. I hope to God he did not suffer so much as he appeared. He became quite suddenly worse. It was Scarlet fever. It was the most blessed relief to see his poor innocent little face resume its sweet expression in the sleep of death.'[6]

He added, 'Poor Emma behaved nobly & how she stood it all I cannot conceive. It was wonderful relief when she could let her feelings break forth.'[7] By the same post, he sent to Hooker, now Assistant Director of the Botanical Gardens at Kew, his 1844 abstract. 'I really cannot bear to look at it,' Darwin confessed.

Panic – Darwin's own word[8] – continued to grip Down as scarlet fever raged through the village. The Darwins made plans to get the children away as quickly as possible, but Etty, though recovering, was too weak to move – 'she has not even put on her clothes',[9] he noted on 6 July.

So it was that when the meeting took place at the Linnean Society in London on 1 July, the news of it was filtered to Darwin through a miasma of domestic pain – 'death and severe illness & misery amongst my children'.[10]

To the scientists who assembled at Burlington House – there were about thirty all told, including two foreign visitors – Wallace's name, if known at all, was that of a commercial purveyor of specimens. Darwin had been elected one of their Fellows in May. Neither Darwin nor Wallace was present.

The proceedings that afternoon began with fulsome praise of the lamented Robert Brown. Then the Secretary read the two papers, Darwin's and Wallace's. Hooker and Lyell were there; Wallace's natural history agent, Samuel Stevens; two of Darwin's friends, William Carpenter and William Fitton. After Darwin's and Wallace's papers had been read, the Fellows heard five other papers on zoological or botanical subjects before adjourning for tea. In his old age, Hooker looked back on the tea-drinking and claimed that Darwin's theory was discussed with 'bated breath', while also thinking that 'the subject was too novel and too ominous for the old school to enter the lists before armouring'.[11] It was equally possible that they had listened with only half an ear, half convinced or half unconvinced. Of those present, Daniel Oliver and Arthur Henfrey became convinced evolutionists while Cuthbert Collingwood would remain sceptical.[12]

When Thomas Bell, President of the Linnean Society, gave his Presidential Address in May 1859, he would opine: 'The year which has passed has not, indeed, been marked by any of those striking discoveries which at once revolutionize, so to speak, the department of science on which they bear.'[13]

Darwin's chief priority, during that week when his theory was launched upon the scientific world, was to get the family away from the village. The number of children in the village who had died had risen to five.[14] The baby Charles was not the only child in the house who appeared to have scarlet fever. Jane, the nursery maid (Parslow's daughter), manifested some of the same symptoms – though she would recover.[15] The Darwins planned to spend a few weeks on the Isle of Wight.

By July they had established themselves at Shanklin. 'This place has evidently sprung up, like a mushroom, & there are three hotels and many villas,' he noted.[16] The toll of family deaths, however, was not complete. Less than a month after the baby's funeral, and just as they were beginning to enjoy the sea air, there came news that Darwin's sister Marianne had died at the age of sixty. To their cousin Fox, Darwin wrote in a spirit of resignation: 'A blessed relief after long continued & latterly very severe suffering'.[17]

On the island, his health momentarily recovered. He went for long walks on his own. Hooker was trying to persuade him, by letter, that he should write an abstract of his theory for the journal of the Linnean Society. Darwin was reluctant at first. 'How on earth I shall make anything of an abstract in 30 pages of Journal I know not.' Yet, as he would add, 'I am extremely glad I have begun in earnest on it.'[18]

What he learnt on his long solitary walks on the Isle of Wight was that he had been waiting for this moment. His natural propensity to hesitate, to hold back from actions and decisions, had been keeping him silent for years – certainly since he had finished the first abstract of the theory in 1844.

Two things were holding him back. One was dread of what the scientific world would make of the theory. The very fact that Darwin was so intensely ambitious – that he wanted to be cock of the walk in the world of Victorian science – made him afraid that his theory would fail to convince, on scientific grounds. For the rest of 1858, therefore, we see his correspondence bulging with scientific inquiries to colleagues. To Asa Gray, for example, at Harvard, we find Darwin asking not only botanical questions directly relevant to Gray's own research, but also questions about the age of the earth and the different temperature of the earth after the glacial epoch.[19] As we shall see, a decade later he would become deeply alarmed by the theories of William Thomson (later Lord Kelvin) about the age of the planet, for the whole Darwinian theory of natural selection depended upon a geological time-span of hundreds of millions of years.

This question of the age of the planet had been the first great breakthrough for the Victorian thinking classes, brought about by Lyell in 1830. The importance of Lyell, as we have seen, was that he made it impossible to believe that Archbishop Ussher's dating of the Bible – placing the creation of the earth in 4004 BC – was sustainable. Chambers, in the anonymous *Vestiges of the Natural History of Creation*, had popularized Lyell's ideas, and those of Lamarck. Darwin, as we have seen, was both appalled by Chambers's book – which seemed to pip him to the post, in terms of enlightening the general public about the advances recently made in science – and aghast at the religious controversy it had excited.

Darwin believed that his own theory, like Lyell's *Geology*, made it impossible to believe in the Bible. By now he had parted company with faith, but he was forever cautious about admitting it, either to his pious wife or to his public.

After the holiday in the Isle of Wight, they returned to Down, and at his desk there Darwin began to write in an unstoppable flow. The thirty-page abstract which had been suggested as a suitable article for the journal of the Linnean Society was turning into a book – *the* book which would make his name. It would contain not a word about the origins of the human race, and not a word about the Bible. Behind its analyses of geology, botany and ornithology, however, shimmer all his preoccupations with these other subjects. The very fact that Darwin was too afraid to mention them overtly – in this book at least – was what made his pages so electrifying. *Vestiges* had, in its slapdash way, presented a compendium of the current state of scientific knowledge and, with journalistic genius, goaded its readers into paroxysms of doubt and anguish. One anonymous best-seller, *Vestiges*, had been followed by another, namely Tennyson's *In Memoriam* (1850), which, in mournful, perfectly musical lyrics, agonized about the ebbing of faith, and tried to clutch at it as it ebbed, 'Believing where we cannot prove'.[20]

On the Origin of Species, when it had been presented to the public, would be the doubt that did not dare to speak its name, and that was one of its best-selling ingredients. The generation immediately following the Victorians – their attitude typified by Lytton Strachey's satirical *Eminent Victorians* – mocked their parents and grandparents for what seemed like hypocrisy, their double standards. Another way of viewing them was that they learnt to live with doubt. They were able to hold two ways of looking at the world in balance.

Looking back from the perspective of 1873, Etty (by then Henrietta Emma Litchfield), a decided unbeliever, wrote to her brother George about 'the evils of concealment about religion, & the cowardliness of much that is written upon scientific religion. But I don't think, since one sentence in the Origin which I have groaned over in Spirit, Father has ever practised anything but what, I consider, the wise reticence of a man who does not care to give his opinions to the world upon a subject which he has not mastered.'[21]

It is hard to be certain what is the one sentence which made Etty groan, but it is presumably the point, in the second edition of *Origin*, where Darwin substituted the word 'Creator' for 'Nature', to play down the impersonality of the natural process he had spent the whole book depicting.

Darwin had come to disbelieve in Christianity. No doubt family bereavements, and above all watching his children die, had confirmed his loss of faith, but it seems to have been one of those slow unravellings with no discernible single cause. Concealment, however, was a habit which came from the very core of his own nature; it came from his tenderness towards Emma, and his personal deviousness, his inability to come clean about religion as about other issues (such as his debts to previous scientists). Etty, when she edited a hundred years of Darwin family letters, remarked upon the fact that the ebbing and flowing of faith was not overtly discussed at Down House. Of Emma, Etty wrote, 'As years went on her beliefs must have greatly changed, but she kept a sorrowful wish to believe more, and I know that it was an abiding sadness to her that her faith was less vivid than it had been in her youth.' To Darwin, at about the time he was writing *The Origin* (the letter is undated, but Etty assigned it to this period), Emma wrote, 'My heart has often been too full to speak,' but she commended to him the text from Isaiah, 'Thou wilt keep him in perfect peace, whose mind is stayed on thee.'[22]

Just as it was Lyell who had organized the manner in which Darwin's theory should be aired at the Linnean Society in the summer of 1858, so it was Lyell – when Darwin's words began to flow from his pen so unstoppably – who arranged a publisher, his own, and the publisher of *The Voyage of the Beagle*, John Murray.

Before he had direct dealings with him over the publication of *The Origin*, Darwin, with a mixture of dread and excitement at the hornet's nest he was about to stir up, asked Lyell,

> Would you advise me to tell Murray that my Book is not more *un*orthodox, than the subject makes inevitable. That I do not discuss origin of man. – That I do not bring in any discussion about Genesis &c., & only give facts, & such conclusions from them, as seem to me fair. –
>
> Or had I better say *nothing* to Murray, & assume that he cannot object to this much unorthodoxy, which in fact is not much more than any Geological Treatise, which runs slap counter to Genesis.[23]

Darwin wrote these words when the book was all but complete – 28 March 1859. This was after an intensive winter of writing. The 'abstract' had swollen to a substantial book. The work on 'horrid species' caused him intense nervous anxiety. The time of health and vigour on the Isle of Wight in the summer of 1858 had been short lived. The 'old severe vomiting' returned. By the time of his fiftieth birthday in February, he was drained of strength.

He sent the book chapter by chapter for criticism by Hooker at Kew Gardens. This was not without its hazards. Hooker told Thomas Huxley,

> I proposed ending the week by finishing Darwin's MS when to my consternation I find that my children have made away with upwards of ¼ of the MS. By some screaming accident, the whole bundle, which weighed over 1lb when it came (Darwin sent stamps for 2lbs) got transferred to a drawer where my wife keeps paper for the children to draw upon – & they have of course had a drawing fit ever since. – I feel brutified if not brutalized for poor D. is so bad that he could hardly get up steam to finish what he did – How I wish he could stamp and fume at me – instead of taking it so good-naturedly as he will.[24]

The consolation for Darwin was that, very slowly, and after fourteen years of expressing doubts, Hooker had been converted to Darwin's theory. Darwin clung to those who believed in him. Huxley would be the most vociferous, Hooker the most loyal, Lyell (when he came round to the theory with at least part of his mind) perhaps the most distinguished. For the most part, although Darwin's book would persuade the thinking world what it had known or suspected since at least 1844 (and the publication of *Vestiges*), that evolution was true, he would have a harder job persuading the scientific academy that one species could evolve into another. In his lifetime, he never did so in Britain, though there were those abroad, as we shall see, especially in Germany, who became *plus royaliste que le roi* in their enthusiasm for the survival of the fittest.

Lyell's choice of Murray as the publisher was the perfect one. John Murray, as an observant low-churchman, remained loyal to the Creator, but as a good 'tradesman' – the word Lyell often loftily applied to this honest Scotch publisher[25] – he could already hear, like the merry peal of church bells, the ringing of cash tills in the bookshops. His father had published Lord Byron and Jane Austen. The younger Murray had an interest in science.

Once Lyell had acted as the go-between, Parslow, Darwin's butler, was sent up to London with the completed manuscript of *The Origin of Species* wrapped in brown paper, to be left at Mr Murray's offices in Albemarle Street: to the very house where an earlier John Murray, scandalized by their contents, had burnt Byron's memoirs in the drawing-room grate.

Murray sent the manuscript to two readers for their assessment. One of them was Whitwell Elwin. It was said that 'No important decision could be arrived at in Albemarle Street without the advice and approval of the Rector of Booton,'[26] a parish in rural Norfolk. (For Elwin was a clergyman. He was intensely conservative politically, and was indeed the editor of the Tory periodical the *Quarterly Review*, which Murray himself published. This was the journal which, under the editorship of Sir Walter Scott's son-in-law, Lockhart, had so disgracefully savaged John Keats.)

Although Elwin's report on Darwin's book has become the object of derision in most of the biographies, it does contain some telling criticisms. As is well known, the general assessment he made, and which he wished to communicate to Darwin himself, was that the book was too diffuse and tried to cover too many areas of science and, in biology alone, too many species, too many examples. He suggested – and this is what provoked the guffaws of posterity – that Darwin should reframe the book and limit himself simply to the study of pigeons. 'Every body is interested in pigeons. The book would be reviewed in every journal in the kingdom, & would soon be on every table.'[27]

The core of Elwin's objections to Darwin's book, however, is that it does not prove the theory which it expounds. 'At every page I was tantalised by the absence of proofs. All kinds of objections & possibilities rose up in the mind, & it was fretting to think that the author had a whole array of facts, & inferences from the facts, absolutely essential to the decision of the question which were not before the reader. It is to ask the jury for a verdict without putting witnesses in the box.'[28] He also believed – and here Lyell endorsed his view – that, by contrast with *The Voyage of the Beagle* ('one of the most charming books in the language'), *The Origin* was not only diffuse but dry. Darwin viewed Lyell's and Elwin's objections as 'impracticable'. He told Murray, 'I have done my best. Others might, I have no doubt, done [*sic*] the job better . . .'[29]

George Frederick Pollock, who had acted as a publisher's reader for John Murray's father, was more positive than Elwin. Although Murray confided in Pollock that he considered the theory as absurd as contemplating a fruitful union between a poker and a rabbit, Pollock assured him the book would be much discussed. And he emphasized one of the book's virtues: it concedes the 'difficulties' of the theory. He admired the way 'Mr Darwin had so brilliantly surmounted the formidable obstacles which he was honest enough to put in his own path'.[30]

Preparing the book for press, rewriting and marking proofs was a nervous strain. The last chapter caused 'bad vomiting' and 'great prostration of mind and body'.[31] Luckily, for a couple of years now, he had discovered peace at the hands of Dr Edward Lane, a hydropathic practitioner who lived in the former home of Sir William Temple, where Jonathan Swift had been the household tutor and chaplain:

Moor Park in Surrey. In this place, Darwin could receive the familiar water cures without the painful reminders of Annie which were excited by Malvern. The Lanes, and Mrs Lane's mother, Lady Drysdale, were congenial company, who gave him his much needed solitude among their wooded walks, decorative lakes and Dutch parterres.

That year, 1859, brought not merely the tensions of work but also the domestic dramas played out by that volatile breed, the governesses. Etty was now sixteen years old, Annie would have been eighteen. Miss Pugh, who had sat at meals with tears streaming, left the house early in the year. Madame Grut arrived from Switzerland. Etty took an instant dislike to her. In a letter to her brother William, Etty reported:

> Solemn events have happened. Mrs Grut is gone for ever. This is how it came about. On Monday at breakfast mama said very civilly that she wanted some alteration in Horace's lessons. Mrs Grut was evidently miffed at that, & then I said I thought *s'eloigner* wasn't to ramble very mildly & that miffed her again & she made some rude speech or other, 'Oh very well if I knew better than the dictionary' . . . Nothing more came of it then, & all went smooth till I went up to my German lesson in the evening. When I came in I saw there was the devil in her face, well she scolded the children a bit & then sat down by me, when I showed her my lesson (a bit of very bad French) she said, if I knew better than she did it was no use her teaching me & so on & so on, till it came to a crisis & she worked herself into a regular rage . . . I left the room then, & went down stairs to tell my injuries . . . When Papa and Mama heard all about it they settled she shd go at once, so Papa wrote a letter telling her she shd have her 33£ and nothing more . . . then Papa was to go upstairs & deliver the letter . . . Papa got *such* a torrent, telling him he was no gentleman, & white with passion all the time, wanting to know what she had done, what he had to accuse her of – telling him he was in a passion – she would give him time to think . . . We had a very flustered tea, & all evening we sat preparing for the worst, what we shd do if she refused to go out of the house, etc. However she did turn out much milder & sent us a letter to say she wd go on Wednesday.[32]

The lachrymose Miss Pugh returned, for a while, to take Madame Grut's place. Then came the Miss Thorleys, who had been with them of old, followed by a Miss Latter, who left shortly after arriving, to become a teacher in a local school. The swift turnover of governesses suggests – though such an explanation never seems to have crossed any of the Darwins' minds – that they were an intimidating group of children. It was with huge pride that Darwin noted the burgeoning entomological interests of Francis (eleven), Leonard (nine) and Horace (eight), and on their behalf he drafted a letter to the *Entomologist's Weekly Intelligencer.*

> *Coleopters at Down.* We three very young collectors have lately taken, in the parish of Down, six miles from Bromley, Kent, the following beetles which we believe to be rare, namely, *Licinus silphoides*, *Panagus 4-pustulatus* and *Clytus mysticus.* As this parish is only fifteen miles from London, we have thought that you might think it worth while to insert this little notice in the 'Intelligencer'. Francis, Leonard & Horace Darwin.[33]

The boys' discovery was as welcome a distraction from proof-correction as the succession of governess crises was unwelcome. In October, as he was finishing his corrections, Darwin was persuaded by Murray to change the title to *On the Origin of Species by Means of Natural Selection, or the Preservation of Favoured Races in the Struggle for Life.*

The day after he had dispatched the last set of proofs to Albemarle Street, Darwin set off to yet another water-cure establishment, this time at Ilkley, at the foot of the moor immortalized in song. The whole family joined him there, the children remembering it later as a time of 'frozen misery'.[34]

Ilkley Wells House was a huge neo-Tudor extravaganza on the edge of Rumbold's Moor, looking over the windswept, sleet-swept Wharfedale. Autumn was blustery and furious that year in Yorkshire. The invalids were carried, a twenty-minute donkey-ride, from the comfort of the house across a rutted track, in pelting rain, to the healing springs and the baths, which had been constructed on a nearby hillside in a brick-built terrace.

While he was here Darwin sprained his ankle, but believed the treatment to be helping him. By 15 October, he could write to Hooker, 'You cannot think how refreshing it is to idle away a whole day, & hardly ever think in the least about my confounded book, which half killed me.'[35]

All through the illnesses of the summer and the autumn, Darwin had busily corrected proofs. There were so many corrections and changes that the printers charged £72 and 8 shillings.[36] Murray proposed printing 1,250 copies, which would yield £240, two-thirds of which he offered to the author – a very generous deal indeed by the standards of twenty-first-century publishers. The stout green-bound volume would sell at fourteen shillings.

Murray had known his business. The first print run of 1,250 copies sold on the first day of publication – 500 going to Mudie's Circulating Library, with its 25,000 subscribers. There was no doubt that, despite its cumbersome title, this was a book for which the world was waiting. Rather than doing a simple reprint, Murray asked Darwin to prepare a second edition – that is, a reprint with corrections and emendations. Almost instantly, from Louise Swanton Belloc (an Irishwoman who had married a Frenchman, and who would become the grandmother of poet and controversialist Hilaire Belloc), there came the request to translate the book into French.[37]

Darwin had expected a repetition of the storm which had greeted the publication of *Vestiges of the Natural History of Creation* fifteen years before. The fear of public exposure, of causing religious hurt to Emma, of falling out with old friends in the world of science – all these fears brought on his symptoms to excess.

The intellectual climate, however, was not the same as it had been in 1844. *Vestiges* had disconcerted Darwin because he feared it would steal his thunder, and because it had revealed, among his scientific mentors, the depth of hostility to evolutionary theory. Rough and ready as *Vestiges* had been, however, it had dug over fertile soil. If the professors of science were not yet ready to believe that species could mutate, the thinking public most definitely *was* ready. *Vestiges* had cobbled together the geological theories of Lyell and Agassiz, the evolutionary doctrines of Lamarck and the mysterious evidence of the fossils. The book was riddled with howlers, but those who read it with an open mind discovered that their world-view had changed. Hitherto, the world had been a place in which every species had been separately placed, rather as the plants in formal gardens are taken out of the hothouse and arranged individually in the soil. As with the plants, so with the animal species. But now palaeontology had shown that whole species arose and were apparently discarded. Versions of species visible upon earth in the nineteenth century, such as the horse, could be found in the old stones, and these fossil versions showed that nature was not in a fixed condition. It was in a state of flux. Nineteenth-century England, with its ever-expanding systems of transportation, its social flux, its population growth, its political upheavals, was ready for a theory of nature which revealed everything in existence to be in a state of becoming, rather than fixed arrival. What was lacking, after *Vestiges*, was a theory which could account for the mechanics of such becoming, the method by which molluscs, insects, fish, fowl, mammals developed into the species discernible by natural history. Darwin, with the authority of a man who had studied so many areas of science, and with the simplicity akin to genius, believed he had supplied an answer to that puzzle.

How the world would respond to his theory, this was to dominate the rest of his earthly existence, and, after his death, this would shape his place in history.

12

Is It True?

Darwin described his book as an 'abstract', which he had been 'urged' to publish, 'as Mr Wallace, who is now studying the natural history of the Malay Archipelago, has arrived at almost exactly the same general conclusions that I have on the origin of species'.[1]

It is perhaps worth a brief recapitulation of the central tenets of *On the Origin of Species*, of what made it so distinctive. The reader who has reached this point in our story will understand that scientists since the end of the eighteenth century had begun to realize that life on earth had evolved, and was evolving. In a notebook of 1837, Darwin had written, 'If we choose to let conjecture run wild, then animals, our fellow brethren in pain, disease, suffering and famine – our slaves in the most laborious works, our companions in our amusements – they may partake [of] our origin in one common ancestor – we may all be netted together.'[2]

Versions of this idea – that we are all descended from a common ancestor, that life is one – could be found in the writings of Goethe, Erasmus Darwin, Lamarck, Cuvier, not to mention the Eastern religious philosophical writings in the Upanishads and the *Bhagavad Gita*. Darwin's distinctive contribution to the idea, shared in part with Blyth and almost completely with Wallace, was, first, to posit a very simple explanation for how and why this evolution from earlier forms occurs; and secondly, to conclude that we may indeed be 'all netted together'. The Linnaean notion of taxonomy has to be radically revised, if not altogether scrapped, since there is not necessarily a hard and fast distinction between the species. Indeed, the origin of one species is to be found in the adaptive changes taking place in its predecessor (so to say) in the evolutionary chain.

So, first – the how and why. Species of plants and animals are very fertile. They produce far more than can ever survive into adulthood. Their population remains, in the opinion of Thomas Malthus who converted Darwin to this way of thinking, unchanged. So there is a fierce struggle for life. The large numbers of eggs, spores, offspring, seeds being multiplied by nature have to fight for limited amounts of nutrition. It follows, therefore, that the stronger and the better fitted to survive, the greater a species' chance of survival. (This was the theory which Spencer would call the survival of the fittest, but Darwin did not use the phrase in the first edition of *The Origin*.)

Now, in the reproduction of species by means of sex, there is variation. No two individuals are ever completely identical. Therefore, in a world of stable populations where each individual struggles to survive, those with the most favourable characteristics will be the ones who survive. Nature chooses the best, and discards the weakest. This is natural selection. Plants and animals will adapt themselves to their environment.

There is another distinctive feature of the Darwinian theory which it is worth mentioning at the outset, because it is the one which caused Darwin very grave difficulties even within the body of the text of *The Origin* in its first edition. That is, for his theory to work, it has to be demonstrated that nature does not proceed by leaps. Every homologue in nature – all the various building-blocks, as we have called them in earlier discussions of this matter – have evolved from some pre-existent form. Those homologues which appear to have arrived from nowhere – cells, hair, features, pentodactyl limbs, eyes, angiosperm flowers – did in fact come into being little by little by little. Somewhere, fossil evidence will one day provide us with evidence of their having crawled into something like their present form by an infinite series of adaptations. This is the contention. For Darwin's theory to be persuasive, it was necessary for him to demonstrate or show to be probable (a) that natural selection occurred by sexual means; (b) that progress in nature took place through struggle, through the strong eliminating the weak; and (c) that what appeared to be adaptations might all be accounted for by progressive adaptation – that some ur-form adapted itself into what we see as a cell, a hair, a feather, an angiosperm plant, an eye and so on. Behind this

Erasmus Darwin, Charles's grandfather, was the pioneer of British evolutionary theory. Charles always played down this fact.

Darwin's father, Dr Robert Waring Darwin, was both a medical practitioner in Shrewsbury and a private banker of prodigious wealth.

Darwin's mother, Susannah Wedgwood, is the young woman on the pony, in this portrait by George Stubbs. The famous potter, Josiah, Charles's other grandfather, sits to the extreme right of the picture.

N° 27 Session 1825–6

University of Edinburgh.

Chemistry and Pharmacy

BY

THOS. CHAS. HOPE, M.D & P.

Mr Charles Darwin

It was at Edinburgh University in 1825 that Darwin first heard evolutionary theories expounded from an academic podium. This class-card gained him admittance to lectures in chemistry and pharmacy.

Robert Grant, who taught Darwin medicine and biology, was the first to open his mind to the evolutionary theories of Jean-Baptiste Lamarck.

Robert Jameson's Edinburgh lectures on the origin of species, like Grant's exposition of evolution, were conveniently airbrushed from Darwin's memory when he tried to assert his own personal originality.

William Darwin Fox became what his cousin Charles was meant to become – a country parson. The two shared a passion for insects.

Darwin told Fox that his 'ardour' for 'insectology' had redoubled as their friendship developed. In this letter to Fox of 1828 (*below*), Darwin comments on the illustration by his sister: 'the insect is more beautiful than the drawing'.

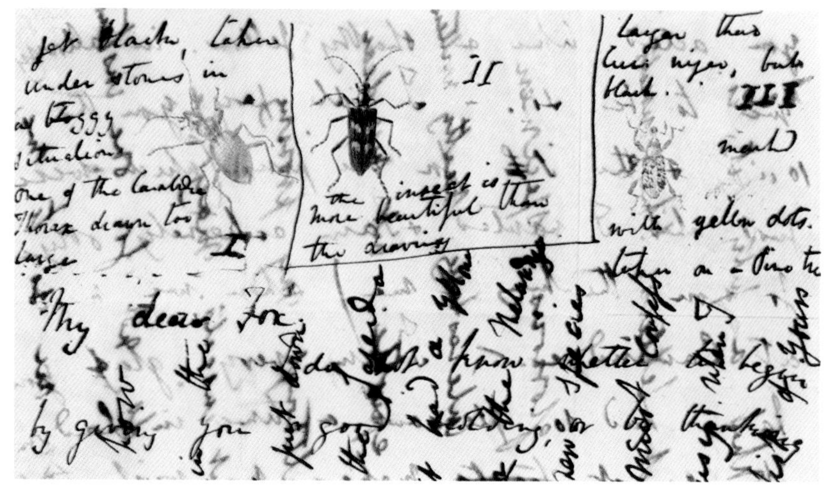

jet black, taken under stones in a boggy situation. One of the Carabidae Those drawn too large

I

II

the insect is more beautiful than the drawing

larger than ... but black.

III

with yellow dots. taken on a Pine tree

My dear Fox

John Stevens Henslow, Professor in turns of Mineralogy and of Botany at Cambridge, was one of Darwin's father-substitutes. He secured Darwin his place as the naturalist aboard HMS *Beagle*.

Adam Sedgwick, Professor of Geology at Cambridge, took the young Darwin under his wing and urged him to read Charles Lyell. Sedgwick was bitterly shocked by Darwin's theory of evolution by natural selection.

Lyell's oft-revised *Principles of Geology* demonstrated that the creation was not a finished thing, and that the processes which led to the earth's formation were still at work – as in the eruptions of Mount Etna in Sicily.

Founder of the Natural History Museum and anatomist of genius, Richard Owen gave a ground-breaking paper on the evolution of the limb in 1849, a full ten years before Darwin's *Origin of Species*. Darwin and Huxley regarded Owen as a bitter enemy.

Thomas Huxley, initially sceptical about Darwin's ideas, became his most vociferous champion. He relished the fisticuffs of public controversy as much as Darwin shrank from them.

Joseph Dalton Hooker, the director of the Botanical Gardens at Kew for many years, was one of Darwin's most loyal and affectionate champions.

When a ship crossed the line of the Equator, the hierarchies in a Royal Navy vessel were inverted. A sailor dressed as Neptune 'baptized' those, such as Darwin, who had not crossed the line before; that is, they were dipped in the ocean, smeared with tar and ritually humiliated.

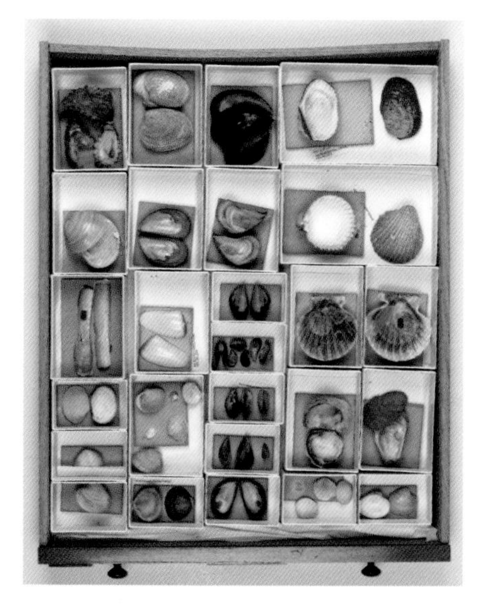

Some of the shells sent back from his *Beagle* voyage by Darwin, the avid collector and classifier.

'The Gaucho is invariably most obliging', Darwin told his journal. Weeks spent in the country outside Montevideo were among the happiest times of his South American travels. Galloping after ostriches on horseback was one of the more amusing activities in Argentina. Darwin noted that their meat 'would never be recognized as a bird but as beef'.

The Zoological Society of London named this creature *Rhea darwinii*, even though it had already been classified by French naturalist Alcide d'Orbigny.

The famous finches of the Galápagos. These are warbler finches. It was John Gould, ornithologist in London, and not Darwin, who first spotted the phenomenon of the finches' beaks adapting to different environmental conditions from island to island.

Modest, unassuming Alfred Russel Wallace who, simultaneously with Darwin, formed the theory of evolution by natural selection.

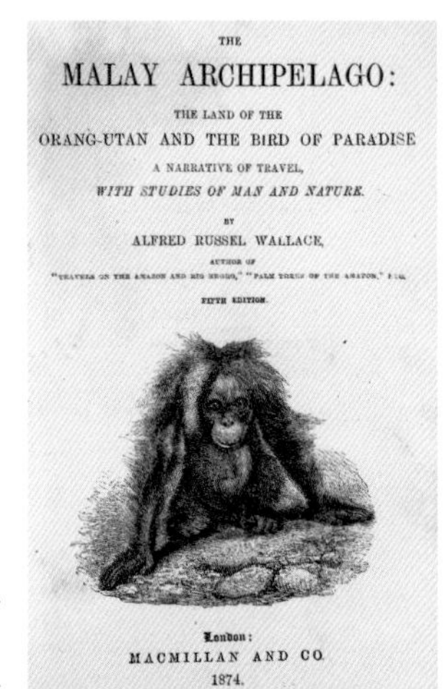

THE

MALAY ARCHIPELAGO:

THE LAND OF THE

ORANG-UTAN AND THE BIRD OF PARADISE

A NARRATIVE OF TRAVEL,

WITH STUDIES OF MAN AND NATURE.

BY

ALFRED RUSSEL WALLACE,

AUTHOR OF

FIFTH EDITION.

London:

MACMILLAN AND CO.

1874.

It was in the Malay Archipelago that Wallace made his momentous speculations. Our cousin the orang-outang looks out at us from Wallace's title page.

Some of Wallace's collection of butterflies, which he deposited in Owen's Natural History Museum.

Marriage to a cousin was almost inevitable for Darwin. He chose wisely in the patient, loving figure of Emma Wedgwood, the mother of his nine children, his nurse and his kindest friend.

Family was all to Darwin and he was a loving father, here depicted with his first-born, William.

Annie Darwin, eldest of his three daughters and the love of his life, died of tuberculosis in 1851, aged ten.

The study at Down House. Note the curtained privy to the left of the fireplace, to which Darwin was often driven by 'stomachic catastrophe'.

Light streams through the study window. Notice his geological hammer, and the microscope through which his searching gaze penetrated the mystery of the barnacle.

The Sandwalk at Down, scene of many a Hamlet-like pacing, as he wrestled with his theories and their implications.

A sucker for the chicaneries of Dr Gully, undergoing the water-cure, as Darwin did, at Malvern Spa.

Darwin's health-obsession was an acquired characteristic inherited by his daughter Henrietta, here seen with a device attached to her nose with the purpose of keeping cold germs at bay. Drawing by her niece Gwen Raverat.

Samuel Wilberforce, Bishop of Oxford. His measured review of *The Origin of Species* raised questions which Darwin did not fully answer; his clumsy joke at Huxley's expense, however, was never forgiven and he goes down in history as 'Soapy Sam'.

The Captain of HMS *Beagle*, in later life, now Admiral Robert FitzRoy, a distinguished meteorologist whose tormented life ended in suicide.

St George Mivart was an early convert to Darwin's theories, but further thought led him to question them. Darwin incorporated many of Mivart's objections into the sixth edition of *The Origin*, thereby leaving the theory threadbare.

Samuel Butler's mischief-making back-to-Lamarck reflections were regarded by Darwin as impertinent and disloyal, particularly from a friend with many family connections.

Darwin's cousin Francis Galton developed the ideas of *The Descent of Man* into a full-blown programme of eugenics with the aim of eliminating the weak and undesirable.

Herbert Spencer, who coined the phrase 'survival of the fittest', supplied Darwin with many of his less plausible ideas, including the 'evolution' of human language.

Edward Blyth, a poor pharmacist from Tooting, was unguarded enough to expound his theory of evolution by natural selection in magazine articles in 1837. Darwin stole these ideas, and covered the evidence of his plagiarism by slicing the relevant pages from his notebooks.

AGITATED TAILOR (to foreign-looking gentleman), *"Y-you're rather l-long in the arm, S-sir, b-b-but I'll d-d-do my b-b-best to fit you!"*

Among the innumerable cartoons which Darwin inspired, this one, in which a tailor measures a client for a new coat, is among the more charming.

This cartoon shows the evolution of the canine household pet, the dog, into a useful household servant, the butler.

Darwin devoted a decade of his life to the study of the barnacle. His monograph on the theme shows him to be a first-class naturalist.

One of Darwin's most attractive books is *On the Expression of Emotions in Man and Animals* which repeatedly reveals his love of dogs. This illustration shows a dog caressing his master.

Darwin's friendship with fellow pigeon-fanciers brought him some of his happiest social pleasures. He always forgot to acknowledge that his cultivation of the birds, such as this fine English Fantail, was in imitation of his parents at The Mount, Shrewsbury.

Darwin's fervent ambition was to be regarded as the English Alexander von Humboldt, the German traveller-scientist and universal genius, shown with his party at the foot of Mount Chimborazo, Ecuador.

One of Darwin's most ardent disciples was the German Ernst Haeckel, who believed not only in human 'cousinage' with gorillas, but also the hierarchy of human beings from mere savages at the bottom to the Aryan race at the peak.

lurked another question, which is perhaps thrown into the highest relief when we ask about the origin of the most primitive building-block of life – the cell. To be fair to Darwin, he would have recognized such a question to be way beyond the scope of *The Origin*, but the failure of science, either in his day or our own, to come up with answers to this question is part of our story.

The first chapter, following Blyth's original article of 1837, traces the patterns of variation in plants and birds and animals under domestication. He writes with authority of pigeon-breeding and of hybrid plants. What will strike any twenty-first-century reader, used to the vigour of contemporary debate about Darwinism, is how tentative he is in advancing his theory. He admits that variability 'is governed by many unknown laws'. Chief of these unknown laws, of course, is genetics, the whole science of which was still waiting to be discovered.

Darwin's next chapter applies the principles of artificial hybrid-breeding under domestication to the life of nature itself. This chapter remains, more or less, an abstract, as he called it. If you read it in conjunction with Chapter Nine, in which he very candidly lists the difficulties of his theory, it posits one which is, on one level, merely academic and on another fundamental to his whole thesis. It is this. 'No one definition has as yet satisfied all naturalists' as to what, in fact, constitutes a species. Linnaeus drew up his classifications, but almost immediately these were called into question, especially by botanists. If you arrive at a definition of a species which you find satisfactory, that species must, by that very definition, be different from other species. And, clearly, those genera which are polymorphic or protean pose particular difficulties for the taxonomist. Darwin cites *Rubus*, *Rosa* and *Hieracum* among plants, and several branchiopod shells, as well as a number of insects, as species which actually change their characteristics in particular circumstances. This, however, is not a difficulty which relates specially to protean genera – if Darwin's thesis is maintained. For if his theory is correct, then every species, including humanity, is protean. We are all, from the smallest mollusc to the most brightly coloured cockatoo, from the spouting whale to the bewhiskered Victorian gentleman in his club, in a state of infinitely slow evolution into something else. This makes taxonomy at best a rough and ready sketch; all natural phenomena are 'works in progress', which makes the task of the Victorian natural historian different from his predecessors, who thought they were describing fixed phenomena.

In Chapter Three, Darwin expounds the Malthusian principle of life on this planet, applying what the economist believed about the human race to every life-form: namely, the 'Struggle for Existence'. The chapter is a classic exposition of what we now call ecology. He cites a patch of land near Farnham in Surrey, extensive heaths with a few clumps of Scotch fir. In the previous ten years, large tracts of this land had been enclosed. Fir trees were beginning to spring up, unsown or planted by human hand, because their seedlings were being chomped by the cattle who grazed there. 'Here we see that cattle absolutely determine the existence of the Scotch fir; but in several parts of the world, insects determine the existence of cattle.' Darwin had come to think that the dependency of one organic being on another was of its essence 'parasitic'. Every organic being, therefore, owes its constitution to that of its near neighbours, and every organic being is either parasitic upon another or in competition with another. If the missel thrush flourishes in some parts of Scotland, it is at the expense of the song thrush. In Russia, the Asiatic cockroach ruled the roost against all the other cockroaches. The chapter ends with one of Darwin's strangest sentences – one to which we must return later in this story. 'When we reflect on this struggle, we may console ourselves with the full belief, that the war of nature is not incessant, that no fear is felt, that death is usually prompt, and that the vigorous, the healthy, and the happy survive and multiply.'[3] This sentence sounds like the sort of thing a father would say to his child, when telling a bedtime story which had become too frightening, and which had made the child cry. 'There, there. It all ended happily ever after.' He had just spent the whole chapter telling his readers that nature *was* in an incessant state of war, only to assure them that it wasn't, not really.

In Chapter Four, Darwin moved to the central platform of his thesis: 'Natural Selection', and how it would work. As often in his writings, Darwin anthropomorphized what he was at pains to demonstrate was an impersonal process. So, he wrote, 'It may be said that natural selection is daily and hourly scrutinising, throughout the world, every variation, even the slightest: rejecting that which is bad, preserving and adding up all that is good; silently and insensibly working, whenever and whatever opportunity offers, at the improvement of each organic being in relation to its organic and inorganic conditions of life.'[4]

What Darwin had to demonstrate was that the process of natural selection was a creative thing; that evolution is a consequence of natural selection acting on the variations within a species. It is easy to see how the process of natural selection would lead to the extinction of species, or certain forms of species. He asserted that 'Natural selection acts solely through the preservation of variations in some way advantageous, which consequently endure.'[5] It follows, according to Darwin, that those less well adapted to their environment, or simply rarer, will eventually die out. This dying-out process is observable in the fossil records. You can actually see species which have become extinct. Did Darwin, however, demonstrate that natural selection can be creative? This is a much harder thing to prove. And does there exist any evidence – any at all – of how natural selection might account for the appearance of the building-blocks, the basic homologues? Until such evidence is demonstrated, Darwin's case remains, at best, not proven. Darwin was hurt to hear, on the grapevine, that Sir John Herschel, the astronomer and Master of the Mint, had called the theory of natural selection 'the law of higgledy piggledy'.

Darwinism is not simply an argument that the forms of nature come into being by common descent. It proposes that every form in nature, every distinctive, taxa-defining novelty has come into being by a gradual process of adaptation and change.

It is crucial to realize – as Darwin's fluidity of mind seems not to have done – that these are two quite separate claims. Alfred Russel Wallace's famous 'Sarawak Law' paper of 1855 had demonstrated that 'Every species has come into existence coincident both in space and time with a pre-existing closely allied species.'[6] It was called the Sarawak Law because Wallace had written his essay, so remarkably like Darwin's theory, in the wet season, in a little house at the mouth of the Sarawak river, at the foot of a mountain in the Malay Archipelago.[7] It is impossible to doubt that descent by modification occurs. For example, the first amphibians which most closely resemble the fish lived in the same geographical region and at the same time (the Late Devonian). This suggests an ancestor–descendant relationship.

What Darwin introduced into the argument – as did Wallace four years after his 'Sarawak Law' paper – is something which has never been demonstrated. This is that the evolutionary novelties which make for a new 'species', and the highly complex homologues which are to be found in the new species, have come about as a result of a slow process.

What we can now see is that while some adaptation, of the kind noted in the Galápagos finches, or of the kind chronicled in Wallace's 'Sarawak Law', takes place within an immutable *Bauplan*, there is no evidence of the homologues themselves having come into being as a process of adaptation. Etienne Geoffroy Saint-Hilaire's picture, in the early nineteenth century, followed to some extent by Owen, was that all animals shared the same basic body-plan – a plan which was fixed. Darwin and the Darwinists believed they had destroyed this idea, and replaced it with the picture of gradual adaptation and change. Evolutionary developmental biology, however, shows that new homologues do indeed appear complete, fixed. For example: some ancestral chordate switched its body-plan from the design previously shared with all animal groups, in which the nerve chord is in a ventral position and the heart and the main blood vessel are placed dorsally, to a design which is the exact reverse: the nerve chord is placed dorsally and the heart is placed ventrally. This has remained invariable in all chordates since. The same genes are involved in specifying the dorsal–ventral axis of chordates and non-chordates. The older nineteenth-century biologists have been proved right, and Darwin wrong.

Michael Denton puts it like this. There is no evidence for all changes in our ancestral evolutionary history having been adaptive or gradual. And if 'the pentadactyl limb or the insect *Bauplan* are also a-functional patterns, like the thirty-four petals of a field daisy or the shape of a maple leaf, then *the whole Darwinian enterprise breaks down*'.[8]

Darwin was honest enough to admit, in his Chapter Six, that there were 'Difficulties on [*sic*] the theory'. For example, 'why, if species have descended from other species by insensibly fine gradations, do we not everywhere see innumerable transitional forms? Why is not all nature in confusion instead of the species being, as we see them, well defined?'

Darwin's way of arguing was one which could be described as slithery. He presented difficulties, but then waved them away rather in the manner of a conjuror covering his top hat with a silk handkerchief. He assured the readers of *The Origin of Species* that the fossil records were imperfect. 'The crust of the earth is a vast museum . . .'[9] Sooner or later the doors of the museum will open. The public – you and I – will be allowed in to view the conclusive evidence provided by the exhibits.

Well, a lot of fossil evidence has been unearthed since 1859. There is *some* evidence of intermediate forms linking major groups of vertebrates. That is, there are *some* 'missing links' – the *Panderichthys*, an intermediate form of fish–amphibian; the *Thrinaxodon*, a mammal-like reptile; the *Pezosiren*, a land-mammal–seacow intermediate. We should expect, though, if Darwin were correct, that there would be hundreds, thousands of examples of such transitions. Fossil evidence is vastly more comprehensive in our day than in the nineteenth century, and it simply fails to endorse the Darwinian theories of gradualism. For Darwin to be right, all novelties in nature have to be explained by gradual mutation. Stephen Jay Gould, one of the most distinguished palaeontologists of the twentieth century, quipped that the absence of transitional forms was 'the trade secret of palaeontology'.[10]

There has never been a coherent explanation of the emergence of highly complex forms. Darwin, again, acknowledged this. 'To suppose that the eye, with all its illimitable contrivances for adjusting the focus to different distances, for admitting different amounts of light, and for the correction of spherical and chromatic aberration, could have been formed by natural selection, seems, I freely confess, absurd in the highest degree.'[11]

Darwin contended that, in spite of the apparent absurdity, a complex organ such as the eye could, little by little, evolve rather than appearing ready-formed. Whatever conclusion a modern reader draws from this old chestnut, it is perhaps only fair to mention the gallant effort made by two Swedish biologists, Dan-Eric Nilsson and Susanne Pelger, who in 1994 published their findings after experiments with a computer model: 'A Pessimistic Estimate of the Time Required for an Eye to Evolve'.[12] By speeding up the whole process on a computer, Nilsson and Pelger have demonstrated that it would take 'only' about 364,000 generations to evolve a good fish eye with a lens. Their 'pessimistic' estimate was so called because they deliberately, in their experiments, stacked the odds against themselves. It would probably take less time for the fish eye to evolve.[13] Perhaps half a million years. As Richard Dawkins wrote, 'And that is a very, very short time indeed, by geological standards. It is so short that, in the strata of the ancient eras we are talking about, it would be indistinguishable from instantaneous.'[14] I have said it is fair to mention Nilsson and Pelger, because they have demonstrated how a fish eye – or any other type of eye, come to that – could evolve. Darwin has to this extent been shown to be propounding a theory which could never have been proved in his lifetime, but which, 150 years later, is demonstrated to be, at least, possible. Having been fair to the Swedish biologists, it is also necessary to be fair to the haddock or the bream, who might, if capable of speech, have wanted to say to Professor Dawkins, 'Half a million years might seem to be "instantaneous" in Oxford, but down here on the ocean bed, when we needed an eye to protect us from predators such as sharks, it seems rather a long time.'

Darwin wrote in complete ignorance of the modern science of genetics, and what he knew of embryology was, by the standards of our times, primitive in the extreme. We know that the same genes control the development of the same organs in a widely different range of species. For example, a gene called Pax6, in the fruit fly *Drosophila melanogaster*, controls and co-ordinates some 2,000 other genes to make its eyes, which have multiple lenses. In the mouse, however, this same gene, Pax6, regulates different genes to make the mouse's single-lensed eyes. Remove the Pax6 gene from a fruit fly and place it into a frog and it will produce frog eyes. These facts are perhaps best explained by the hypothesis that this wide range of species carry genes which have been conserved across the centuries of evolutionary history from a common ancestor. Once we know about the existence of these genes, however, the ponderous and improbable theory of Darwin – that the complex mechanism of optic nerves came into being over hundreds of thousands of years – becomes completely unnecessary. The gene is already present, and will organize and control, here a frog eye, here the eye of a newt, here the eye of Galileo looking through his telescope.[15]

For Darwin's original readers, however, the puzzle of the evolution of the eye would be only one of the phenomena which assailed the imagination. For its author, the book might have been only an abstract. For most readers, even for most scientists, it must have seemed, like Humboldt's *Kosmos* in an earlier generation, a book which was in effect about everything. The author's sheer range still has the capacity to dazzle. The book takes us to every corner of the known world. It leads us through the unfolding succession of geological eras, as previously explored by Lyell. It looks at fossils and is honest enough to say that, as far as proving the theory of natural selection goes, the evidence is inconclusive. From woodpeckers and pigeons to whales and bees, from the strange evidence, and non-evidence, of fossilized trees and extinct species, it appears to consider every species which had ever existed – every one, that is to say, except humanity. About the origins of species generally, Darwin was, in his hesitant, modest tone, expansive and eloquent. About humanity he was, as yet, silent. Knowing the kind of reaction he could expect from those who still equated the belief in the fixity of species with theology, he ended on a lyrical but also unapologetic note. It was a clarion call for a new way of viewing existence itself. 'There is grandeur in this view of life, with its several powers, having been originally breathed into a few forms or into one; and that, whilst this planet has gone cycling on according to the fixed law of gravity, from so simple a beginning endless forms most beautiful and most wonderful have been, and are being, evolved.'[16]

It is almost certain, if you, the reader of *The Origin of Species* in the twenty-first century possess a copy of the book, that you have read the first edition, unless you read the old Everyman edition (published by J. M. Dent & Sons Ltd, 1928), which reproduced

Darwin's sixth edition with all his doubts and qualifications. The two copies which sit on my table as I write are, on the one hand, a paperback, edited for Penguin Classics by J. W. Burrow, and on the other, edited with an introduction by Edward O. Wilson for W. W. Norton & Company, New York, with the self-explanatory title, *From So Simple a Beginning: The Four Great Books of Charles Darwin*. Burrow explains, 'The edition here is the first edition. There were six editions published in Darwin's lifetime, and a large number of changes were made . . . Not only does the first edition possess a unique historical interest . . . but it also presents in many ways a more clear-cut and forceful version of Darwin's theories than the later editions, in which Darwin weakened his argument in an attempt to meet criticisms.'

Edward O. Wilson, introducing what he calls 'the greatest scientific book of all time', also chooses to reproduce the first thoughts of its author, rather than his last. Yet, as Peter J. Vorzimmer wrote in 1970, 'the story of Darwin's handling of his theory of natural selection after 1859 is one of documented qualification and nagging doubt'.[17] In some senses, the story of Darwin's life, from the time of *The Origin*'s publication to the time of his death, was the story of these doubts, so that to read only the first edition, as though it were an unalterable sacred text, is to distort the story. The Botany School Library in Cambridge (England) contains Darwin's collection of learned offprints and articles by other scientists, nearly all reflective upon his evolutionary theory. Darwin's annotations and marginalia to these printed works amount to over a hundred thousand words, and reflect twenty-three years of doubt. He never withdrew his theory, but he made many emendations to it, based upon the scientific objections which he discovered in the writings of others. Darwin, even more than Chambers in *Vestiges*, taught the world to see that nature is not in a fixed or still condition. It is on the move. So, too, is the mind of any serious scientist. Some of the moves which Darwin made, in response to critiques of his work, were away from the truth. Hamlet took over from the confident scientist. What the biographer notes is that Darwin himself, doughty as a warrior for his theory, nevertheless had many moments either of doubting it or (which is different) of not seeing how it could be defended. The young man who wished to be another Humboldt, a great scientist who would explain everything, had, like a child in a fairy story, been punished by being granted his wish. It was by no means always a happy experience. The rest of our story, then, is of Darwin tacking this way and that in stormy seas, allowing his allies (especially Huxley) to fight his battles, while he retreated into the calmer joys of pure natural history, devoting, for example, to earthworms the patient attention which might have been theirs had his life followed a different path, had he never set sail on the *Beagle* and had he taken orders and become a naturalist-parson in the Gilbert White mould. The stockade of family life would, through these last decades, keep him safe and sane, but the storms without were violent. Some accounts of Darwin's life speak as if the storms were theological; as if, confronted with the impersonal processes of natural selection, it was from Church and clergy, all challenged to take leave of their God, that Darwin suffered his most cutting assaults. As any perusal of the various editions of Lyell's *Geology* would show, or as you would learn from the *Westminster Review* in its many editions, or as the translations from the German of Mary Ann Evans (aka George Eliot) would make plain, Victorian intellectuals had been wrestling with religious doubt for years before Darwin published *The Origin*. His book might have confirmed some in their unbelief, but it was not essentially atheistic in texture. After all, the belief in a Creator is distinct from a theory about how the creation came into being. The notion that 'the Church' was the first to attack Darwin came about because some of the more established academic scientists were at the older universities; and Fellows of colleges in Oxford and Cambridge in those days were clergymen. Many of Darwin's most enthusiastic supporters, however, were Christians. Embarrassing as Darwin might find the disapproval of the cloth, and painful as he found his religious difference from Emma, what was professionally troubling to him, was his difficulty in persuading his fellow *scientists*.

13

The Oxford Debate and its Aftermath

THE BRITISH ASSOCIATION for the Advancement of Science (BAAS) held its meetings in a variety of venues, and the meeting for the summer of 1860 was scheduled to happen in Oxford. Darwin planned to attend, together with Lyell and Hooker. He even made inquiries about taking lodgings in the town,[1] and he looked forward to walks round college gardens with Hooker.[2] It was somehow inevitable, however, that from Darwin's great scene, the Oxford debate, the chief player should be absent. Etty became ill in the course of the summer, and this provided the perfect excuse for him not to attend the meeting. When she showed signs of improvement, it was necessary for his own body, with all its familiar psychosomatic manifestations, to come to the rescue. To Lyell, on 25 June, he wrote, 'I have given up Oxford; for my stomach has utterly broken down & I am forced to go on Thursday for a little water-cure, to "Dr Lanes Sudbrook Park, Richmond Surrey", where I shall stay a week.'[3]

So it was that Darwin never came to meet the man who had written an especially hostile review of *The Origin of Species* for the *Quarterly Review*, Samuel Wilberforce.

Wilberforce's review highlighted two problems with the theory of natural selection, as expounded by Darwin. The first concerned the analogy Darwin wished to draw with the selective breeding of domesticated species. Darwin envisaged natural selection, a sort of impersonal deity, 'daily and hourly' scrutinizing species over the space of entire geological epochs. The problem with the analogy, Wilberforce said, was that domestic breeders do not, in fact, create new species – they merely modify existing species – and the wild descendants of domesticated types, rather than continuing to 'develop', in fact revert to the original type. If anything, therefore, the behaviour of animals under domestication disproved rather than proved the Darwinian thesis.

Wilberforce's second accusation was that Darwin, if not misrepresenting Lyell, misused him. Lyell's *Geology* shows that there is no geological evidence which proves the existence of transitional forms, of one species turning into another. Darwin acknowledged 'gaps' in the geological evidence, but appeared to be enlisting Lyell for his argument. In fact, there were no 'gaps', simply insufficient evidence. Darwin acknowledged that Wilberforce's argument was 'uncommonly clever' and that he 'picks out with skill all the most conjectural parts, and brings forward well all the difficulties'.[4]

As far as I know, no Darwinian has ever given a satisfactory answer to Wilberforce's two points. The significance of Wilberforce, however, from a mythic perspective, was not that he was a scientist, but that he was a bishop, the Bishop of Oxford no less. As a young man, an undergraduate at Oriel College, he had sat at the feet of the 'Noetics', those dons who in the 1820s were considered to be dangerously in the vanguard of progressive thought. Never a professional scientist, Wilberforce was nevertheless a clever man, with a first-class degree in mathematics, who kept up his scientific interests while being successively a parish priest in the country, Dean of Westminster and Bishop of Oxford. In 1850, he attended Owen's entire course of lectures on anatomy. 'I could give the Bishop of Oxford a certificate for most regular attendance,' wrote Owen.[5]

Wilberforce, his father William canonized by history as the philanthropist whose tireless zeal led to the abolition of the slave trade in the British Empire, was doomed to be one of history's losers – the man who pompously and facetiously belittled the theory of evolution. In a sense, whether he did so *qua* bishop or *qua* supporter of the scientific orthodoxy of his day no longer matters. He could not know, as he prepared for the Oxford conference, that for almost everyone in future times he would be 'Soapy Sam', the man who set himself up against scientific truth.

When the BAAS met in 1845, Herschel had devoted his Presidential Address to denouncing *Vestiges* for religious heresy. It was a measure of how that book had changed the atmosphere that, by 1860, no mention at all was made of *The Origin* in the Presidential Address. It was, nevertheless, on everyone's mind. Two separate papers, on different days of the meeting, promised, however, to devote themselves to Darwinian questions. On Thursday 28 June, the Oxford Professor of Botany, Charles Daubeny, presented a paper, 'On the Final Causes of Sexuality of Plants with Particular Reference to Mr Darwin's Work "On the Origin of Species by Natural Selection"'. He broadly speaking supported Darwin's theory. Both Owen and Huxley were present. When Daubeny had finished, Huxley was asked if he would like to make a contribution. He declined, stating – quite strangely, given the later significance with which he invested the BAAS meeting – that he did not think a public venue of this kind was a suitable place to discuss Darwin's theory, because 'sentiment would unduly interfere with intellect'.

A few speakers had their say, and Owen then rose to describe some of the facts which he considered useful to those members of the public who wanted to come to terms with Darwin's theory. He proceeded to repeat his views on the vast differences between the human brain and that of the gorilla. This 'sentiment' was more than enough to inflame the 'intellect' of Huxley – or perhaps the other way around. Huxley bounced to his feet to tell the audience that there was no time to explain to them the many similarities between the brains of apes and people; but he wished to say that Owen's position was based on fear that humanity might lose its privileged place among the earth's species. It is difficult for us, at this distance of time, not to see that this was true, and that this was what underlay the second and much more explosive public airing of the Darwinian question during that Oxford meeting. This was the Saturday session of Section D – that is, the section of science which contained zoology, botany and physiology.

Huxley was, at this stage of the meeting, so bored by the proceedings that he had decided not to attend the Saturday meeting, intending to leave Oxford for Reading, where his wife's brother-in-law had a house. He was persuaded to stay by, of all people, Robert Chambers. It really did seem, as the audience gathered for Saturday's session, as if all the living figures who had played an important role in Darwin's intellectual journey had gathered at the newly built Natural History Museum in South Parks Road.

The building itself was a parable. The Honour School of Natural Science had started in Oxford only in 1849. The Natural History Museum was seen as the home of Oxford science, and of Oxford's approach to science. The architects, Thomas Deane and Benjamin Woodward, made something which was midway between a North German *Rathaus* and the aisle of a Gothic cathedral temporarily set aside for the exhibition, in glass cases, of skeletons, fossils and other natural history specimens. According to John Ruskin's friend Henry Acland, University Reader in Anatomy, Gothic was the appropriate style for science – by contrast with the fairly recent Cambridge Museum, which was classical. Richard Cresswell, one of the founder members of the Ashmolean Society, saw natural science as 'a kind of religious contemplation'. Cresswell, a clergyman like so many dons, saw the collection as a celebration of the wonders of God's creation.[6]

It was in this cathedral of contemplative science that the great debate was to take place. It had been finished only months earlier. Each day, the architects would turn up with plants borrowed from the Botanic Gardens so that stone-carvers could copy their likeness in the capitals of the columns. The wrought-iron roof was similarly decorated with motifs from nature, finished by the firm of Skidmore, of Coventry. In the spandrels could be seen interwoven branches, with fruits of lime, chestnut, walnut, sycamore, palm and other exotic trees such as Darwin and Huxley might have seen on their voyages. And there, beneath these reminders of exotic tropical botany, was none other than Admiral FitzRoy himself. Here was Darwin's old friend and mentor Henslow, who had only a year to live. Here was Owen again. Lyell and Hooker were present. The word had somehow got about that this was to be a momentous meeting. Although Oxford was already deep in the Long Vacation – so the young were not present in any great numbers – here were the local clergy, en masse, with their wives; here was 'town' as well as 'gown'. Long before the meeting was due to start, 700 people were in attendance. The lecture hall was soon filled to capacity, and it was necessary to move the audience to the (as yet unfinished) library.

The meeting began with a talk by John Draper, who had been Professor of Chemistry at the University of the City of New York since 1838. The room became hot. The American was slow-spoken and prolix. Hooker, who was present, felt that he was merely echoing things which had been written, and printed, by the likes of Herbert Spencer and Henry Thomas Buckle. Draper's theme was 'On the Intellectual Development of Europe, Considered with Reference to the Views of Mr Darwin and Others, that the Progression of Organisms is Determined by Law'.

Draper was a determinist, and a progressivist, like Herbert Spencer. He argued that the progress of civilizations was determined, as was the history of organisms. One of the only sentences recalled by one of the witnesses (Isabel Sidgwick) was, in Draper's American voice, 'Air we a fortuitous concourse of atoms?' Taking the audience through five phases of ancient Greek history, Draper suggested that all civilizations, including that of the Victorians, inexorably followed these patterns, just as plants and birds and mammals are determined, in their development, by the processes of natural selection, as described by Mr Darwin. It is fascinating to note that, on this its very first public airing, Darwinism, which began as a kind of metaphor (nature imitating Malthus), was interpreted as a social metaphor by Draper. Those who gathered in the library in South Parks Road were not discussing who was right, Darwin or Wilberforce, in the matter of whether animals, artificially bred in domestication, can change species. They were not discussing who was right, Darwin or Owen, in the assertion that there was, or wasn't, proof for the transmutation of one species into another. Instead, they were immediately confronted, in Draper's elongated, ponderous words, with the question of what Darwinism said about humanity.

When Draper had finished, it was Richard Cresswell who was first on his feet, vehemently denying that you could force an analogy between the history of nations and that of organisms. How could there be any parallel drawn between the intellectual progress of man and the physical development of the lower animals? Then Sir Benjamin Brodie arose – a royal surgeon who had attended upon William IV and the present Queen – simply to deny Darwin. That was what they were there for. Why not take the gloves off and start the fight? Several anti-Darwinians spoke, and Henslow, who was in the chair, asked Huxley if he would respond in Darwin's defence. Huxley – for the moment – declined. Then Samuel Wilberforce rose. He was a good-looking man, and an experienced public speaker. This was what the crowds had come to hear. Wilberforce gave the audience a rather dry précis of his review of *The Origin of Species*. He repeated his argument that a comparison between the breeding of hybrids under domestication and of animals in the wild actually disproved, rather than proved, natural selection. The fossil evidence had 'gaps' because there was none.

Had the Bishop of Oxford left his argument there, he might well have been deemed the victor in the debate that morning. But having scrutinized Darwin's inductive methodology for about half an hour, the Bishop could not resist disobeying Henslow's injunction that speakers should keep the discussion on a scientific footing. Christianity, he stated, offered a nobler view of life than Darwinism. The Bishop shuddered to think of a world where Darwinian evolution would be adopted as a creed. He rejoiced that the 'greatest names in science' had already rejected Darwin's theory, which, he believed, was 'opposed to the interests of science and of humanity'.

Even now Soapy Sam, in spite of having spoken for too long, could have sat down covered with honour. He had the audience on his side, however, and their excitement went to the Bishop's head. He could not resist a little quip. He turned to Huxley who was, he patronizingly said, 'about to demolish me' and inquired, 'Was it through his grandfather or his grandmother that he traced his descent from an ape?'

Huxley, who was sitting beside Sir Benjamin Brodie, murmured, 'The Lord hath delivered him into mine hands.' Brodie 'stared at me as if I had lost my senses'. Soapy Sam's quip was a frivolous response to a serious scientific question. As such, it antagonized some of the audience. It was also – and here it especially cut Huxley to the quick – a snob remark. While the Bishop was merely rebutting the thought that humanity had actually descended from apes, Huxley felt the wounds of social humiliation. Standing among the Oxford professors, he became once more the son of the impoverished school usher, who had only enjoyed minimal formal education until enlisting as a surgeon's assistant. Not for Huxley the luxury of a college education or a university degree. While Brodie, and Wilberforce, and Acland, and these other learned men looked on, with their condescending expressions, Huxley was the raw man of the suburbs. The new science was not the metaphorical expression of new politics, exactly speaking; but for those, like Huxley, who had no time for the formularies of theology, the Established Church, with its incomprehensible doctrines and its Thirty-Nine Articles of Religion, and its hold on the universities, was an embodiment of everything to which a free spirit must instinctively feel itself opposed.

With great restraint, Huxley replied to the Bishop that he was 'unable to discover either a new fact or a new argument' in his speech, 'except indeed the question raised as to my personal predilections in the matter of ancestry'. Only two days before, baiting Owen, Huxley had raised the matter of human kinship with the apes, but on this occasion, for rhetorical effect, he expressed surprise that the Bishop should have brought up such a topic in a serious discussion. 'If, then, the question is put to me would I rather have a miserable ape for a grandfather or a man highly endowed by nature and possessed of great means and influence and yet who employs those faculties for the mere purpose of introducing ridicule into a grave scientific discussion – I unhesitatingly affirm my preference for the ape.'[7]

Henry Fawcett, later Professor of Economics at Cambridge, would say that no one present could ever forget the impression made by Huxley's riposte. (Fawcett, after the debate, was heard saying in a loud voice that he did not believe the Bishop had even read Darwin. Wilberforce, who was near by, was about to pitch in to defend himself when he saw that Fawcett was blind.)[8] The audience, who had roared with amusement at Soapy Sam's joke, now clapped Huxley. Lady Brewster – wife of Sir David Brewster, the astronomer, and keen ally of Owen – actually fainted, perhaps at Huxley's impudence, perhaps from sheer excitement.

This was not the end of the drama by any means. No sooner had Huxley resumed his seat and the applause died out than another figure arose. Huxley, by now the self-appointed 'Bulldog' of Darwin and Darwinism, was the newest friend of the reclusive sage of Down. But here stood one who had once been his friend and was no more. Hook-nosed, beetle-browed, still with a fine head of wavy brown hair, here at the age of fifty-five stood the tall figure of Rear Admiral (as he was by now) Robert FitzRoy, clutching in his strong capable hand the book which he considered to be imperilled by the Darwinian thesis: the Holy Bible. FitzRoy wanted to put it on record that he regretted ever giving Darwin the chance to sail around the world, thereby beginning the train of thought which undermined the word of God. FitzRoy was a clever man, a Fellow of the Royal Society, and now, in his retirement from the Royal Navy, chief of the meteorological department of the Board of Trade. He was a pioneer of storm warnings for the help of those at sea, and a skilled meteorologist. His speech against Darwin, however, was not a clever speech, either rhetorically or intellectually. It left the impression that the theory of natural selection was an uncomfortable truth which he regretted Darwin having unearthed. What seemed to distress him was not the possibility of Darwin's being wrong, but the possibility of Darwin's being right. Henslow, in the chair, and a friend to Darwin, though by no means convinced by the theory, now turned to Hooker and asked him if he had anything to say.

For many who attended the meeting, it was Hooker's speech, rather than Huxley's, which really carried the day for Darwin. Hooker spoke calmly and reasonably. He told the audience that he had been privy to Darwin's hypothesis for the previous fifteen years, and that he had applied Darwin's views 'to botanical investigations of all kinds in the most distant parts of the globe, as well as to the study of some of the largest and most different Floras at home'. He explained that he had originally believed that species were 'original creations', but that he could no longer subscribe to this belief, since it did not explain the evidence at hand. He rebutted Wilberforce's description of Darwin's views as a 'creed', and said, rather, that the theory 'offers the most probable explanation of all the phenomena presented by classification, distribution, structure, and development of plants in a state of nature and under civilization'. Hooker said that he was ready to abandon a belief in evolution by natural selection but only if and when a better hypothesis appeared. Until then, Darwinian evolution offered 'the best weapon for future research'.

Here was an expanded version of something which, a little earlier in the debate, had been advanced by Darwin's friend and neighbour John Lubbock. The debate had at last been anchored (as Henslow had implored from the beginning that it should be) in science. When he heard about the debate, Darwin felt heartily glad not to have been there. 'I would as soon have died as tried to answer the Bishop in such an assembly.'[9]

When it passed into myth, the Oxford debate became the occasion when – to quote one of the most vociferous atheists of recent decades, Christopher Hitchens – Huxley 'in front of a large audience cleaned Wilberforce's clock, ate his lunch, used him as a mop for the floor, and all that'.[10] At the time, however, views were more nuanced, and more balanced. The *Athenaeum* definitely regarded Hooker's as the most impressive speech. The *Press* considered the debate not as a battle in the grand war between God and science, but, by contrast, as evidence of 'wide and wise toleration', showing a spirit of co-operation between the 'Christian' and the 'scientific'.[11] Undoubtedly, there was a dimension in the debate – introduced by Wilberforce and painted in lurid colours by FitzRoy – of religious paranoia, the fear that by adopting a theory as to how nature operates they would somehow have discarded God Himself. But a much bigger factor than the theological element to the debate was that of an academic orthodoxy under threat. The great majority of scientists, especially in Britain since the Napoleonic Wars, had rejected evolutionary theory as continental claptrap. They clung to 'original creations' because it suited them to do so. Most science, particularly in the setting of university faculties, finds it easiest to operate within a status quo.

There is a paradox here, of course, because scientific progress depends, not on the reaffirmation of old ideas, but on the testing of new ones; and some of those new ideas will, inevitably, overthrow orthodoxy. James Secord called the debate 'a minor incident later raised to mythical dimensions by the need to make the Darwinian debate look more heated than it actually was'.[12] And Ian Hesketh rightly adds the question, 'Isn't that precisely what makes the debate significant?'[13]

Soapy Sam, *pace* Hitchens, was not eaten for lunch or used as a floor-mop. In terms of victory or defeat, the debate was in fact a dead heat. Despite the best endeavours of Henslow and Hooker, however, it was not possible to limit this discussion to a simple question of science.

The Darwin family recalled the period when *The Origin of Species* was published as a time of 'frozen misery', as we have seen. When the book had actually appeared, the whole family had left the icily uncomfortable lodgings at Ilkley to return to Down House. Darwin felt himself to be in the dock as the reviews appeared. In the intervening months, the children watched the reaction of Charles and Emma Darwin to the stream of letters and reviews. Emma read most of the reviews to the children, but she concealed, even from Etty, the denunciation by Adam Sedgwick.[14] Although we know, from the written evidence left behind, that Emma was deeply troubled by the religious implications of Darwin's theories, she gave little indication, if any, of such worries to her children. One biographer has likened the terseness of Emma's letters to her children to the laconic military dispatches of the Duke of Wellington.[15]

> My dear Lenny. You cannot write as small as this I know. It is done with your crow-quill. Your last letter was not interesting, but very well spelt, which I care more about. We have a new horse on trial, very spirited and nice-looking, but I am afraid too cheap. Papa is much better than when Frank was here. We have some stamps for you: one Horace says is new Am, 5 cent. Yours, my dear old man, E.D.[16]

That was written in November 1863 to her thirteen-year-old. 'We are cool fish, we Darwins,' Lenny said many years later.[17] Illness, as Darwin himself demonstrated to them on an almost perpetual basis, was one way of thawing the cool and attracting affection and cherishing. Etty, who in her mid- to late teens seemed worryingly prone to unspecified illness, noted that 'Both parents were unwearied in their efforts to soothe and amuse us whichever of us was ill; my father played backgammon with me regularly every day and [Emma] would read out to me.'[18]

Indeed, it was Etty's illness, even more than Darwin's own, or the Oxford debate about *The Origin of Species*, which dominated the late summer and early autumn of 1860. They took her to Eastbourne, from whence he reported, in a letter to Henslow about botany, that 'I am glad to say she has benefited decidedly from sea-air.'[19] In October, she had 'such a week as I did not know man could suffer. My daughter grew worse and worse, with pitiable suffering, so that all the Doctors thought we should lose her.'[20] In fact, she rallied, and would live until 1927 – reaching the age of eighty-four. It was not the Darwin family, but the Huxleys who would lose a child that autumn. On 20 September, Huxley wrote in his journal:

> our Noel, our first-born, after being for nearly four years our delight and our joy, was carried off by scarlet fever in forty-eight hours. This day week he and I had a great romp together. On Friday his restless head, with its bright blue eyes and tangled golden hair, tossed all day upon his pillow. On Saturday night the fifteenth, I carried him here into my study and laid his cold still body here where I write. Here too on Sunday night came his mother and I to that holy leave-taking.[21]

Darwin wrote to Huxley:

> I know well how intolerable is the bitterness of such grief. Yet believe me, that time & time alone, acts wonderfully. To this day, though so many years have passed away [since Annie's death], I cannot think of one child without tears rising in my eyes; but the grief is become tenderer & I can even call up the smile of our lost darling, with something like pleasure.[22]

'Children are one's greatest happiness,' he wrote to Asa Gray, 'but often still a greater misery. A man of science ought to have none – perhaps not a wife; for then there would be nothing in this wide world worth caring for, and a man might (whether he could is another question) work away like a Trojan.'[23]

In this phase of life, when one anxiety or another about his family was never far absent, and when his own health was always precarious, he worked like a Trojan indeed. There was the everlasting need, occasioned by reviews and correspondence, to emend *The Origin of Species*. Murray wanted a third edition; work on this began in November 1860 and was completed by March 1861. There was work on *Variation of Animals and Plants under Domestication*, published in two substantial volumes in 1868 – in effect, an expansion of the first few chapters of *The Origin*, work which occupied him for most of the 1860s. And there was also his work on orchids. During the extended summer holidays of 1860, when he and Etty had both been ill, and when excitements and fears about the Oxford meeting of the BAAS threatened, the Darwins stayed in Hartfield, Sussex, where Emma's sisters, Elizabeth Wedgwood and Charlotte Langton, had houses next to one another. Looking for native orchids in the boggy Sussex hollows, he began a special study of the sundew, *Drosera rotundifolia*, whose small, round, reddish leaves, covered in sticky hairs or tentacles, not unlike a sea anemone, flexed and bent to ensnare insects. In the course of that autumn, he wrote detailed, happy letters about them to Henslow, who responded with advice, and seeds and specimens, to help him with his labours. This was what Darwin did best, and which caused him most delight: close observation of natural history. Everything, of course, in the end, referred back, in his mind, to the Theory. That relentlessly logical figure later in the century, Sherlock Holmes, in the story called 'The Naval Treaty' expressed the belief that the gratuitous beauty of flowers argued for the existence of a beneficent Creator.[24] Orchids were surely nature's version of art for art's sake. Not so, demonstrated Darwin. The apparently meaningless ridges and horns in the sundew and other orchids were weapons in the struggle for survival. It was only by the arrival of insects in their tentacles that these beautiful plants could cross-pollinate.

Sherlock Holmes would seem to be in tune with the facts of the case. Angiosperms (that is, flowering plants which produce seeds with a female reproductive organ, a carpel) provide one of the most devastating challenges to Darwin. Once again, as with the feather, as with the pentadactylic limb, the homologue appears, as it were ready formed, some 140 million years ago. From the original *Bauplan*,[25] over 250,000 species have been evolved or adapted. There are no transitional forms in the fossil evidence. Michael J. Sanderson, in an article 'Back to the Past: A New Take on the Timing of Flowering Plant Diversification',[26] reminds us that it is not simply the origin of angiosperms which poses an insoluble mystery. 'A number of much more recent clades of angiosperms could be interpreted as mini-abominations: that is, they have poor fossil records at their base, *novel innovations with unclear transitional forms among related taxa*' (my italics).

The correspondence with Henslow limited itself to natural history, and steered clear of theology, as the two men always, very wisely, had done. Their friendship was old, as Darwin fondly remembered. Henslow 'was deeply religious'.[27] As for Darwin's theory, he respected the younger man's point of view, but he regarded it as not proven. 'God does not set the creation going like a clock, wound up to go by itself,' he told his brother-in-law Leonard Jenyns.[28] Then, at the age of only sixty-four, Henslow, who lived at Hitchin, suffered a stroke and it was obvious that he was close to death. Hooker and his wife (Henslow's daughter) immediately made the journey from Kew. The elderly Sedgwick, by now frail, tottered over from Cambridge, arriving in time to whisper words of faith into Henslow's ear, not knowing whether his friend heard them or not.

Hooker, knowing of Henslow's special fondness for Darwin, told his friend of the crisis. He happened to know that, in the previous couple of weeks, Darwin had been up to London for a Philosophical Club dinner, and had made a rare speech at the Linnean Society. He had also entertained Mary Butler at Down House, one of his fellow patients from the Ilkley water cure, who made an unsuccessful attempt to demonstrate spirit-rapping and table-turning. Yet, when it came to attending the death-bed of his mentor, Darwin pleaded illness as an excuse to stay away. '[I]f Henslow . . . would really like to see me, I would of course start at once,' he wrote to Hooker. His sole reason for not offering to come,

> was that the journey, with the agitation, would cause me probably to arrive utterly prostrated. I shd be certain to have severe vomiting afterwards, but that would not much signify, but I doubt whether I could stand the agitation at the time. I never felt my weakness a greater evil. I have just had a specimen, for I spoke a few minutes at Linnean Society on Thursday & though extra well it brought on 24 hours vomiting. I suppose there is some Inn at which I could stay, for I shd not like to be in the house (even if you could hold me) as my retching is apt to be extremely loud.[29]

The journey from King's Cross station in London to Hitchin takes approximately half an hour. Darwin could easily have undertaken the journey within a day from Down, allowing himself time to make loud retching noises in wayside inns and railway lavatories. After Henslow had died, Darwin felt eaten up with guilt. Unable to speak to him about this, Emma wrote him a letter, urging him to pray.

> When I see your patience, deep compassion for others, self command & above all gratitude for the smallest thing done to help you I cannot help longing that these precious feelings should be offered to heaven for the sake of your daily happiness. But I find it difficult enough in

> my own case. I often think of the words, 'Thou shalt keep him in perfect peace whose mind is stayed on thee'.[30]

Darwin wrote 'God bless you' in the margin of this letter, but he had long ago left religious belief behind. When, years later, in 1871, Etty married Richard Litchfield, Darwin's letter of advice and blessing contained the sentences, 'I have had . . . a happy life, notwithstanding my stomach; and this I owe entirely to our dear old mother, who, as you know well, is as good as twice refined gold. Keep her as an example before your eyes, and then in future years, Litchfield will worship and not only love you, as I worship our dear old mother.'[31]

Some readers of such words, aware that Emma was *not* Darwin's dear old mother, would find it hard not to see a direct correlation between the motherless Darwin's need for his lost mother's love and Emma's constant willingness to pander to his psychosomatic whims. It is interesting that in her letter to him suggesting prayer she treasured his slavish gratitude to her.

Or was there a simple physical explanation? We have already dismissed (p. 127) the possibility of Darwin having contracted Chagas's disease in South America. A more convincing explanation was offered by Anthony K. Campbell and Stephanie B. Matthews, who argued in a 2005 paper, 'Darwin's Illness Revealed', that Darwin suffered from systemic lactose intolerance.[32] The symptoms of this condition fit Darwin's case. He seems to have suffered his worst bouts of flatulence and vomiting about two hours after eating. This is the time it takes for lactose to reach the large intestine. Dr Gully's regime included the removal of lactose from the diet, and the time Darwin underwent the water cure seems to be the only time he was free of his symptoms. Emma's recipe book, preserved in the University Library at Cambridge, reveals Darwin's sweet tooth and love of rich food, especially for creamy puddings. As well as puddings made of egg yolks and cream, Darwin, when ill, frequently took to his bed with bread and milk, the last thing he should have been doing if Campbell and Matthews are right in their diagnosis. Theirs certainly seems to be the sanest explanation yet for Darwin's symptoms, but it also remains true that there was a psychosomatic element to the condition. Symptoms became much worse when Darwin was under stress, or when he was in danger of having to break his self-absorbed and highly domestic routines.

The death of Henslow removed a father-figure, whom Darwin had revered. 'Poor dear & honoured Henslow. He truly is a model to keep always before one's eyes . . . I fully believe a better man than Henslow never walked this earth.'[33] So Darwin to Hooker. It probably made it more painful to contemplate Henslow's virtue – and Emma's virtue – during the 1860s when he cogitated the next big step, his book on *The Descent of Man*. Henslow, who never doubted so much as one of the Thirty-Nine Articles, and Emma, who urged Darwin to pray his way through his troubles, could scarcely endorse the views of humanity which he entertained: views which he had kept secret, but which he was now preparing to share with the world. In addition to this, he was revising and revising *The Origin of Species*, as the corrections and criticisms of the early version continued to cascade on to his desk at Down House. No wonder that the early 1860s corresponded to his worst yet spells of ill health. In September 1863, Emma persuaded him to return to Dr Gully's clinic in Malvern. They took the Villa Nuova in Malvern Wells and Darwin attended the clinic as an outpatient. Dr Gully was ill and unable to supervise Darwin's treatment. Dr James Ayerst was put in charge of the case and Darwin had no confidence in him. By now, Darwin's skin was so raw with eczema that he could scarcely endure the rubbing and slapping and sousing which the 'cure' entailed. He was suffering memory loss – a fact about which Emma was in denial, but it was noted by Etty.[34] He seemed to be on the verge of seizure, perhaps epilepsy. An especially painful aspect of the Malvern visit was that they at first supposed that Annie's gravestone had been moved or stolen. Darwin wrote to his cousin William Fox:

> Emma went yesterday to the church-yard and found the gravestone of our poor child gone. The Sexton declared he remembered it, & searched well for it & came to the conclusion that it has disappeared. He says the churchyard, a few years ago, was much altered, & we suppose that the stone was then stolen. Now, some years ago, you with your usual kindness visited the grave & sent us an account. Can you tell us what year this was? I was so ill at the time & Emma

> hourly expecting her confinement that I went home & did not see the grave. It is not likely, but will you tell us what you can remember about the kind of stone & where it stood; I think you said there was a little tree planted. We want, of course, to put another stone. I know your great & true kindness will forgive this trouble.[35]

In fact, the gravestone was merely hidden by some undergrowth and Emma eventually found it. While they were in Malvern, the Darwins heard from Hooker that he had lost his six-year-old daughter. It was a painful example of how frequent child-death was in Victorian England.

Darwin returned from Malvern iller than when he had arrived. He could now scarcely walk. From now onwards, he moved quite surely from being a middle-aged man, who was occasionally struck down by his mysterious symptoms, into being an old man, a full-time invalid, much of the time bed-bound.

'For 25 years', he wrote in 1865, supplying notes to yet another medic, Dr John Chapman, 'extreme spasmodic daily & nightly flatulence; occasional vomiting, on two occasions prolonged during months . . . All fatigue, especially rocking [he means the rocking of a carriage or a railway] brings on the head symptoms . . . cannot walk above ½ mile – always tired – conversation or excitement tires me most.'[36]

The whiskers, which had been sprouting from English male cheeks since the 1840s, crept round during the Crimean War (1854–6) to cover the chin. The difficulty of obtaining shaving soap and razors – and the even greater difficulty of shaving in the wintry blizzards above Sevastopol – led to a relaxation of military discipline. Before the war, to be fully bearded in England implied that you were either an eccentric sage like Thomas Carlyle or a working-class labourer (also like Carlyle), or deranged. After the war, a full beard was a sign of being a war hero. Huxley, who had been smooth of chin during the Oxford debate with clean-shaven Bishop Wilberforce, by 1863 had a throat which was as furry as an orang-outang, and he continued to grow his beard until it reached his chest. So did Anthony Trollope. Alfred Russel Wallace returned from Singapore thickly bearded. Lord Tennyson, Poet Laureate, Sir Henry Cole, the mind behind the Great Exhibition of 1851, John Ruskin, the future Prime Minister Lord

Salisbury, they all sprouted long beards, so Darwin's trademark Father Christmas growth was nothing unusual. What was unusual, however, as his white beard lengthened over his coverlets, was how old, how very old, he now appeared to be. From the age of fifty onwards, he had become an ancient, resembling the wounded king Amfortas in the Parsifal legend – which Richard Wagner sketched out in April 1857, though he would not write the music drama until the late 1870s, early 1880s. Without suggesting that Huxley and friends were the Grail Knights, there was something of the mythic, or allegorical, about Darwin's hermit-existence from now onwards. He had started the nineteenth century on an imaginative journey for which a dreadful price seemed to be being paid.

'I suppose your destiny is to let your Brain destroy your Body,' his cousin Fox told him.[37] From Darwin's point of view, work was the only thing which made him forget his pain. Yet, by paradox, 'I know well that my head would have failed years ago, had not my stomach always saved me from a minute's over work.'[38] Francis Darwin would look back and remember his father's inability to sleep once he was fired up by an intellectual problem, or disturbed by churning emotions. (He once woke up his son William in the small hours to apologize for a remark he had made the previous day, and which his mind, in the watches of the night, had magnified out of all proportion.)

The illness now dominated. When Hooker questioned his claim that he had vomited dozens of times during a day, and asked, 'Do you actually throw up or is it retching?', Darwin admitted that 'I seldom throw up food, only acid & morbid secretion.'[39]

Perhaps the simplest way of living with the condition was to accept that the Darwins and the Wedgwoods, multiply intermarried cousins, were a hypersensitive tribe, much given to nervous disorder. It was a question of heredity: like everything else apparently.

14

Adios, Theory

IN 1856, WHILE Darwin had still been at work on *The Origin of Species*, a jolly, cigar-smoking Augustinian friar, Gregor Mendel, began a series of experiments in the garden of St Thomas in Brno – in what is now the Czech Republic, but was Brünn in the Austro-Hungarian Empire. Mendel deliberately kept his experiment very simple. He worked on the ordinary garden pea. It was ideally suited for the study of the problem of inheritance, since it is short lived and could be rapidly studied through several generations. Also, it might be either self-fertilized or cross-fertilized, so each mating could be controlled.

Mendel crossed yellow-seeded peas with green-seeded. The hybrid seeds were all yellow, so he named the yellow-seeded character the *dominant* and the green-seeded the *recessive*. He allowed 258 of the yellow-seeded hybrids to seed themselves, thereby obtaining 8,023 seeds, of which 6,022 were yellow and 2,001 were green. So there was a three to one proportion for every pair of characters involving a dominant and a recessive. The recessive trait-determiner – what the world now knows as the *gene* – re-expressed itself in the second hybrid generation whenever it was by chance combined with another recessive. Mendel had discovered the modern science of genetics.

His experiments with the peas allowed him to draw two conclusions. The first principle was the law of segregation. Genes come in alternative varieties known as alleles which influence phenotypes – such things as the shape of a seed if you are a pea, or the colour of your eyes if you are a human being. Two alleles are inherited, one from each parent. If different alleles are inherited, one is dominant and the other is recessive. The recessive gene can be passed on (as Queen Victoria passed on haemophilia) without the carrier having any of its characteristics.

Mendel's second principle was the law of independent assortment. The inheritance pattern of one trait does not influence the inheritance pattern of another trait. So the genes that encode the shape of a pea's seed do not encode its colour. Each Mendelian trait is passed on in the ratio of three to one, according to the dominance of the pattern of the genes involved.

These laws of Mendel were both, in fact, slightly inaccurate. Not all characteristics follow the simple patterns of dominance which he suggested. Only after the structure of DNA was discovered in the 1950s could the full science of genetics be developed. Mendel, however, had made a discovery of a truly Newtonian or Copernican dimension. Up to this point, the study of inheritance, and the mystery of how characteristics are passed on through the generations, was shrouded in muddled theories. Mendel moved out of theory into something which could be demonstrated, over and over again. All that was required, after Mendel, was for his discoveries to be refined. He read his paper on inheritance to the Natural History Society of Brno. In 1868, he was elected abbot of his monastery and gave up scientific research. He died in 1884. The organist at his funeral was the young Janáček.

Mendel's paper about the peas lay unremarked until, in and around 1900, various scientists, including Hugo de Vries, Carl Correns and Erich von Tschermak, who were working on what could be learnt about the mechanism of evolution by hybridization, rediscovered him and recognized his importance. William Bateson, in Cambridge (England), an evolutionary biologist, immediately saw the importance of Mendel's work and coined the term 'genetics'. Not long afterwards two American scientists – Edmund Beecher Wilson at Columbia University and Nettie M. Stevens at Bryn Mawr – concluded, independently, that the determination of sex, including the one-to-one sex ratio, was caused in Mendelian fashion by the segregation and reunion of the X and Y chromosomes.

Darwin was working in complete ignorance of what science would know as genetics. One reason, perhaps, for his failure to come anywhere near understanding the process of inheritance was that he was casting his net so wide, writing off to India to get Blyth's views on every creature from ants to elephants, dissecting barnacles and insect-eating orchids and, of course, mixing with the pigeon-fanciers, whose genuine knowledge of their subject he lapped up. The result was that he assembled a collection of facts so enormous that it would have been hard to know where to focus. Although he believed there was one simple explanation for how inheritance works, he thought to test out the idea by a Humboldtian investigation of *everything*. Mendel stuck to peas and observed one simple process at work. The old saying about the hedgehog and the fox comes to mind. Having grasped the principle which was expounded by Mendel, it became possible, by the mid- to late twentieth century, for biology to see how densely complicated, how various, how simply massive is the amount of information stored, conveyed and passed on through the double helix, which 'packs over a hundred trillion times as much information by volume as the most sophisticated computerized information system ever devised'.[1]

Mendel's theory, confirmed by the genetic researches following Watson and Crick's stupendous discovery of the double helix and its structure, is really lethal to Darwinism. As mentioned, by the beginning of the twentieth century, Mendel's theory had been largely forgotten, and it was only when it was revived, and the science of modern genetics began, that some scientists devised what has been called neo-Darwinism, or a supposed synthesis between the two ideas – Mendel and Darwin, genetics and gradualism. In 1937, the experimental zoologist Theodosius Dobzhansky wrote *Genetics and the Origin of Species* in which he argued that it was possible to reconcile Darwinism, with its commitment to a gradualist evolution, and genetics, with its factual evidence that the gene is passed down like a hard, unvarying particle, like a piece of shot. His views were adopted by Thomas Huxley's grandson Julian Huxley in his *Evolution: The Modern Synthesis* (1942). The ideas of this synthesis were applied to palaeontology by George Gaylord Simpson in *Tempo and Mode in Evolution* (1944). From these three books stem the fusion of Darwinian theory and genetic knowledge which is orthodoxy in most academic schools of biology and which has been popularized, for example, by Richard Dawkins. It attempts to blend population genetics with the old Darwinian doctrines: so we see randomly produced genetic variations helping individuals in a species population better adapted to compete for resources in their environment. The favourable genes are then, supposedly, inherited in greater numbers in the gene pool, the total number of genes in any particular species, until the new species emerges. The trouble with this is that mutations in genes tend not to be minor; nor are they obviously always to the advantage of a species – witness the many mutations responsible for diseases.

It has been suggested that Darwin owned a copy of Mendel's paper. In Michael R. Rose's *Darwin's Spectre* it is stated that 'an unopened copy of Mendel's crucial paper on inheritance in peas was found among Darwin's files after his death'.[2] This, however, turns out not to be true. Investigations in the Cambridge University Library, where the bulk of the Darwin archive is stored, reveals no work by Mendel among Darwin's books or papers, and Francis Darwin writing in *More Letters of Charles Darwin* says, 'It is remarkable that, as far as we know, Darwin never in any way came across Mendel's work.'[3] Peter Gautrey, in his paper written with Robert Olby, 'The Eleven References to Mendel before 1900', points out that Darwin is known to have had copies of two works which did quote from Mendel's paper.[4] Darwin made a number of annotations in his copy of one of these, by Hermann Hoffmann, but not on the page which refers to Mendel; the other, *Die Pflanzen-Mischlinge*, by Wilhelm Olbers Focke, was acquired by Darwin in November 1880 but the pages which refer to Mendel's work were never cut.

Darwin was aware, almost before the first edition of *The Origin of Species* had been published, that there was something defective about his inheritance theory.[5] He devoted only one chapter of *The Origin* to the question, but he had accumulated so much material on the subject that it was his intention to work up at least three more chapters in a later book. And the book published as *The Variation of Animals and Plants under Domestication* represents his attempt to come to grips, not only with the critics of *The Origins*, but also with the core scientific problem of his book – a problem, incidentally, with which science had been wrestling certainly since 1799, when Thomas Andrew Knight published a paper for the Royal Society on . . . none other than that revealing genus the *Pisum*: a foreshadowing of Father Mendel in his monastery garden. Knight was more naturalist than theorist, so he observed *how* peas bred, and *how* hybridization affected them more than he asked *why*. Darwin, having come to see that variation was passed on, but having accepted (for some decades anyway) that Lamarck's theory was wrong, had to find a way of explaining how variation was handed on from one generation to the next. Knight crossed grey peas with white peas and noted the phenomenon (though he did not call it that, of course) of the dominant gene. Only grey peas were produced. The white pea was recessive.[6] Darwin supposed that when two parents mated, their characteristics were *blended*. Lyell, never fully a convert to Darwin's idea of natural selection, had weighed the arguments, in *Principles of Geology* (volume two), in favour of the transmutation of species. 'The entire variation from the original type, which any given kind of change can produce, may usually be effected in a brief period of time, after which no farther deviation can be obtained by continuing to alter the circumstances, though ever so gradually – indefinite divergence either in the way of improvement or deterioration, being prevented, and the least possible excess beyond the defined limits being fatal to the existence of the individual.'[7]

In Darwin's copy of this book, he has underscored the words 'improvement or deterioration' and added, 'If this were *true*, adios *theory*.'[8]

This was the point upon which Wilberforce seized, both in his speech at the BAAS in Oxford and in his thirty-page review of *The Origin of Species*. Had he limited himself to the discussion of variation in species, and avoided the cheap gag about Huxley's grandmother, Wilberforce's place as a footnote in the history of science would be more glorious. The same point was made by the Scottish engineer Fleeming Jenkin, in his review of the fourth edition of *The Origin* in a periodical called the *North British Review*, in 1867. Jenkin was taking a break from his work together with James Clerk Maxwell on the calibration of standard coils of wire, which would eventually perfect cable communications and regularize the use of telegraphy. He was, therefore, one of the real pioneers of the modern world, bringing to pass the marriage of technology and political expansionism which made Britain the predominant power during these decades.

Jenkin's review was immensely long. Given its date – written just a decade after Mendel's discovery – it is tantalizingly close to Mendel. He can see the problem which Mendel solved, without – of course

– knowing the solution. 'A given animal or plant appears to be contained as it were within a sphere of variation; one individual lies near one portion of the surface, another individual, of the same species, near another part of the surface; the average animal at the centre . . .'; Darwin envisaged a slow process of variation. But, said Jenkin – here following Wilberforce's argument, and agreeing with what Agassiz also said – for Darwin's theory to work it is necessary for time to 'fix' the variety. He imagined each species, or ur-species, in a 'sphere', and for Darwin's theory to work variations within that sphere would have to get beyond its own limit. He acknowledged, as any countryman or rose-breeder or pigeon-fancier might do, that slight improvements might be discernible within a species. But there remained Wilberforce's puzzle: how come artificial hybrids revert in the wild? Jenkin can see, in other words that there is a *something* which determines how species inherit their characteristics – what we know to be genes. Darwin could see it too, but he could not answer either Jenkin or Wilberforce or Agassiz. 'Fleeming Jenkin has given me much trouble, but has been of more real use to me that any other essay or review.'[9] If he could not answer Jenkin, it did look very much as if he would, after thirty years' work, be obliged to say adios, theory.

Hooker, such a kind friend to Darwin and such a conscientious scientist, could see that the theory on which Darwin's reputation was based was a theory of variation: how are variations passed on through inheritance? And do such variations account for the 'origin' of species – that is, do species evolve from variations into other species? In 1862, he and Darwin had exchanged letters on the subject, and Darwin had written that he had, in effect, moved back to the position of Lamarck. 'I hardly know why I am a little sorry but my present work is leading me to believe rather more in the direct action of physical conditions. I presume I regret it, because it lessens the glory of Natural Selection, and is so confoundedly doubtful. Perhaps I shall change again when I get all my facts under one point of view, and a pretty hard job this will be.'[10]

Hooker, six years later, broke it to Darwin that he was going to devote his Presidential Address at the BAAS to 'the fact that Darwin's theory had failed'.[11] (So much for the 'victory' of Huxley in the 1860 debate!) Darwin's crestfallen reply is painfully revealing, since it suggests that – at the time of writing at least – he had abandoned the theory in favour of Lamarckianism. He was saying that the *Origin* triumph was not that it was true in detail, but that it made people believe in evolution in general.

> I am glad to hear that you are going to touch on the statement that the belief in Natural Selection is passing away. I do not suppose that even the *Athenaeum* would pretend that the belief in the common descent of species is passing away, and this is the more important point. This now *almost universal* belief in the evolution (somehow) of species, I think may be attributed in large part to the *Origin* . . . If you agree about the non-acceptance of Natural Selection, it seems to me a very striking fact that the Newtonian theory of gravitation, which seems to everyone now so certain and plain, was rejected by a man so extraordinarily able as Leibnitz. The truth will not penetrate a preoccupied mind.[12]

This is a sad letter not merely because he sees a trusted friend publicly disputing his theory, but because, in the space of only a few sentences, he appears to be saying two quite contradictory things. First, he says that the theory of natural selection is *not* true, but that at least it made people believe in 'the evolution (somehow) of species'. Then, appalled by the thought of saying adios to his theory, he compares it, not for the first time, to the discovery by Newton of how the universe holds together by gravitational force.

Darwin felt isolated. Wallace had returned to Britain in 1862, his own health – though not as poor as Darwin's – badly diminished by living in the Malayan Archipelago. Wallace was taken up by Huxley, and introduced to Lyell's rather grand drawing-room. But salon-life was not for him, and this wraith-like, gentle, bearded man felt out of his depth in London society. He made visits to Down, but there was an awkwardness about the relationship between the two discoverers of natural selection. Wallace remained firm in his faith in the theory. He was a purer 'Darwinian' in this sense than Darwin. The Duke of Argyll, who would be Secretary for India in Gladstone's first government, wrote a review of Darwin's orchid book, maintaining the Sherlock Holmes argument that these flowers showed the Creator's hand, creating beauty for beauty's sake. The Duke followed it up with a book of his own, *The Reign of Law* (1867), in which he developed his theme: 'Ornament or beauty is in itself a purpose, an object, an end.' He cited the hummingbird with its topaz crest. This, surely, was an example of beauty having been created for its own sake? Darwin amazed Huxley by being rather impressed by this argument. He was shocked by Huxley's perky class chippiness in calling the Duke 'the Dukelet' and the 'little beggar'. 'I have always thought the D. of Argyll wonderfully clever,' Darwin countered. 'As for calling him a little beggar, my inherited instinctive feelings wd. declare it was a sin thus so speak of a real old Duke.'[13] His own view of how the humming-bird got its crest began to modify. 'Natural' selection was not sufficient to explain so extravagant a manifestation. Sexual selection was more to the point. Wallace urged Darwin simply to stick to his guns.

The combination of ducal piety on the one hand and the scientific rigour of Jenkin on the other forced Darwin to come up with some 'it', some phenomenon which could explain how variation was passed down through the processes of inheritance. He could see, in other words, the hole in the story, which *we* can see being filled by Mendel's genetics. Not knowing the Mendelian solution – and, so frustratingly for us, it was sitting there, waiting for Darwin to read about it, in an article which he possessed but did not read! – Darwin came up with an explanation which was simply wrong. He called his mysterious X, his explanation, 'pangenesis'. The science of genetics, as pioneered by Mendel and expanded since the discovery of the double helix, does not require, even in the metaphorical sense used by Darwinians, any *motivation* as an explanation of why some forms are handed down. What genetics reveal is *how* the process works. Darwin could see characteristics being handed on, by orchids, by pigeons, by members of the Darwin family – for the illnesses which beset his children in the nursery all tormented him, not only as a parent, but as a scientist who feared that he and Emma had bred too many of their neurotic family ailments. Pangenesis was a rather feeble combination of Lamarck and his own musings. He proposed that every tissue, cell and living part of an organism produced unseen germs, gemmules or granules which carried on inheritable characteristics. There were gemmules which decided the size or shape of your feet, the colour of your hair, and so forth. These characteristics were, he believed, mixed up by sexual congress. For Darwin, the granules, or whatever they were, were a sort of melange produced by sex; but also – and here he added a dash of Lamarck to his rather desperate concoction – some limited effects from the environment were embedded in each individual's constitution, and could be handed on to the offspring, through the gemmules. Wallace urged him to stick, simply, to the idea of natural selection – as the majority of Darwinists do in our own day. When Huxley first read Darwin's attempt to articulate pangenesis, he playfully said that he had donned his 'sharpest spectacles and best thinking cap'. 'Somebody rummaging among your papers half a century hence will find *Pangenesis* and say, "See this wonderful anticipation of our modern theories and that stupid ass Huxley prevented his publishing them".'[14] It was Huxley's kind way of telling him not to publish the pangenesis idea. The advice was not heeded.

Someone who came much closer to a Mendelian solution of the problem of inheritance was a writer who was not in the narrow sense of the word a scientist at all – Herbert Spencer.

> Just as, during the evolution of an organism, the physiological units derived from the two parents tend to segregate, and produce likeness to the male parent in this part and to the female parent in that: so, during the formation of reproductive cells, there will arise in one a predominance of the physiological units derived from the father, and in another a predominance of the physiological units derived from the mother. Thus, then, every fertilized germ, besides containing different *amounts* of the two parental influences, will contain different *kinds* of influences – this having received a marked impress from one grandparent, and that from another. Without further exposition the reader will see how this cause of complication, running back through each line of ancestry, must produce in every germ numerous minute differences among the units . . .
>
> From the general law of probabilities it may be concluded that while these involved influences, derived from many progenitors, must, on the average of cases, obscure and partially neutralize one another; there must occasionally result such combinations of them as will produce very marked divergences. There is thus a correspondence between the inferable results and the results as habitually witnessed.[15]

The Danish biologist Søren Løvtrup, citing this passage, adds, 'Evidently, Spencer was a Micromutationist insisting on the necessity of macromutations, which he assumed to arise through particular combinations of micromutations. Surely, if Darwin had paid a little more heed to Spencer, he could have saved himself much trouble.'[16]

When Spencer met Darwin, before the publication of *The Origin*, and heard what the theory was, he quipped that it might well be termed 'the survival of the fittest' – a tag which Darwin adopted, and which stuck. Spencer was eleven years younger than Darwin. Like any human being, Darwin found Spencer's voluminous writings hard going. 'His conclusions never convince me: and over and over again I have said to myself, after reading one of his discussions – "Here would be a fine subject for half-a-dozen years' work."'[17] On other occasions, however, he deemed Spencer 'wonderfully clever, and *I dare say mostly true*' (1866 to Hooker), and by the end of the decade he even suspected 'that hereafter he will be looked at as by far the greatest philosopher in England; *perhaps equal to any that have lived*' (1870 to E. R. Lankaster).[18]

Every now and then there is a brave attempt by an academic to remind us of Spencer's very existence.[19] He has largely evaporated from the consciousness of thinking or reading people. Whereas some in the twenty-first century have read Carlyle – and many have read Darwin, Newman, Mill – Spencer is, of all the nineteenth-century British intellectuals, the one whose significance has been most utterly lost. It is easy to mock him.[20] His hypochondria alone was far more absurd than Darwin's: his insistence, for example, upon taking his pulse at particular times of day and, if the pulse rate was too fast, turning at once for home; his refusal to wear conventional dress (caps rather than silk hats, no evening clothes – though he relaxed this rule when invited to dine with Monckton Milnes – Lord Houghton – to meet the King of the Belgians); his 'love affair' with George Eliot, which was never consummated; and the sheer volume of what he wrote – *A System of Synthetic Philosophy* (1860), *First Principles* (1862), *The Principles of Biology* (two volumes, 1864–7), *The Principles of Psychology* (1870–2), *The Principles of Sociology* (1876–96), *The Data of Ethics* (1879), not to mention the thousand-page autobiography. He was probably the most famous 'philosopher' of the day, easily as famous as in their day Sartre or Lévi-Strauss. When he died, the Italian parliament observed a minute's silence. They did not go quite so far in the US Congress, but he was hugely revered in America, where audiences put up with his strange habit of accepting engagements and then refusing to speak to them. It was enough to see the Master. His works were translated into languages as different as Chinese and Mohawk. Progressive-minded Indians absorbed him, as did the French and the Germans.

Experts in any of the fields which he entered might find his prolix lucubrations to be slapdash, while academic logicians or metaphysicians might find his use of language to be too loose to be durable. If, however, you want to know what a thinking person in the mid- to late nineteenth century thought, then take a tour of Herbert Spencer's mind. The very inconsistencies, and changes of mood, in his vast output are characteristic of his age. His early work seems to a later reader almost insane in its optimism – its belief that the Great Exhibition of 1851 represented a high point not only of free trade but of human achievement and human happiness. This was followed by a nervous breakdown, in which Spencer felt that his nonconformist childhood in Derby had been a hell. Very much like Mill, he had been brought up to believe in work, work, work. He could not shake off the habit, and absurdly worked himself into the ground in the second half of his life extolling the virtue of leisure; and, while leading a life of celibacy, advancing a theory of happiness based on passion. He was above all – and this was why he became a Victorian best-seller – a clever man who banished the theocentric view of life. He was, however, anti-materialist. Like Huxley, he was agnostic rather than atheist. In early life – during his days of hanging around with George Eliot before she became George Eliot – he was a worshipper of what he called the Unknown, and he believed there was something essentially mysterious and sacred about Life with a capital L. But he also believed that such metaphysical musings were reinforced by science and its study. As Søren Løvtrup's tribute shows, Spencer's biology was well informed. He always remained, however, a popular metaphysician-cum-higher journalist; science was not his primary interest. He was explaining the nineteenth century to itself, explaining how it had to come to terms with the scientific outlook while not losing its soul. Huxley, who came from a very similar provincial lower-middle-class background (both men had unsuccessful impoverished schoolteachers as fathers) read the proofs of Spencer's *First Principles* (1862) and wrote to him: 'It seems as if all the thoughts in what you have written were my own and yet I am conscious of the enormous difference your presentation of them makes in my intellectual state. One is thought in the state of hemp yarn and the other in the state of rope. Work away thou excellent rope maker and make us more rope to hold on against the devil and the parsons.'[21]

This is the key to Spencer's importance, not only in Huxley's inner world, but in that of all his Victorian contemporaries. Today his ashes rest in Highgate Cemetery, just opposite the gigantic memorial to Karl Marx, and a few yards from George Eliot. They were all in their way ardent warriors against the parsons: George Eliot, who in her incarnation as Mary Ann Evans had translated Ludwig Feuerbach and David Strauss, undermining the philosophical basis of Christianity and shaking the foundations of belief in the Bible; Marx, whose dialectical materialism took atheism for granted, and replaced the justice and providence of the Hebrew Scriptures with a vision equally prophetic, equally deterministic, that the victory of organized labour over capital would reorder human society; and Herbert Spencer, who believed in the 'survival of the fittest' even before Darwin had seen how aptly it described his own view of life.

The debate at Oxford, and the denunciation of *The Origin of Species* by a bishop (albeit a scientifically minded bishop), hardened, almost mythologized, the battle-lines. Darwin had not set out on his Humboldtian journey of being a great scientist solely in order to wage war on religion. Having Huxley as his champion, however, and Spencer as a friendly philosopher, now determined the direction in which his theory would take him. The assaults on *The Origin of Species* which worried him were scientific assaults on its veracity. With each revision of the book, through six editions, he discarded more and more of its central theory. To some extent, the assaults which were based upon its supposedly heretical content provided a screen for the book's scientific errors. When, by the end of the century, 'Darwinism' had been all but put to sleep, and science had moved on, other reasons could be found for unbelief – especially in the pages of Spencer. Fascinatingly, when neo-Darwinism revived, from the mid-twentieth century onwards, it awoke with all its mid-Victorian anti-religious trappings. It is hard to think of any other branch of modern science – quantum theory, for example, or discoveries in electromagnetism, neuroscience or astronomy – whose proponents spend as much time talking about the errors of theology as of the truth of their own area of expertise.

We'll come back to Spencer, but it is time to mention another figure in the story – one of the villains, if you take a partisan neo-Darwinian viewpoint. St George Jackson Mivart (1827–1900) was the son of a successful hotel-owner in London, who studied law at Lincoln's Inn while also devoting himself to the study of biology. He was a student of Huxley's and knew Owen, and through these distinguished mentors he managed to get a post as a teaching biologist at St Mary's Hospital Medical School in Paddington. He knew Darwin slightly, but revered him deeply. He also knew Wallace. He was a convinced evolutionist and never wavered from this position. A convert to Roman Catholicism at the age of sixteen, he was eventually excommunicated from that Church by Cardinal Vaughan for his refusal to deny evolution.

Mivart had read *The Origin of Species* with enthusiasm when it was first published. Darwin, Huxley and friends, when Mivart became their enemy, denounced him as if he were some kind of mouthpiece for his Church, whereas the truth was that, however much he loved the Church of his adoption, he loved scientific truth more. One of his difficulties with Darwin's mode of expression in *The Origin*, in fact, was his way of speaking as if natural selection was purposive; for Mivart, both as a scientist and as a Catholic, the process of nature (whether or not to be understood as set in motion by a divine creative inbreathing) was to be seen as just that, a process – and, in this sense, as impersonal.

> It is easy to complain of one-sidedness in the views of many who oppose Darwinism in the interests of Orthodoxy; but not at all less patent is the intolerance and narrow-mindedness of some of those

> who advocate it, avowedly or covertly, in the interest of heterodoxy. This hastiness of rejection or acceptance, determined by ulterior consequences believed to attach to 'Natural Selection', is unfortunately in part to be accounted for by some expressions and a certain tone to be found in Mr Darwin's writings. That his expressions, however, are not always to be construed literally is manifest. His frequent use metaphorically of the theistic expressions, 'contrivance', for example, and 'purpose' has elicited from the Duke of Argyll and others criticisms which fail to tell against their opponent, solely because such expressions are, in Mr Darwin's writings, merely figurative – metaphors and nothing more.[22]

It may be hoped, then, that a similar looseness of expression will account for passages of a directly opposite tendency to that of his theistic metaphors.

> Moreover, it must not be forgotten that he frequently uses that absolutely theological term 'Creator' and that he has retained in all the editions of his 'Origin of Species' an expression which has been much criticized: he speaks of 'life with its several powers, having been originally breathed by the Creator into a few forms, or into one'. This is mentioned in justice to Mr Darwin only, and by no means because it is a position which this book [Mivart's riposte to *The Origin*, consisting of a number of articles gathered in book form under the title *On the Genesis of Species*] is intended to support. For, from Mr Darwin's usual mode of speaking, it appears that by such Divine action he means a supernatural intervention, whereas it is here contended that throughout the whole process of physical evolution – the first manifestation of life included – *supernatural* action is not to be looked for.

Christian writers, over the ages, have differed on the question of whether the age of miracles was restricted to the era of Christ, or whether miracles are still possible. Mivart was merely stating that to believe in a Creator is not a belief in One who intervenes in the created order when that order is in process. To this extent, the question of 'creationism' is irrelevant to the purely scientific question of how species evolve. It was this, Darwin's specific theory of evolution by natural selection, which Mivart had come to doubt. He summarized his objections to *The Origin of Species* thus:

> That Natural Selection is incompetent to account for the incipient stages of useful structures.
>
> That it does not harmonize with the co-existence of closely similar structures of diverse origin.
>
> That there are grounds for thinking that specific differences may be developed suddenly instead of gradually.
>
> That the opinion that species have definite though very different limits to their variability is still tenable.
>
> That certain fossil transitional forms are absent which might have been expected to be present.
>
> That some facts of geographical distribution intensify other difficulties.
>
> That the objection drawn from the physiological difference between 'species' and 'races' still exists unrefuted.
>
> That there are many remarkable phenomena in organic forms upon which 'Natural Selection' throws no light whatever, but the explanations of which, if they could be attained, might throw some light upon specific origination.[23]

Consider the giraffe. Put simply, macro-mutationists think that there is not much future in being a giraffe if your neck does not grow as high as the nearest tree-foliage. Otherwise, you might have stayed like the short-necked oxen or antelopes who inhabit similar parts of southern Africa. For a macro-mutationist, a giraffe is a creature with a neck long enough to reach leaves on trees. For a Darwinian, a giraffe is the result of a long history of struggle in which a number of weak, short-necked would-be giraffes have been panting to reach those leaves, but without success. Darwin tried to answer this in the sixth edition of *The Origin*: 'Those individuals which had some one part or several parts of their bodies rather more elongated than usual, would generally have survived. These will have been intercrossed and left offspring, either inheriting the same bodily peculiarities, or with a tendency to vary again, in the same manner; whilst individuals, less favoured in the same respects, will have been most liable to perish.'[24]

Mivart also asked, if natural selection is so potent, and if high browsing is so advantageous, why did not all the creatures in southern Africa develop long necks? Darwin's answer to this conundrum is one of his least persuasive analogies. 'The answer is not difficult,' he wrote,

> and can best be given by an illustration. In every meadow in England in which trees grow, we see the lower branches trimmed or planed to an exact level by the browsing of the horses or cattle; and what advantage would it be, for instance, to sheep, if kept there, to acquire slightly longer necks? In every district some one kind of animal would almost certainly be able to browse higher than the others; and it is almost equally certain that this one kind alone could have its neck elongated for this purpose, through natural selection and the effects of increased use. In S. Africa the competition for browsing on the higher branches of the acacias and other trees must be between giraffe and giraffe, and not with the other ungulate animals.[25]

There seem to be two flaws in this strange (almost surreal) illustration. One is that it would not be the sheep but the horses or cattle which had to grow longer necks in order to reach the trees (in which they had shown no interest anyway); and secondly, if the 'nascent giraffe' had a neck long enough to reach foliage, it would not need to 'compete' with other nascent giraffes in order to reach the tree. Either it was tall enough to reach the tree, or it wasn't, and if it wasn't, then it could not have existed. It could not have eaten. One is reminded of Darwin's belief that the highly complex optic nerve 'developed' over time, and of the modern research which has demonstrated by computer that it takes 'only' half a million years or so for a haddock to develop an eye. There was another fish conundrum to which Mivart drew Darwin's attention. The flatfish or *Pleuronectidae* have asymmetrical bodies, but their eyes are placed on the upper side of the head. During early youth, the eyes stand opposite to one another thereby making the body symmetrical. Mivart wrote that a sudden 'spontaneous transformation in the position of the eyes is hardly conceivable'.[26] But what if it were gradual? Then 'how such transit of one eye a minute fraction of the journey towards the other side of the head could benefit the individual is indeed far from clear'.[27]

Darwin answered this objection in the most surprising way. He appealed to a Lamarckian ichthyologist called Malm, published in 1867. According to Malm's theory, the young *Pleuronectidae* cannot retain a vertical position for long, owing to the excessive depth of their bodies, the small size of their lateral fins, and their having no swimbladder. Exhaustion therefore causes them to drop to the bottom of the ocean on one side, but while they lie there on the bed of the sea, they twist, like restless sleepers, so that the lower eye is looking upwards. 'The forehead between the eyes consequently becomes, as could be plainly seen, temporarily contracted in breadth.'[28] Darwin did not appear to notice that Malm, and Dr Albert Günther, 'our great authority on fishes', both advocated an essentially Lamarckian explanation for the positioning of the flatfish's eye.

The truth was that Darwin was horribly discountenanced by Mivart's objections to his theory. To Wallace, he wrote, 'Mivart's book is producing a great effect against Natural Selection, and more especially against me.'[29]

Moreover, Mivart was not the only scientist who appeared not merely to modify but to destroy the theory. William Thomson (later first Baron Kelvin of Largs, 1824–1907) was the child of a professor of mathematics in Belfast, who later moved to Glasgow. Thomson was a physicist of prodigious energy and brilliance, who, having studied at Glasgow, entered Peterhouse, Cambridge, and made his mark, not only academically (second wrangler, and a string of learned papers before he was twenty-one), but also a keen oarsman and co-founder of the Cambridge University Musical Society. He became Professor of Natural Philosophy in Glasgow at the age of twenty-two. In his first four years as professor he published no fewer than fifty papers, some of them in French. His great subject was the transformation of heat, and he developed a theory that you could date the age of the planet by assessing the length of time it had taken to cool from its initial boiling lava. (This research was carried out alongside his practical skills as an engineer and his pioneering interest in telegraph.) It was his experiments in thermodynamics, however, which convinced him that the earth was far younger than the time span needed for Darwinism to be true. A hundred million years, by Thomson's calculation, was long enough for the whole story to be told, from its first fiery beginning to its end. Thomson was a great scientist, but his calculations were wildly wrong. In 1903 – four years before he died – it was discovered that radioactive elements constantly emit heat. A year later, Ernest Rutherford suggested that the ratio of the abundance of radioactive elements to their decay products provided a way to measure the ages of rocks and the minerals containing them. By 1911, Arthur Holmes was using uranium/lead measures to estimate the ages of rocks from the ancient Pre-Cambrian period, and calculating their age as 1,600 million years old. Isotopes were discovered in 1913, and in the 1930s the modern mass spectrometer was developed. It was now possible to see that the earth was at least 4,000 million years old, perhaps 5,000 million. Then, in 1956, the American physicist Clair Cameron Patterson (a man) compared the isotopes of the earth's crust with five meteorites. On the basis of these calculations, he was able to work out that the earth is about 4,550 million years old. 'All subsequent estimates of the age of the earth have tended to confirm Patterson's conclusion.'[30]

Darwin did not know any of this, of course. For him, Thomson's dating of the earth was 'an odious spectre': 'Thomson's views of the recent age of the world have been for some time one of my sorest troubles.'[31] In fact, as we now realize, Darwin, in the first edition of *The Origin*, underestimated the age of the earth, by about 4,200 million years; but at least his underestimate gave time for natural selection to have occurred. Thomson's mere 100 million was not enough. Darwin told his son George that this was the single most intractable point levelled against the theory. He implored George the Cambridge mathematician to make alternative calculations. George loyally obliged and proposed a modification of Thomson's figures which might make the theory tenable. Darwin meanwhile racked his brains to see if he could somehow speed evolution up, or somehow force the facts to fit the theory. He had abandoned the scientific theory in favour of propaganda technique. Everyone mocks Edmund Gosse's father for his pious belief that fossils had been placed in the rocks by God to test our faith, though they were really only as old as Archbishop Ussher had calculated. Darwin's revision of *The Origin of Species* was worked through in a comparable state of mind, believing where he could not prove. In fact, there was a strange double-think going on. With one hand, he constantly revised *The Origin*, diluting the theory to the point where it made no sense, even by its own terms. On the other hand, he could not acknowledge that his original theory had turned into a convoluted version of Lamarckianism. The wider the dissent, the more doggedly he believed in 'his' theory ('our' theory when writing to Wallace) and the more resolutely he basked in Huxley's histrionic public defences of it. The revisions of *The Origin* attracted small public notice. Only 311 people bought the fifth revision, for example, and sales for *Variation of Animals and Plants under Domestication* were comparably low.

A survey of the reviews of the first edition of *The Origin of Species* could not make happy reading for its author.[32] Two rejected the idea of evolution outright (Sedgwick and Agassiz). Three (Hooker, Hutton and Wright) accepted Darwin's thesis but with deep reservations. The remaining eleven – all of them, however, respectful – suggest that micro-mutation is an implausible explanation for the origin of species, and propose some form of macro-mutation. Some of the reviewers also point out similarities between Darwin and Lamarck and between Darwin and *Vestiges* which he was unwilling to acknowledge himself. The book had confirmed in the reading public the hunch which it had had since reading Lyell and Chambers years before: namely that evolution is true. As a specific explanation for the origin of species, it had failed, and Darwin was gored, wounded, by the result. The side of him which was the faithful naturalist would acknowledge this, and was preparing to withdraw from the fray. The other side of him – the side which had believed in 1844 that he was guilty of murder, for having murdered his wife's God – was preparing the development of his biological theory into a metaphysical treatise. *The Origin of Species* had not said a word about the origin of the human race, although it was patently obvious that this was what made the central argument a matter of controversy. Why else would Bishop Wilberforce have been so foolish as to ask Huxley on which side of his family he was related to a monkey? Having remained taciturn, except in his letters, and in implicit revisions to *The Origin*, about this vital matter, Darwin was now at work on the book which really spelled out, not only the logical conclusion to natural selection, but also the first principles that had enabled him to form the theory in the first place. Darwin, after all these years of caution, was about to come clean about the central question of all: about who he thought we human beings are.

The Origin of Species had not ventured an opinion about the origin of the human species. Not everyone, perhaps, had paid sufficient attention to Chapter Three of that book, 'Struggle for Existence', in which the Malthusian doctrine was applied 'with manifold force to the whole animal and vegetable kingdoms'.[33] Malthus's central idea – that populations increase proportionately to the amount of food available to them, and after that perish – was, of course, an idea which he applied exclusively to the human race. It was Darwin (and Wallace) who applied it to plants and fish and birds. When he came to apply it back again to humans, Darwin blurred the distinction between two sorts of 'struggle' for existence: namely, the 'struggle' of seeds, or incipient life, to come into being at all, and the struggle, when alive, to be able to eat. So he spoke of American forests being scenes of perpetual warfare, with different tree-types vying with one another for space, different insects battling it out.

The implication, though he did not state this openly in 1859, was that the human race was exactly comparable to Californian redwoods, beetles, cockroaches and the rest – all fighting with one another, as were Father's spermatozoa and Mother's eggs, to be conceived in the first instance, and then, when born, scrambling violently for the porridge bowl, the rice or the mock-turtle soup. A moment's reflection would show us just how untrue this is, as a picture of actual human behaviour at any stage of history: 'each species [he did not except humanity], even where it most abounds, is constantly suffering enormous destruction at some period of life, from enemies or from competitors for the same place or food'.[34] Even the slow breeders, like elephants or human beings, would, by the Malthusian principle, eventually breed so excessively that there would not be enough food for them to eat.

Darwin here was *not* talking about the 'struggle' for eggs to be fertilized by spermatozoa. He was talking about the life of living beings – human included – after they have been born. Many more individuals are born, he maintained, than can possibly survive. This is plainly not true. Even in the darkest days of mid-nineteenth-century England, where child mortality was painfully high, it was still the case that the huge majority of human beings survived childhood. The most extreme and apparently 'Malthusian' phenomena in Darwin's lifetime – the Irish potato famines – were not caused by there being too many Irish people. They were caused by the failure of the potato crops, the forcible restraint of the population who were trying to get their hands on the abundant wheat harvests, which were exported by landlords for profit under the very eyes of the starving, and the Malthusian beliefs of the Liberal government who thought in the first instance that the famine was 'inevitable'.

Darwin's proposition that infant mortality is 'very high', and that only the 'fittest' could survive such a struggle, simply bore no relation to the facts of the case. By speaking of child mortality existing on the same scale as, say, the death of seedlings in a pine forest, he was positing – though not quite openly stating – a child mortality rate which was comparable, say 70 or 80 per cent. So far as I know, none of his first readers pointed this out, and it was only the philosopher David Stove[35] who makes clear the full absurdity of the Darwinian analogy.

You might suppose that when he came to write about the descent of man directly Darwin would have come to revise this view, but this was not the case. The Malthusian doctrine is retained in all its nonsensical plentitude in Chapter Four of *The Descent*, where he says that overpopulation in Britain was prevented only by 'the greater death-rate of infants in the poorer classes'. As for the 'savages', although it would seem that they are 'less prolific than civilised people', they too would easily breed more children than they could feed were it not for the fact that their numbers are 'by some means rigidly kept down'.[36] A sinister phrase here, which is not fully explained.

Long ago, in 1820, William Godwin had objected to Malthus that, if his theory had been true, the English would have become 'a people of nobles', the 'fit' surviving and the feckless unwashed poor dying of starvation. This did not happen. Contemporaries of Darwin as various as Wallace himself, as W. R. Greg – in *Enigmas of Life* – and as Darwin's cousin Francis Galton all addressed themselves to the puzzling fact that the exact opposite of the Darwinian–Malthusian process seemed to be at work. As Greg, for example, a retired mill-owner, put it, 'The careless, squalid, unaspiring Irishman multiplies like rabbits: the frugal, foreseeing, self-respecting, ambitious Scot, stern in his morality, spiritual in his faith, sagacious and disciplined in his intelligence, passes his best years in struggle and celibacy, marries late, and leaves few behind him.'[37]

David Stove was particularly clever at capturing Darwin's very distinctive way of dealing with an argument. Greg, Dalton and Wallace were all lined up in opposition to Darwin's views. If Darwin was correct, and natural selection was busy at work discarding feckless, lazy or unintelligent people, or people of feeble physique, while preserving the muscular and the upright, then surely the 'people of nobles' would already have been in existence. Here is David Stove:

> He discusses at length the relevant writings of Greg, Wallace and Galton . . . Yet he somehow manages to do so without ever once betraying the faintest awareness that what he is dealing with is an *objection* to his theory. Well, that was Darwin's way. He was temperamentally allergic to controversy, and would always, if he could, either ignore or else candidly expound a criticism of his theory, as a substitute for answering it. The result might be, and often was, that his own position became hopelessly unclear, or else clear but inconsistent. But then, he did not mind *that* at all![38]

Readers of *The Descent of Man* might very well find it a disappointment. It sold well. Murray sent Darwin a cheque for £1,470. 'You have produced a book wch. will cause men to prick up what little is left them of *ears* – & to elevate their eyebrows . . . it cannot fail, I think to be much read.'[39] There are, indeed, parts of *The Descent of Man* which still make the eyebrows soar. The book does not, however, quite give its money's worth. Any reader might hope that it would contain an account of *how* the human race is descended from some semi-hominid ape and, if so, which primates are our closest cousins. About these matters Darwin is both brief and vague. He gives a far less lucid account of the human descent from apes than is to be found among his twentieth-century exponents, such as John Maynard Smith in *The Theory of Evolution* (1958, revised 1975 and 1993) and Julian Huxley in *Evolution in Action* (1953). Indeed, I suspect that if we had not read some such book first, *The Descent of Man* would be only semi-intelligible as a fulfilment of our expectations. It is a wide-ranging ramble over the field of evolution in non-human species; and these ramblings are interspersed with reflections upon his fellow human beings which make Darwin, when placed beside even the most reactionary or fascistically inclined readers of the twenty-first century, seem simply monstrous. For here in all its fullness is an exposition of his belief in the survival of the fittest, by which he meant the white races of the globe in preference to the brown-skinned races; and, among the white-skinned races, the supremacy of the British; among the British, the class to which Darwin happened himself to belong; and among that class, the Darwin family, and himself, in particular. The grand end of the struggle for life was to allow the rentier class to live in comfort while lower ranks toiled. 'I feel no remorse from having committed any great sin,' Darwin added in a note to his *Autobiography*, 'but have often and often regretted that I have not done more direct good to my fellow creatures.'[40]

In *The Origin of Species*, he had softened the cruelty of his view of life – of the endless struggle and fight which it involved – by adding, as we observed in an earlier chapter, 'we may console ourselves with the full belief, that the war of nature is not incessant, that no fear is felt, that death is generally prompt, and that the vigorous, the healthy, and the happy survive and multiply'.[41] Anyone who has seen a zebra chased and mauled by a lion or a bird quivering in the mouth of a cat will question the idea that there is no fear in nature. But Darwin's best-sellers were based, perhaps like many best-sellers, on mythic contradiction. The vigorous happy and healthy who bought books, and rode in carriages, and sent their sons to Rugby and Haileybury and Eton were consoled to realize that, although it required much destruction and death to allow them to exist, the inferior and discarded breeds felt no pain as they died out. Darwin had, for the moment at least, left biology behind and ventured into the popular realms of mythology. He had given the Victorians a myth which reflected their own self-made achievements as the summit, not merely of recent history, but of the whole impersonal order of nature. It was a consoling myth, not least because, unlike the myth of Genesis, the Man and the Woman who walked together in the Garden would never be called to account. Adam and Eve walked naked until they heard the voice of Jehovah and hurried to cover themselves with leaves. Mr and Mrs Victorian Bourgeois, already swathed in bonnets, crinoline, frock coats and stove-pipe hats, felt no comparable reproach when they contemplated the inexorable and impersonal force of Nature. God was moral. Natural selection appeared to endorse all their baser selfishness.

★

It is open to question how many Europeans ever took the story of Adam and Eve in the Garden of Eden literally. It was first written down as a myth, and it remained – remains – a myth: that is, a story by which human beings understood their origins, and their place in the world. At some point, perhaps in the seventeenth century, the myth came to be understood as what we should call history, perhaps because this was a period when modern historiography itself knew its beginnings. Clearly, this literalism would need to be corrected, and the story of evolution, as it had come to be explored from the closing decades of the eighteenth century onwards, provided this corrective.

It is probably fair to say that nowadays in the West – with the exception of some biblical literalists – the evolutionary picture, of humanity having emerged from a long ancestry of lower primates, is now the dominant myth. It has replaced Adam and Eve in the Garden. 'Every human being, the first man marching ahead of the endless ape armies of prehistoric times, you and I, no less than our descendants, are the heirs of all the ages, poised on the perilous brink of time,' as an English poet (born 1892) once phrased it.[42]

The most extended metaphysical exploration of the myth is perhaps to be found in the writings of the French Jesuit palaeontologist Pierre Teilhard de Chardin, who in *The Phenomenon of Man* (English edition 1959)[43] saw the entire natural order beginning with the stuff of the universe itself and unfolding on earth over millions of years, from the first manifestations of life, through protozoa, coelenterates and chordates leading to fish, amphibia, reptiles, birds, mammals and eventually primates, to the dawning of human consciousness: to what he called the hominization of the species. Thomas Huxley's grandson Julian Huxley commended the work to English-speaking readers after Teilhard had died. (In his lifetime, his religious superiors forbade him to publish his work.) Above all, as a humanist, Julian Huxley commended the Christian priest's sense that biology commanded us to be aware of our common humanity and our kinship with the natural order.

If Teilhard's name is not so prominent in the public mind as it once was, he and Julian Huxley's generation surely were influential on later generations. The Green Movement and our sense of the interrelatedness of all species on the planet, and of our ecological co-dependency, all surely stem from a profound reading of evolutionary ideas. So too, in the time that the evolution myth has had to penetrate our collective consciousness, has evolution worked to break down racial barriers. Our knowledge that the human race almost certainly derives from hominids who left Africa 1.7 to 1.8 million years ago would, you might suppose, make racism intellectually and emotionally impossible. Human beings, however, are not purely logical, and even scientists are capable of using what they suppose to be scientific facts to bolster what are in fact simple prejudices. Darwin was no exception, and his book *The Descent of Man*, published by John Murray in February 1871 – is a work bristling with the late Victorian high-bourgeois mindset. It represents humanity's climb from hirsute higher primate to frock-coated member of the Athenaeum Club as a mirror of the social climbing which had enabled such ascents as his own grandfather's family from the dreaded depths of trade into the work-free existence of the educated rentiers: 'Man may be excused for feeling some pride, at having risen, through his own exertions, to the very summit of the organic scale; and the fact of his having risen, instead of being aboriginally placed there, may give him hopes for a still higher destiny in the distant future.'[44] In something of the same way, a Victorian bourgeois gentleman, who owed his leisure to ancestors in trade, might hope to marry a daughter into the ranks of the aristocracy. Moreover, Darwin showed the classic British – perhaps it would be more accurate to say English – love of animals, a love which is stronger than his disdain for brown-skinned and 'savage' people. 'For my own part I would as soon be descended from that heroic little monkey, who braved his dreaded enemy in order to save the life of his keeper . . . as from a savage who delights to torture his enemies, offers up bloody sacrifices, practises infanticide without remorse, treats his wives like slaves, knows no decency, and is haunted by the grossest superstitions.'[45] There are moments, however, when he admits to finding it difficult to accept that he is, indeed, part of the same species as the savage, 'even if we compare the mind of one of the lowest savages, who has no words to express any number higher than four, and who uses no abstract terms for the commonest objects or affections, with that of the most highly organised ape'.[46]

The 'savages' whom he had encountered as a young man in Tierra del Fuego flicker across the sixty-year-old Darwin's brain like figures in a nightmare, and he compares their reactions to life, and their levels of intelligence, with his own domestic pets. Once, when a gust of wind moved an open parasol on the lawn at Down House, Darwin's dog, 'a full-grown and very sensible animal', growled and barked, believing the movement to have been caused (Darwin supposed) by 'some strange living agent'. He compares the dog barking with the hostile reaction of the Fuegians when the surgeon on the Beagle went duck-shooting. The 'savages' believed it would bring bad luck: 'much rain, much snow, blow much'.[47] In his *Voyage of the Beagle*, he had already noticed that the language of the Fuegians 'scarcely deserves to be called articulate'.[48] Only a few decades after the *Beagle* had made her short visit to Tierra del Fuego, the parents of Lucas Bridges went to live there as missionaries. His father compiled a dictionary of the Yaghan language, far from complete, which revealed that they had a vocabulary of over 32,000 words. In *The Descent of Man* Darwin was still repeating the claim that the Yaghans ate human flesh and maltreated their children. Bridges, who lived among them for years, encountered no instance of cannibalism, and found them to be intensely devoted to their children. Darwin told readers of *The Descent of Man* that 'we could never discover that the Fuegians believed in what we should call a God'.[49] (Interesting use of the word 'we' incidentally.) Bridges found the Yaghans to be intensely religious – as, *pace* Darwin, every anthropologist has found every 'primitive' people on the face of the planet to be.

The notion that 'savages' do not really have 'languages' in 'our' sense of the word bore no relation to any study of the language of 'savages', nor to any knowledge, so far as we know, of even the simplest facts about what in modern parlance would be called linguistics. His book reflects what Herbert Spencer and other intellectuals of the time supposed about this interesting and complicated branch of study. Because Spencer believed that everything was on an upward spiral of progress, he imagined that language, like plant forms, or the necks of giraffes, was perpetually improving itself. With (so far as one can work out from his *Autobiography*) either no Greek or only the most basic knowledge of Greek, he decided that Homer wrote in a 'primitive' language which was obviously less sophisticated than that of poets in Spencer's own day like Tennyson or William McGonagall. (If this analogy strikes you as unfair, Spencer actually believed that *The Iliad* is more 'primitive' than Victorian three-decker novels, such as the masterpieces of Mrs Henry Wood.) Hebrew, in Spencer's view, is further back in the evolutionary scale than English. 'If we compare, for instance, the Hebrew Scriptures with writings of modern times, a marked difference of aggregation among the groups of words is visible.'[50]

If it is doubtful whether Spencer knew Greek, it is almost certain he knew no Hebrew, and it would be interesting to find any Hebrew-reader who considered, let us say, the Book of Job, with all its parallelism, its ambiguities and its poetic sweep, to be less linguistically sophisticated than Spencer's *First Principles*.

So influential was Spencer, however, and the habits of mind to which he was prone in the 1860s, that he managed to persuade even those who were experts in their academic fields. Professor Max Müller is a case in point; an accomplished linguist who was also much loved by Queen Victoria for his supposed resemblance to the Prince Consort. Darwin quotes with approval in *The Descent of Man* some words of Max Müller, printed in the periodical *Nature* in January 1870. 'A struggle for life is constantly going on amongst the words and grammatical forms in each language. The better, the shorter, the easier forms are constantly gaining the upper hand, and owe their success to their inherent virtue.'[51] Sad words to read for the more mellifluously long-winded of Max Müller's contemporaries such as George Meredith or the young Henry James – if they ever did read them. Quite how a short word could be viewed as more 'virtuous' than a long word is difficult to define, though Spencer, who seems to have invented this particular and distinctive approach to human vocabulary, castigates the languages of North America for their polysyllabic primitiveness. The Ricaree and Pawnee languages were, for example, obviously 'uncivilized', because they persisted in saying *ashakish*, three wasteful syllables, when they could be saving useful time by using the manly English word 'dog'.[52] Darwin, who was a poor linguist – never getting very far with those early German lessons and relying on a children's governess to write his letters to German correspondents – swallowed all this stuff quite seriously and absorbed it into *The Descent of Man*.

The Descent of Man is a longer book than *The Origin of Species*. This is chiefly because, even more than in the previous book, Darwin makes little or no attempt to follow a tightly argued structure, preferring to flood his chapters with information about matters which have, at most, an indirect bearing on his supposed subject.

Most of the second part of the book is not concerned with the human race at all. Chapter Ten, for example, concerns the 'Secondary Sexual Characters of Insects'; Chapter Eleven continues the story, narrowing his theme to *Lepidoptera*, but without the suggestion that human beings derive from butterflies; Chapter Twelve concerns 'Fishes, Amphibians, and Reptiles', and the next four chapters are devoted to birds. Much of it is fascinating, but it scarcely throws any light on the supposed subject of the book, namely the descent of man. About this question, Darwin was disappointingly reticent, concluding one of the lamer chapters, on the manner in which human beings might have developed, with the sentence, 'we do not know whether man is descended from some comparatively small species, like the chimpanzee, or from one as powerful as the gorilla'.[53]

In 1858, the discovery of flint artefacts and animal bones in Brixham Cave – in Devonshire – by the Scottish palaeontologist Hugh Falconer and the English geologist and schoolmaster William Pengelly, established what Roderick Murchison, of the Royal Geographical Society, called 'a great and sudden revolution in modern opinion'[54] about the antiquity of the human race. The artefacts were deemed to be coeval with the bones of the extinct animals. By 1872, Gabriel de Mortillet had organized the surviving artefacts, some found in the Awirs Cave, near Engis in Belgium (1829), some in the Neander Valley in Germany (1856), into a plausible sequence. Almost complete Neanderthal skeletons would be found in Belgium in 1886. A little later, in Java (now Indonesia), the young Dutch doctor Eugène Dubois found a primitive skull cap, and the world was introduced to *Pithecanthropus erectus*. The debate could now be conducted, among informed scientific opinion, about what connection, if any, these discoveries had with later humanity. It used to be supposed, for example, that our ancestors eliminated the Neanderthals, but it is now thought by *some* scholars that the earliest humans coexisted with the Neanderthals for thousands of years.

To none of these scientific questions was Darwin, who wrote before the major palaeontological discoveries, able to give his mind. Nor is there much evidence that his work prepared any useful groundwork for the actual, scientific investigation into 'the descent of man'. What his book of that title does offer, on the other hand, is a series of highly contentious generalizations and assertions about the *metaphysical* nature of humanity.

An example which leaps immediately to mind, in his chapter 'On the Development of the Intellectual and Moral Faculties', is his explanation of the success of the British Empire, and how 'a nation which produced during a lengthened period the greatest number of highly intellectual, energetic, brave, patriotic and benevolent men, would generally prevail over less favoured nations'.[55]

Why, for example, have the British in the nineteenth century been so much more successful, in the previous couple of centuries, than the Spanish – when once the Spaniards were such successful colonists? The answer is that the British were Protestants. Just as the Spanish Empire was getting under way, 'all those given to meditation or culture of the mind' were lured into the celibate priesthood. Meanwhile, the 'freest and boldest men' were all burnt at the stake 'at the rate of a thousand a year'.[56] Small wonder, then, that the British ended up with a more prosperous and extensive empire than the Spaniards.

Darwin perhaps cannot be blamed for believing that the Spanish Inquisition killed a thousand people a year, since this story was aired in various learned journals, reflecting the visceral anti-Catholic prejudice of the Victorian Establishment.[57] Modern historians think that the Inquisition between 1480 and 1520 in fact killed some forty people a year, not a thousand. It is hard to see how this level of judicial murder, however immoral, could be held responsible for reducing the number of adequate Spanish colonists. Many of those killed by the Inquisition were executed not for heresy but for a practice of which Darwin himself hotly disapproved, both on evolutionary and bourgeois principles, namely sodomy. (Darwin was probably too innocent to realize that many of the most successful soldiers, colonial administrators and merchants in the British Empire were in fact homosexual. He lived too late to know the life story either of Roger Casement or Cecil Rhodes, or to read the dream diaries of Robert Baden-Powell.)

The Descent of Man is, moreover, riven with self-contradiction. On the one hand, Darwin clung to the Malthusian view that the secret of life is a struggle for existence. The being which can push, elbow or kill its way to the most nutritious fodder will come out on top. On the other hand, Darwin was a kindly, high-minded liberal, whose father-in-law had been elected to Parliament as one of the first wave of Liberal MPs after the passing of the Reform Bill. His Wedgwood grandfather had campaigned against slavery, and, whatever Darwin's view of 'savages', he never stooped to regarding them as a saleable commodity. So it was that Darwin wanted to conclude his *Descent of Man* on a tone of moral uplift. Here is a paragraph towards the close of his treatise:

> The moral nature of man has reached the highest standard as yet attained, partly through the advancement of reasoning powers and consequently of a just public opinion, but especially through the sympathies being rendered more tender and widely diffused through the effects of habit, example, instruction and reflection. It is not improbable that virtuous tendencies may through long practice be inherited.[58]

A very typical late-Darwin sentence, that. He puts 'It is not improbable' at the front of it, and expects the group of words to sound plausible, even scientific. *If* 'virtuous tendencies' are hereditary, presumably he is speaking in a Lamarckian sense – that we learn certain patterns of behaviour, which could then be passed on, not through instruction to children already born, but actually through inheritance. We'll return in a moment to this, for it turns out to be one of Darwin's most lasting and influential ideas. For the moment, though, let us stay with the moral sense.

> With the more civilised races, the conviction of the existence of an all-seeing Deity has had a potent influence on the advancement of morality. Ultimately man no longer accepts the praise or blame of

> his fellows as his chief guide, though few escape this influence, but his habitual convictions controlled by reason afford him the safest rule. His conscience then becomes his supreme judge and monitor: nevertheless, the first foundation or origin of the moral sense lies in the social instincts, including sympathy; and these instincts no doubt [here we go again, with 'no doubt'!] were primarily gained, as in the case of the lower animals, through natural selection.[59]

In spite of Darwin putting 'no doubt' into that sentence, the difficulty of accepting it as reliable will be immediately apparent. Our minds inevitably go back to the celebrated third chapter of *The Origin of Species*. 'We behold the face of nature bright with gladness, we often see superabundance of food; we do not see, or we forget, that the birds which are idly singing round us mostly live on insects or seeds and are thus constantly destroying life or we forget how largely these songsters, or their eggs, or their nestlings, are destroyed by birds and beasts of prey . . .'[60]

The principle of natural selection, as defined by Darwin – inspired by Malthus – is a struggle for existence. 'It is the doctrine of Malthus applied with manifold force to the whole animal and vegetable kingdoms.'[61] It is not logically possible to believe in this 'Malthusian' principle and at the same time to maintain that natural selection leads to the growth of 'virtuous tendencies', 'social instincts including sympathy'. The process which has now taken Darwin's own surname as an epithet – Darwinian – is a ruthless process. 'In the state of nature', Huxley maintained in *Evolution and Ethics*, '[human] life was a continual free fight.'[62]

David Stove, in a posthumously published essay called 'Darwinism's Dilemma', places the problem very amusingly and forcefully. It is obvious that human life could not be sustained if Darwin's view of it were actually true. Human life simply isn't a free fight and a constant struggle. The existence of hospitals, nurseries, poorhouses in the old days, unemployment relief in our own day, shows that struggle and fight are not everlasting, and that societies do not survive by allowing the weak to go to the wall. Apart from anything else – and this is not an argument Stove advances, it is my own addition – the weakness of human babies and children positively demands levels of unselfishness on the part of their parents which are at variance with the Darwinian 'struggle'.

Stove argued that there were three ways out of Darwinism's dilemma. One route was what he called the Cave Man theory. Maybe we are kindly and civilized now, because we are nineteenth-century liberals; but in 'the state of nature' human beings really were like that. 'Life was a continual free fight, and beyond the limited and temporary relations of the family, the Hobbesian war of each against all was the normal state of existence.'[63] Stove asks what this selfish cave man was doing with a family *at all*. 'Any man who had on his mind, not only his own survival, but that of a wife and child, would be no match for a man not so encumbered.'[64]

If you abandon the Cave Man arguments, you are left with two choices, in Stove's judgement. One is to be a Darwinian Hard Man. No one could ever be a completely consistent Hard Man. Indeed, if you were a serious Hard Man you would probably have to say that not only should there be no hospitals, orphanages, care homes for the old, and so on, but that there are none – there could not be any, if 'the struggle for existence' was still in full storm. Herbert Spencer came close, in his philosophy, to being this Hard Man, developing as he did, in *The Man against the State*, the view that the existence of hospitals and the rest weakens the natural process. In a healthy society, Spencer came to believe, communities, and certainly states, would simply let the weak go to the wall.

The third option, of being a Darwinian Soft Man, is the option favoured by most of those who say they believe in Darwin. Darwin himself, for most of the time, was such a Soft Man. The Soft Man 'believes all the right things'[65] – in welfare programmes, kindliness and so forth – but realizes that in today's world it would seem bizarre to admit to not believing in *The Origin of Species*. Soft Man simply fails to think the thing through.

Many of Darwin's contemporaries were not content with this double-think. And before we pour too much scorn on those older critics of Darwin, such as Sedgwick, who failed to grasp the fundamental fact of evolution, we should perhaps allow them the credit for failing for an honourable reason. The picture of *humanity* presented by Darwin was an ugly and an unrealistic one.

★

The problem, in part, went right back to the difficulty of defining a 'species'. Bertrand Russell was perhaps never more Victorian, indeed almost toppling over into the world of Lewis Carroll, when he wrote, 'If men and animals have a common ancestry, and if men developed by such slow stages that there were creatures which we should not know whether to classify as human or not, the question arises: at what stage in evolution did men, or their semi-human ancestors, begin to be all equal?' All readers who open *The Descent of Man* for the first time would consider this to be one of the central questions which the book would answer; in fact it is silent. Russell went on,

> Would *Pithecanthropus erectus*, if he had been properly educated, have done work as good as Newton's? Would the Piltdown Man have written Shakespeare's poetry if there had been anyone to convict him of poaching? A resolute egalitarian who answers these questions in the affirmative will find himself forced to regard apes as the equal of human beings. And why stop with the apes? I do not see how he is to resist an argument in favour of Votes for Oysters.[66]

The passage loses none of its force, even if it was written before the exposure of the Piltdown Man – the supposed 'Missing Link' – as a forgery in 1953. The problem was put even more succinctly by Sedgwick, who wrote in the margin of *Vestiges of Natural Creation*, 'But why do not monkeys talk?'[67]

For the first generation of British scientists who tried without success to come to terms with the fact of evolution, human kinship with the apes was the sticking point – it was for Sedgwick, for Wilberforce, for Henslow, for many of them. In some ways, however, it is for the subsequent generations, down to our own, who can see that there is, and must be, some sort of kinship with the higher primates, and who can fully see that evolution is a fact, for whom the questions raised about the distinctiveness of human nature seem all the sharper, and Darwin's answers seem even less adequate than they did for believers in the historicity of Adam and Eve. All the Adam and Eve-believers were saying, in effect, was something which seems on the face of it inescapable, that human beings are different from their ape cousins. Even from their Neanderthal cousins (or mates) in the mists of time, for though the Neanderthals are now thought to have produced artefacts and tools, Russell's reductionist joke applies. We cannot envisage a Neanderthal Newton or Shakespeare.

One of the first publicly to dissociate himself from Darwin's view of humanity was Wallace. Although he considered *The Descent* to be 'on the whole wonderful', and 'a marvellous contribution to the history of the development of the forms of life',[68] he could not agree about Darwin's account of human evolution. Wallace was a diffident man, vegetarian, socialist and given to many of the gentler fads of the age. What he was really saying was that the book was admirable when it spoke of fish, butterflies and birds, but unconvincing when it came to trying to write about the subject of the book's title. He expressed his reservations with such courtesy, however, that Darwin was largely appeased. Wallace, though unwavering in his belief in natural selection, was a religious believer. He believed that man, with all his faculties of language, conscience and spiritual sensibility, had not merely evolved but was, in effect, a new creation by the Almighty.

Mivart thought the same. He recollected the moment he had come clean about the matter to Huxley.

> After many painful days . . . I felt it my duty first of all to go straight to Professor Huxley and tell him all my thoughts, feelings and intentions in the matter without the slightest reserve . . . Never before or since have I had a more painful experience than fell to my lot in his room at the School of Mines on that 15th of June, 1869. As soon as I had made my meaning clear, his countenance became transformed as I had never seen it. Yet he looked more sad or surprised than anything else. He was kind and gentle as he said regretfully that nothing so united or severed men as questions such as those I had spoken of.[69]

By the time he had published his own book, *The Genesis of Species*, in 1871, Mivart was generous in acknowledgement that 'we are mainly indebted to the invaluable labours and active brains of Charles Darwin and Alfred Wallace' for addressing the problems of how natural laws produce new species and new characters. 'Nevertheless, important as have been the impulse and direction given by those writers to both our observations and speculations, the solution will not . . . ultimately present that aspect and character with which it has issued from the hands of those writers.'[70]

Darwin in fact incorporated some of Mivart's more devastating corrections into the sixth edition of *The Origin of Species*, but he was not a good loser. Unable to refute Mivart, and lacking the courage to engage in public controversy himself, Darwin unleashed a pamphlet war which even his admiring biographer Janet Browne admits was 'nasty'.[71] An American admirer of Darwin – though it would be truer to call him a worshipper, since on a visit to Down House the young mathematician-philosopher described meeting the great man as an almost religious experience – was Chauncey Wright, then in his early twenties. Appalled by Mivart's temerity in daring to question the Master, Chauncey Wright had written a lengthy critique of Mivart's *Genesis of Species*, and Darwin paid for this to be reprinted in Britain as a pamphlet. Meanwhile, 'Darwin's Bulldog', Huxley, was trundled into the lists to attack Mivart. Darwin had considered Mivart's article denouncing *The Descent of Man* in the *Quarterly Review* as 'most cutting'. Huxley obliged with some cuts in riposte, though he returned neat rapier thrusts with blows from a blunt hatchet.

Mivart's review had been devastating but polite. Apart from itemizing specific errors in Darwin's book, he committed (in the eyes of Darwin-worshippers) two cardinal sins. One was he attributed the theory of natural selection to Wallace, as well as to Darwin, knowing of course that while Wallace accepted the theory as an account of how plants and animals have evolved, he did not see it as an adequate theory to explain the existence of human consciousness, the moral sense, the aesthetic sense and so on. Secondly, Mivart drew attention to Darwin's use of phrases such as 'we have every reason to believe', 'no doubt', 'unless we wilfully close our eyes' to introduce propositions which by no means command assent.

Already, among the Darwin-worshippers, it had become bad form to attribute too much of the glory to Wallace for his simultaneous proposal of the natural selection theory. Wallace, in their version, was analogous to Trotsky in the story of the Russian Revolution as seen through Leninist eyes – someone whom it would eventually be possible to airbrush out of the story. Huxley was in St Andrews on holiday when he read Mivart's detailed review, and decided to pen a reply. In the course of a detailed scientific critique of Darwin, Mivart made one very small mistake which covers nine lines of his entire essay. He attributed to a Spanish Jesuit called Suárez the view that St Augustine had been right: that nature had evolved, that the divine creation of humanity had been a matter of breathing life into what Huxley called 'the manlike animal's nostrils'. In fact Suárez had rejected this view and believed in something more like a modern-day creationist – namely that humanity was created entire, body and soul, as it were in one creative act.[72]

Mivart had blundered in attributing a view to a Renaissance-era Jesuit which he had not in fact held. That was all. It was not wrong, however, to say that a belief in God and a belief in evolution were quite compatible. Thousands of his contemporaries, including Charles Kingsley, Wallace himself, the mathematician Baden Powell (father of the Scouts' founder)[73] – even the diehard Dr Pusey in Oxford – were evidence of this. Yet for Huxley's son Leonard this was 'one of the most deadly [sentences] in the history of controversy'.[74] Mivart mildly stated, 'It was not without surprise that I learned that my one unpardonable sin – the one great offence disqualifying me for being "a loyal solider of science" – was my attempt to show that there is no real antagonism between the Christian revelation and evolution.' For this, as has been stated, he was excommunicated from his Church. Darwin and his friends also vindictively blackballed him from membership of the Athenaeum Club. Hooker read Huxley's malicious attack on Mivart and wrote to Darwin, 'What a wonderful Essayist he is, and incomparable critic and defender of the faithful.' Darwin himself was beside himself with malicious glee. 'I laughed over Mivart's soul till my stomach contracted into a ball, but that is a horrid sensation which you will not know . . . It quite delights me that you are going to some extent to . . . attack Mivart.'[75]

The Descent of Man and Huxley's assaults on Mivart were demonstrations that Darwinism had now ceased to be a merely scientific theory and was the new religion. Huxley was, as Hooker described him, 'the defender of the faithful'. Moreover, when the Darwinians spoke about eliminating the opposition, they were not dealing just in metaphor. *The Descent* had ended with a nod towards Darwin's cousin Francis Galton (1822–1911), the coiner of the term 'eugenics', whose book *Hereditary Genius* had been published in 1869. Seeds of the eugenic idea are planted at the end of *The Descent*. Just as man rose from the lower animals, so the class to which the Darwinians belonged had risen from a lower class, and they were determined to kick away the ladder from beneath them. By 'selection' man can 'do something not only for the bodily constitution and frame of his offspring, but for their intellectual and moral qualities. Both sexes ought to refrain from marriage if they are in any marked degree inferior in body or mind . . .'[76]

As has been justly said, the majority of Darwin's human contemporaries in Britain were not engaged in a competition for life, but in a competition for early death through alcohol.[77] Before they achieved that end, however, those 'inferior in body or mind' tended to breed in greater numbers than the class to which Darwin himself belonged. Had Darwin's application of the theory of natural selection to sociology been true, it would have been his own class – strong in intellect, talent and enterprise – which eventually came to outnumber the feckless, the weak and the stupid. Paradoxically, it was the very fact that the Darwinian theory of humanity was untrue which would eventually cause Darwinians to embrace eugenics. Precisely because the 'unfit' were outbreeding the 'fit', it was thought necessary to help 'natural' selection a little by artificially eliminating the 'unfit'. Darwin, in the second edition of *The Descent of Man*, would spell out the need for this.

> With savages, the weak in body or mind are soon eliminated; and those that survive commonly exhibit a vigorous state of health. We civilized men, on the other hand, do our utmost to check the process of elimination; we build asylums for the imbecile, the maimed and the sick; we institute poor laws; and our medical men exert their utmost skill to save the life of every one to the last moment. There is reason to believe that vaccination has preserved thousands, who from a weak constitution would formerly have succumbed to small-pox. Thus the weak members of civilized societies propagate their kind. No one who has attended to the breeding of domestic

animals will doubt this must be highly injurious to the race of man . . . excepting in the case of man himself, hardly anyone is so ignorant as to allow his worst animals to breed.[78]

★

In 1871, the first of Darwin's children was married. Etty was their only daughter to take this step, Bessy remaining a spinster. Richard Litchfield, ten years older than his wife, was a lawyer who worked for the Ecclesiastical Commissioners. Janet Browne makes the unavoidable, and surely just, supposition that 'he may have seemed like another version of her father'[79] – though Litchfield, unlike the recluse of Down, was much given to good works at the Working Men's College in London. Etty met him in June, had become engaged in July and was married in August. Litchfield's friend Arthur Munby, whose taste was for working-class women, and who was secretly married to a maid of all work called Hannah Cullwick, was dismayed by Litchfield's wedding which, he said, 'takes away the last but one of my unmarried friends'. Of the twenty-seven-year-old bride, Munby told his diary, 'I have yet dwelt all day on his words, after speaking of her intellect and power – "She does not believe in a personal God". How wide we have to stretch our sympathies now a days!' Munby noted that Etty was 'subdued and self-possessed', finding something in her features which was 'not unlike the photographs of her father'.[80] There is a clue here, incidentally, to the very modern nature of Darwin's fame. In that, the first age of widespread photography, it was possible to be a very famous hermit, and your face to be quite familiar to millions.

Etty and Litchfield, like Francis Galton, and like her brother Leonard, the most fervent eugenicist of the Darwin siblings, would be childless.

Tiny and cantankerous, Etty would lead a life which, by the standards of women of a later age, was eventless almost beyond imagining. In the twentieth century, she once told her niece that, when her maid Janet was away for a few days, '*I am very busy answering my own bell.*'[81] She never posted a letter for herself, never sewed on a button and never travelled anywhere without a maid. Litchfield, exuberantly bearded, usually clad in black coat, striped trousers and egg-shaped waistcoat, was a little taller than she, but not much. It was the survival of the unfittest. Both of them enjoyed ill health, though Etty, a true Darwin, easily beat her husband at the hypochondriacal game. After her 'low fever' aged thirteen, she had been encouraged by the doctor to take breakfast in bed. She never got up for breakfast again for the rest of her life. Her long-suffering maids were asked to place a silk handkerchief over her left foot as she lay in bed, because this, of the two feet, was the one which got colder. If anyone else in the house was suffering from a cold, Etty would wear a sort of gas mask of her own invention, a wire kitchen-strainer stuffed with antiseptic cotton-wool and tied on like a snout with elastic round her ears.

The Litchfields resided in Kensington Square, near the houses of Sir Hubert Parry, John Stuart Mill and Lady Ritchie (Thackeray's daughter). The house was decorated with Morris wallpapers and curtains, blue china, peacock feathers and Arundel prints.[82]

Etty's siblings had lives slightly more crowded with incident, but not dramatically fruitful when viewed from a genetic viewpoint. William, the eldest, who became a banker, married but was childless. Leonard, who succeeded to the presidency of the Eugenic Education Society after the demise of his cousin Francis Galton in 1911, married but also had no children. Francis, who became a botany don at Cambridge after his father's death released him from secretarial and lavatorial-assistant duties at Down House, married and had one daughter, the poet Frances Cornford. George, the Cambridge maths don, had four children, one of whom was Gwen Raverat; and Horace, founder of the Cambridge Scientific Instrument Company, and Mayor of Cambridge in 1896–7, had two daughters, and one son killed in action in 1915.

Of Darwin's nine children, therefore, only three had issue. Indeed, one of the reasons that the Victorian upper-middle class died out was that it did not breed with anything like the fecund abandon of the aristocracy or the lower orders. Samuel Butler's exposition of the sheer hellishness of family life in *The Way of All Flesh* found an echo in many young readers' bosoms. Rose Macaulay and Ivy Compton-Burnett were not unusual in their generation, in being born into enormous upper-middle-class families not one of which bred. Macaulay (1881–1958) was one of seven. Compton-Burnett (1884–1969) was the seventh of twelve children, two of whom died in a suicide pact, one of whom was killed on the Somme. Although Darwinism, if true, should have ensured, by a process of natural selection, the triumph of Darwin's social class and the natural extermination of the hordes of Pooters crowding the new-built suburbs, the vast army of improvident toilers, who filled their slum-dwellings with offspring they could scarcely afford to feed, this did not happen. There were not enough intelligent rich people in proportion to the others whose rowdy lives must be a source of distress to those upper-middle-class people who were kind enough to observe them. Clearly, the more public-spirited members of Darwin's class did their best to extend education to the masses, but it would obviously be better if such troubling individuals did not exist. Although George Darwin's academic expertise lay in the fields of geodesy and of mathematics (he became Plumian Professor of Astronomy at Cambridge in 1883), he shared his cousin Francis Galton's enthusiasm for what would eventually be termed eugenics. In the years after he coined this term (1883) Galton would campaign politically for tax breaks to encourage intelligent people to have large families and to sterilize the 'unfit'.[83] Long before this campaign got under way George Darwin, developing the ideas of his father's *Descent of Man*, had written a proposal 'on beneficial restrictions to liberty of marriage' in 1873.[84] The article appeared in the *Contemporary Review* and was a classic exposition of the 'eugenic' idea, viz. that those deemed by the Darwins to be defective should be forbidden to breed. In July 1874, an anonymous essay appeared in the *Quarterly Review* discussing works on primitive man by John Lubbock and Edward Burnett Tylor. It included an attack on George Darwin's paper as 'speaking in an approving strain . . . of the encouragement of vice in order to check population'. The anonymous author was St George Mivart. Today, 'liberal' opinion in the West deplores eugenics, not least because of the enthusiasm with which it was adopted in Germany in the period 1933–45. It would only be among conservative Christians, however, that you would be likely to find those who believed contraception or medically advised abortion to be immoral. Mivart was, it is true, a Roman Catholic, albeit a convert who had been excommunicated for his belief in evolution. In 1873–4, however, he would probably have been in the huge majority of Victorians in believing contraception to be morally questionable and abortion positively criminal. George Darwin had not even ventured into the notion, which was a commonplace in the entourage of Bertrand

Russell (heterosexual), Lytton Strachey (gay) and the Bloomsbury Set in the 1920s, that homosexuality was another good way of limiting the population explosion. (The Stracheys were a typical example of the upper-middle-class tendency to abandon breeding on the Victorian scale – Lytton was the eighth child of ten, few of whom bred.)

When Charles Darwin read the attack upon his son in the *Quarterly*, he was enraged, and advised him 'that it wd be a good plan to lay the case before an *eminent* Counsel, not necessarily for Prosecution of the author, but that he shd compare the Review with your Article'.[85] George had pointed out (correctly) that in the laws of the German Commune 'Prostitution was not merely tolerated but was secretly promoted as a check to overpopulation, as in Japan at the present day.' (Charles Darwin had darkly alluded to this supposed vice in Japanese society in *The Descent*.) They did not commence legal proceedings, either against Mivart or against the *Quarterly*, which, as Charles Darwin reminded his son, was published by his own publisher, Murray. They considered sacking Murray, but George worried that it would be 'a great annoyance to go to a new publisher'.[86] The spat, however, had awoken uncomfortable feelings. Ever since, in his first notebooks M and N, Darwin had begun to meditate upon the mystery of human descent, he had been aware that by embracing Malthusianism he was overturning conventional Christian morality. Human beings, rather than seen as made in God's image and likeness, were reduced to mere 'surplus population'. It has been rightly said that 'Darwin did not first formulate natural selection and then apply it to human beings: he drew the theory directly from contemporary (and ideologically loaded) assessments of human behaviour and afterwards concealed its implications for over three decades, until *The Descent of Man*.'[87] Eugenics was the natural consequence of Darwin's Malthusianism. Sterilization programmes, state-encouraged prostitution, contraception and abortion were the logical rational measures by which to fulfil his idea of society. He was, of course, unable fully to admit this to himself, still less to 'come out', when an old man, as an advocate of such policies. Safer to consult a lawyer and see if Mivart could not be bullied into silence.

Darwin, in the conclusion to *The Descent of Man*, had endorsed Galton's views that, if the prudent avoid marriage while the reckless marry, the inferior members tend to supplant the better members of society.[88] It was not long before this translated itself in Britain into a fear of 'race suicide'. Sidney Webb, for instance, one of the leading leftist social engineers of his generation, founding father of the *New Statesman and Nation* and of the London School of Economics and one of those who drafted the constitution of the Labour Party, feared that Britain was 'gradually falling to the Irish and the Jews', owing to their high rate of reproduction. Webb, in common with H. G. Wells, George Bernard Shaw and Winston Churchill, shared the view that middle-class women would be 'shirking in their duty' if they did not have families which outnumbered those of the feckless poor. Opponents of eugenics are able to see that, not only were their views morally repellent, they were based upon false science, mistakes which derived directly from Darwinism.[89] Karl Pearson (1857–1936), who became the first Professor of Eugenics at the London School of Economics in 1904, decried Mendelian genetics, advocating Galton's theory of inheritance through continuous blending and variation, 'a view', as Lucy Bland and Lesley Hall accurately remark, 'which became increasingly untenable with the rise of genetics in the interwar years'.[90] It was not long before Professor Pearson was advising two Royal Commissions, one on the Care and Control of the Feeble-Minded (1904–8) and another on Venereal Diseases (1913–16).[91] It is not difficult to see a direct correlation between these noxious ideas and the social programmes of the Third Reich, culminating in the Holocaust. Michel Foucault remarks on the 'mythical concern with protecting the purity of the blood and ensuring the triumph of the race'.[92] Less than thirty years would elapse between boring little Sidney Webb's expressing the fear that his country would fall to the Irish and the Jews and another European country, Germany, enacting the Reich Citizenship Law, the Blood Protection Law, the Marital Health Law and the Nuremberg Laws for racial segregation, all based on bogus Victorian science, much of which had started life in the gentle setting of Darwin's study at Down House.

Darwin, almost from the first, was wildly popular in Germany. One of his most enthusiastic adepts was Ernst Haeckel (1834–1919), who was quick, having embraced the doctrines of the Master, to postulate an evolutionary tree beginning with 'ape men', splitting to a 'primitive' branch, which included 'Negroes', 'Kaffirs', 'Hottentots' and 'Papuas'; then to the 'higher' branch – Finns, Magyars, Japanese and Chinese; and, highest of all, the Indo-Germanic races, the Aryans. Darwin treasured the support and penfriendship of Haeckel, who had received his medical degree from the University of Berlin in 1857 before becoming a lecturer in comparative anatomy at the University of Jena in 1861 and Professor of Zoology, and Director at the German Zoological Institute. In September 1874 Darwin exclaimed to Haeckel, 'What grand success your books have had, & I am delighted to see your name continually quoted in all parts of the world with high approbation.'[93] From Jena, Haeckel could write back with pride and enthusiasm, 'In Deutschland hat der Darwinismus im letzen Jahre sehr grosse Fortschritte gemacht, besonders unter den *Philosophen*'[94] ('Darwinism has made great progress in Germany in the last year, especially among *philosophers*.')

In other words, the Germans, rightly, saw Darwinism less as a purely scientific hypothesis and more as a world-outlook, a way of being modern, a way of reordering society in the wake of rapid industrialization, urban growth, family size and structure, and sexuality. Racial hygiene was a key ingredient in this from early days. (The phrase was coined not by Haeckel but by Alfred Ploetz, but it exactly echoed Haeckel's views.)[95]

Haeckel read Darwin in the German translation of H. G. Bronn. Although considered in his own time to be a great scientist, it is probably fair to say that, today, 'historians of science find him of more interest than biologists do'.[96] Haeckel was not a proto-Nazi. In fact, the actual Nazis dismissed his philosophy as 'a direct violation of Nazi *völkisch* commitments'.[97] Nevertheless, it remains hard to read his reflections on race without the toes curling.

Darwin believed that Haeckel was 'one of the few who clearly understands Natural Selection'.[98] Haeckel had come to England in 1866, at the invitation of Huxley, to attend the British Association for the Advancement of Science, convening in that year at Nottingham. While in England, he took the opportunity of visiting Lyell in London, and, of course, he made the Haj to Down House.

> As the coach pulled up to Darwin's ivy-covered country house, shaded by elms, out of the shadows of the vine-covered entrance came the

> great scientist himself to meet me. He had a tall, worthy form with the broad shoulders of Atlas, who carries a world of thought. He had a Jupiter-like forehead, high and broadly domed, similar to Goethe's, and with deep furrows from the habit of mental work. His eyes were the friendliest and kindest, beshadowed by the roof of a protruding brow. His sensitive mouth was surrounded by a great silver-white full beard. The welcoming warm expression of his whole face, the quiet and soft voice, the slow and thoughtful speech, the natural and open flow of ideas in conversation – all of this captured my whole heart during the first hours of our discussion. It was similar to the way his great book on first reading had earlier conquered my understanding by storm. I believed I had before me the kind of noble worldly wisdom of the Greek ancients, that of a Socrates or an Aristotle.[99]

Etty took a more satirical view of the visit, noting that the shy Jena professor, 'one of Papa's most thoroughgoing disciples', was 'so agitated he forgot all the little English he knew, & he & Papa shook hands repeatedly . . . Talking of dining in London – "I like a good bit of flesh at a restoration."'[100]

For Haeckel, Darwin's theory was something much more than a scientific hypothesis. It was a vision of life itself. Haeckel had lost a beloved wife, Anna. Even before this tragedy, he had written, in 1863, 'I am now convinced that a great future lies before this theory, and that it will slowly but surely loose us from the bonds of a great and far-reaching prejudice. For this reason I shall dedicate my whole life and efforts to it.' When Anna died as a young woman, of what appears to have been a combination of pleurisy and appendicitis, Haeckel embraced the pitilessly tragic quality of the Darwinian theory as a spiritual credo, colouring it with the emotional tones of German Romanticism. Darwin in fact bore no resemblance to Goethe, but it was revealing that Haeckel, on his visit to Down, believed that he did. For Haeckel, eternal life meant life on earth, 'nature, an eternal life, becoming and movement'. Darwin's evolution and Goethe's becoming (*Werdend*) fused in Haeckel's passionate soul. In the *Evolution of Man* he stated, 'All my readers know of the very important scientific movement which Charles Darwin caused fifteen years ago, by this book *The Origin of Species*. The most important direct consequence of this work, which marks a fresh epoch, has been to cause new inquiries to be made into the origin of the human race, which have proved the natural evolution of man through lower animal forms.'[101]

In his defence of Darwin from his lecture podium at Jena, Haeckel – even before *The Descent of Man* had been published – saw Darwinism as a law of progress (*Gesetz von Fortschritts*).[102] He believed one saw the processes of natural selection and the struggle of existence at work in societies and in human history. On one level, those scientists of our own day who believe that Haeckel misinterpreted the Master in certain scientific details are only addressing the scarcely relevant question of what is factually true. Haeckel, of course, believed that Darwin's theory would be borne out by the facts. Inspired by Haeckel, Eugène Dubois (1858–1920) would set out to Java and discover what he believed to be the Missing Link, *Pithecanthropus erectus*, the earliest ape-man. The authenticity or otherwise of the bones discovered by Dubois (some scientists claimed the Missing Link's skull cap had come from an ape and the femur from a human being)[103] was beside the point – of no more capacity to dislodge the Darwinian faith than a sceptical dating of the Turin Shroud or a Relic of the True Cross. Haeckel, more eloquent and more Romantic than many of his generation, had found in Darwinism a new religion. Richard Goldschmidt, the great Berlin geneticist who migrated from Germany to Berkeley in the 1930s, recalled reading Haeckel as an adolescent.

> I found Haeckel's history of creation one day and read it with burning eyes and soul. It seemed that all problems of heaven and earth were solved simply and convincingly; there was an answer to every question which troubled the young mind. Evolution was the key to everything and could replace all the beliefs and creeds which one was discarding. There were no creation, no God, no heaven and hell, only evolution and the wonderful law of recapitulation which demonstrated the fact of evolution to the most stubborn believer in creation.[104]

(The law of recapitulation, now largely disbelieved, was Haeckel's refinement of Darwinism: namely that an embryo, through the process from fertilization to gestation, goes through all the stages of evolution of its remote ancestors.)

Darwin, too, in his quiet way, saw his theory as an alternative religion which put paid to Christianity. His natural diffidence, however, would have made him shrink from making any such public declarations on the subject as Haeckel's. Speaking to the Society of German Natural Scientists and Physicians shortly after Darwin's death in 1882, Haeckel saw Darwin as in essence an anti-religious figure who had bravely hacked through the jungle of religiously overgrown biology following in the footsteps of great Germans – Lessing, Herder, Goethe and Kant. Just as Martin Luther who, 'with a mighty hand tore asunder the web of lies by the world-dominating papacy, so in our day, Charles Darwin, with comparable overpowering might, has destroyed the ruling error doctrines of the mystical creation dogma and through his reform of developmental theory has elevated the whole sensibility, thought, and will of mankind on to a higher plane'.[105]

He went on to quote a private letter which Darwin had sent to a young Russian nobleman, Nicolai Aleksandrovich Mengden, who had written to Darwin about the theological difficulties aroused in his mind by the evolutionary theory. Darwin had replied, 'I am much engaged, an old man, and out of health, and I cannot spare time to answer your questions fully – nor indeed can they be answered. Science has nothing to do with Christ . . . For myself, I do not believe that there ever has been any revelation.'[106]

There are one or two other examples of Darwin's candour in this matter, as he grew older, in letters to total strangers. With Emma, though not with his children and his intimates, he continued to fudge the issue – anything for a quiet life. Yet, though Haeckel's full-blown Romanticism was not Darwin's style, in *belief*, the two men were as one.

15

Immense Generalizations

WHEN LORD SALISBURY became Chancellor of Oxford University, in succession to the 14th Earl of Derby, it was in his gift to nominate the honorary doctors for the first year of his office (1870). He nominated Darwin, Huxley and John Tyndall, superintendent of the Royal Institution and one of the great men of science of his day. Dr E. B. Pusey, the noted Arabist and diehard chieftain of the High Church party, objected, but Salisbury was firm. He shared Pusey's religious convictions, but believed that 'it is not desirable that the Church and those who represent her should condemn scientific speculations when they are only inferentially and not avowedly hostile to religion'.[1]

In the end, in deference to the Puseyite scruples, Huxley's name was removed from the list of honorands, but Salisbury, as so often, was being quite cunning. No doubt he genuinely admired Darwin's industry and range; but in conferring upon him an honorary doctorate (of Civil Law) he was publicly declaring that the Church of England could withstand the winds of Darwinism very easily. Besides, Darwin was by way of being a kinsman of Salisbury's wife – Georgina Drewe, daughter of Caroline Allen of Cresselly, having married Sir Edward Alderson; among their issue was a daughter, Georgina, later Lady Salisbury. Oxford – which since the 1860 debate between Huxley and Soapy Sam had been synonymous in some minds with anti-scientific reactionism – could now be seen as bestowing blessings upon scientific research. Darwin, for his part, was happy to be a member of the Establishment. As it happened, the degree was a purely notional honour. It was impossible to award honorary degrees *in absentia*; and Darwin was, of course, too ill to make the journey to Oxford to take part in a ceremony.[2]

Meanwhile, with *The Descent of Man* complete and published,

Darwin was at work on his final and most popular, evolutionary text, *The Expression of the Emotions in Man and Animals*.

Work had progressed slowly. The burden of Darwin's correspondence was enormous. (The volumes of the Cambridge edition covering the years 1871–3, when he was finishing *Expression* and revising *Descent*, occupy over 2,400 printed pages.) He was more than usually tired, and beset by his symptoms. Aged only sixty-two he could write, 'I feel as old as Methuselah.'[3]

Darwin was one of the great accumulators of evidence, collector of examples. If he belonged to a category of literary history it would be with the *Etymologies* of Isidore of Seville, or with Burton's *Anatomy of Melancholy*, those bulging compilations of information, quotation and cross-reference. The best bits of Darwin's book on the expression of the emotions consist of anecdotage. One can imagine, when reading them, why his friends and family record him as a charming man, within his circle of intimates, and a good talker.

> Horses scratch themselves by nibbling those parts of their bodies which they can reach with their teeth; but more commonly one horse shows another where he wants to be scratched, and they then nibble each other. A friend whose attention I called to the subject, observed that when he rubbed his horse's neck, the animal protruded his head, uncovered his teeth, and moved his jaws, exactly as if he were nibbling another horse's neck, for he could never have nibbled his own neck . . .[4]

'Let us now suppose that the dog suddenly discovers that the man whom he is approaching, is not a stranger, but his master; and let it be observed how completely and instantaneously his whole bearing is reversed . . .'[5] (He wrote more tenderly about dogs than about any other species.) Some of the observations seem almost seventeenth century in their quirkiness: 'The whole body of the hippopotamus . . . was covered with red-coloured perspiration whilst giving birth to her young. So it is with extreme fear; the same veterinary has often seen horses sweating from this cause; as has Mr Bartlett with the rhinoceros; and with man it is a well-known symptom.'[6]

The book is both a wonderful treasure-house of detail – barn owls swelling their plumage and hissing in anger, lizards extending their throat pouches for the same reason and erecting their dorsal crest, expressions they also demonstrate during courtship, snakes and chameleons inflating themselves when irritated, cats arching their backs in terror – and a work of ideology, for he wished to remind his readers every few pages that all species, including the human race, were loosely related. The human passages are richly enjoyable, particularly the evidence of the insane, supplied by the psychiatrist Dr James Crichton-Browne, who ran the Wakefield Asylum in Yorkshire. 'I have been making immense use almost every day of your manuscript,' Darwin wrote to Crichton-Browne. He used the psychiatrist's written evidence, but tenderly eschewed the photographic, with the exception of a woodcut of a madwoman's hair bristling like that of a frightened animal.

The effect of reading *The Expression* however, cannot be what Darwin intended. In the passages which concern, for example, weeping or blushing, he can find no examples in the animal kingdom which match the two activities. (We may take the weeping horses, mourning Patroclus in *The Iliad*, to be a fiction.) There is the standard expression of an Englishman's superiority to the 'savage' peoples of the globe, but he does not find a cockatoo or a chimpanzee who weeps.

> A New Zealand chief 'cried like a child because the sailors spoilt his favourite cloak by powdering it with flour'. I saw in Tierra del Fuego a native who had lately lost a brother, and who alternately cried with hysterical violence, and laughed heartily at anything which amused him. With the civilized nations of Europe there is also much difference in the frequency of weeping. Englishmen rarely cry except under the pressure of acutest grief; whereas in some parts of the Continent the men shed tears much more readily and freely.[7]

The empathetic Darwin, who wept so copiously over the death of Annie, or who bathed with tears a letter from his wife expressing her religious belief, has turned into a stiff-upper-lip Victorian. Likewise, as he disclosed to his *Autobiography*, the boy who had loved reading Byron, Thomson's *Seasons* and Scott had developed into a man who had 'wholly lost . . . all pleasure from poetry of any kind'.[8]

Blushing, he wrote, in one of the least convincing pieces of speculation, towards the conclusion of his book, 'originated at a very late period in the long line of our descent'.[9] It is interesting, and surely wrong, when he identifies the sole, or chief, cause of blushing in human beings as anxiety about their appearance.

Those who opposed Darwin's central contentions must often seem to later generations extraordinarily obtuse, their identification of the fixedness of species, as defined by Linnaeus, with ancient religious orthodoxy simply bizarre. Their questioning of Darwin's view of humanity, however, does not always seem so strange.

In the tenth edition of Lyell's *Principles of Geology*, the geologist announced – it was in 1869 – his total conversion to Wallace and Darwin's theory of evolution by natural selection. It is a good marker of how far, and how quickly, things had come. Only ten years before, even Huxley had been unconvinced by any idea of mutation of species. Lyell was moving in the direction of accepting it, but having seen the flaws in Lamarck he had been slow to accept the new theory; and while accepting the phenomenon of *evolution*, Lyell was slow to accept Darwin's theory about how it happened.

Where opinions began to divide – once evolution itself became accepted in the scientific world – was over the question not only of how human beings originated, but of what they were.

Adam Sedgwick's scribble in the margin of *Vestiges of Creation* – 'But why do not monkeys talk?' – has never been satisfactorily answered by Darwinians, though Stephen Pinker has made formidably impressive attempts to do so. Noam Chomsky, whom Pinker attempted to answer in his work, remains a convincing analyst to many when he posits the apparent phenomenon, unique to human beings, of the 'Language Acquisition Device'. The Darwinian attempt to link this to the 'languages' of dolphins, parrots or chimpanzees only emphasizes the gulf between the signs animals give to one another by their 'languages' and the sheer complexity of the 'Language Acquisition Device', which appears to suggest that babies are 'hard-wired' to build up an understanding of high grammatical complexity. In this, as in their tears and, later in life, their blushes, they are simply different from other primates. Nor does there exist any evidence of human languages 'evolving' from 'simple' to 'complex' in the manner supposed by Herbert Spencer (and swallowed whole by Darwin in *The Descent*).

No wonder, then, that there were so many Victorians, such as Wallace, who resisted the easy way in which Darwin had begun to write as if it were a mere given that humanity's complex linguistic, moral and social capacities had simply 'evolved' from some earlier form of ape-life.

'Here then, we see the true grandeur and dignity of man . . . he is, indeed, a being apart, since he is not influenced by the great laws which irresistibly modify all other organic beings.'[10] When Darwin read these words, in Wallace's review of Lyell's *Principles of Geology* (the tenth edition) he was horrified. 'No!!!' he wrote in the margin, underlining it three times. 'I hope you have not murdered too completely your own & my child . . . If you had not told me, I should have thought that [your remarks] had been added by someone else . . . I differ grievously from you, and I am very sorry for it.'[11] Wallace was led, partly because he could never resist 'an uphill fight in an unpopular cause',[12] into the investigation of, and acceptance of, spiritualism.

Spiritualism, as a phenomenon of the nineteenth century, was a natural reaction to the materialist-rationalist position of Benthamite economists, scientific atheists and others. Of course, shamans and mediums had been summoning up the dead since at least the time when the Witch of Endor called back Samuel in the Bible. The Victorian mania for 'spiritualism', and the number of highly successful 'society mediums', such as Daniel Dunglas Home – the model for Browning's 'Mr Sludge the Medium' – was, in a way, very unspiritual. Even its manifestations, such as 'ectoplasm' flying out of the heads of psychics when they were in a trance, or the tapping of tables, or the vibrations of strange voices from the Other Side – had a physical aspect. The adepts of the cult wished to demonstrate, as you might demonstrate a scientific fact, that human beings were immortal spirits.

Those, chiefly but not entirely from Cambridge, who first convened the British Society for Psychical Research in February 1882 included Henry Sidgwick, Frederic Myers, Edmund Gurney, Lord Tennyson, Leslie Stephen, Mark Twain and Alfred Russel Wallace.[13] Their aim was to collect evidence, to investigate the plausibility or otherwise of the phenomena of spiritualism – seances, spirit-writing and so forth. Wallace had been converted to a belief in spiritualism by Mrs Marshall, who had tapped out a message to him from his dead brother Herbert, which he considered incontrovertibly authentic. The Society still exists today, and has accumulated an enormous archive of 'evidence', but nothing which would be considered a proof to the unconverted.

Poor Mrs Marshall, incidentally, unhappily married to Frederick Myers's cousin Walter, tried to cut her throat and then, when she botched the task, walked bleeding into the shallows of a Swiss lake, and then out, further and further.[14]

Myers, a sceptical would-be believer – it was he who had that famous conversation with George Eliot in the gardens of Trinity Cambridge about God, immortality, duty – was a friend of Darwin's brother-in-law Hensleigh Wedgwood. In the decade before the SPR was actually established, there were many attempts either to debunk or to investigate the spiritualist fad.

Ras Darwin, Charles's brother, arranged for a seance in his house in 1874 conducted by a paid spirit medium. Present were Darwin and Emma, Etty and Richard Litchfield, George Lewes and George Eliot, Hensleigh and Fanny Wedgwood, Frederic Myers, Francis Galton and Ras's friend Mrs Bowen. Lewes and Ras spent a couple of hours beforehand checking under tables and chairs for the apparatus which might be used to move the furniture or create the effect of a psychic phenomenon. When Charles Williams began, however, he insisted upon the room being completely dark, upon which the Leweses and the Charles Darwins left. Darwin's son George, however, remained, as did the others. 'We had grand fun one afternoon,' Darwin wrote,

> for George hired a medium, who made the chairs, a flute, a bell, a candlestick, and fiery points jump about in my brother's dining room, in a manner that astounded every one, and took away all their breaths. It was in the dark, but George and Hensleigh Wedgwood held the medium's hands and feet on both sides all the time. I found it so hot and tiring that I went away before all these astounding miracles, or jugglery, took place. How the man could possibly do what was done passes my understanding. I came downstairs and saw all the chairs, etc. on the table, which had been lifted over the heads of those sitting around it. The Lord have mercy on us all if we have to believe in such rubbish.[15]

Ras – whimsical, more charming than Charles, and far more social – became no less social with the years. He liked to carry a small kitten in his waistcoat in case he met a child who enjoyed playing with it – presumably, only on occasions when kittens could easily be found. His nephew Frank Darwin was another cat lover, and would send up kittens in a basket by train to Uncle Ras. Frank married, in 1874, Amy Ruck, and they set up house in a village cottage at Downe, so that the young man could act as his father's secretary. Ras, with his round-rimmed black hat and black cloak, could have been a mad clergyman, even down to the detail of his entirely sharing his brother Charles's lack of belief. Emma's life of prayer and Bible study had small effect upon her offspring. Gwen Raverat, daughter of George Darwin, recalled how her sister Frances – in Cambridge, where else? – 'took me to a very private place under the wooden bridge on the Little Island and told me there, in confidence, that it was not at all the thing to believe in Christianity any more. It simply wasn't done.'[16] She went on to assert, of her Darwin father and his brothers: 'They were quite unable to understand the minds of the poor, the wicked, or the religious.'[17]

Frank, as the secretary at Down House, had a heavy workload. The fact that *The Origin of Species*, by the time it reached its sixth edition, had been emended so many times as almost to have contradicted its original sunny confidence, in no way diminished the enthusiasm with which, all over the world, those whose imaginations had been opened to the fact of evolution attributed their enlightenment to Darwin. In the year after *The Descent of Man* had been published, Darwin wrote over 1,500 letters, and, although the numbers diminished somewhat, much of his time, for the rest of his life, would be devoted to the fan club. At 9.30 each morning he would come into the drawing-room at Down for the letters, rejoicing if there was only a small postbag. If there were family letters, he would lie on the sofa and hear them read aloud. Then, at half past ten, he would return to the study, with its makeshift privy behind a curtain, and work on his public correspondence until midday.

The retchings, flatulence and bowel disorders were worse on some days than on others. In the lavatorial fug of the study, the patient Frank would sit and help his father cope with the admiring correspondents, from Europe and from America. What the younger German evolutionists had in common was a sense that evolution had finally got rid of religion, disproved it once and for all. Darwin in turn so much admired Haeckel that he said, if he had read the German's *Natürliche Schöpfungsgeschichte* – published in German in 1868, but only translated into English two years later – he would not have bothered to write *The Descent of Man.*[18] The title of Haeckel's work is self-explanatory: the story of creation is to be explained *naturally*. Man evolved. There is no need for a Creator to have started the impersonal process of evolution.

Karl Marx sent Darwin a copy of the second German edition of *Das Kapital* in 1873. Darwin thanked him politely for it: 'Though our studies have been so different, I believe that we both earnestly desire the extension of knowledge & that this in the long run is sure to add to the happiness of mankind.'[19] Janet Browne pointed out that Marx's book was still on the shelf at Down House, but that the pages had not been cut.[20]

There are certain paradoxes about the relations between Marx and Darwin. Darwin's moment of enlightenment about the struggle for survival came to him while reading Malthus, an economist Marx especially deplored. Ideology, however, is seldom logical, and though Darwin's social and political thought was an extension of Malthus, and though the struggle for survival could plausibly be seen as a pseudo-scientific justification of the exploitation of the working classes by the idle bourgeoisie, Marx evidently felt he should admire Darwin. When Marx was buried in Highgate Cemetery on 17 March 1883, with the faithful gaggle of a dozen mourners, they heard Friedrich Engels declare, 'Just as Darwin discovered the law of evolution in human nature, so Marx discovered the law of evolution in human history.'[21] Edward Aveling, the spurious English doctor who helped Sam Moore translate *Das Kapital* into English, had been drawn to Darwin before he became a Marxist.[22] In the early 1880s, he would address working-class evening classes on Darwinian 'philosophy', and published such penny tracts as *The Student's Darwin* and *Darwin Made Easy*. Aveling – the model for Shaw's Dubedat in *The Doctor's Dilemma*, who in effect murdered Marx's daughter by persuading her to enter a suicide pact and then failing to keep his side of the bargain – was probably influential in persuading Engels that Darwinism could be fed into the Communist world-view. Darwin and Marx, according to Aveling, looked alike. (Yes, they both were tall and had white beards, but here the resemblance rather fades.) 'That which Darwin did for Biology, Marx has done for Economics,' Aveling told Engels. 'Each of them by long and patient observation, experiment, recordal, reflection, arrived at an immense generalisation – a generalisation the likes of which their particular branch of science had never seen.'[23]

Probably there has never been a phase in the intellectual history of the West to compare with our own times, in the early decades of the twenty-first century; when 'immense generalisations' have lost their appeal. It is easy to see why, given the vacuum left by the apparent collapse of Christianity among intellectuals, the 'theories' or 'generalisations' of Marx and Darwin made their metaphysical appeal. The decade in which Marx and Darwin died was, by contrast, hungry for a new faith. John Ruskin, writing *Praeterita* – his formidably impressive autobiography – during the same decade, looked back to his young manhood and his agonized agnosticism. In the mid-1840s, he had visited Pisa and had what amounted to an epiphany in the Campo Santo: 'The total meaning [of Christianity] was, and is, that the God who made earth and its creatures, took at a certain time upon the earth, the flesh and form of man'.[24] Ruskin rather fudges whether he actually believed this doctrine, after his Pisa moment of enlightenment, or whether he believed it at the time of writing in the 1880s. He was right, however, to see it as central to the way that Western humanity had viewed the world until the Enlightenment. Marx and Darwin between them constructed alternative Grand Narratives. After such mainstream figures as Lyell had come round to the belief in evolution – in some shape or form – it is impossible to doubt that evolution would have become a scientific given even if Charles Darwin had never been born. Once the connection in respectable English minds had been dissolved between Lamarck and the tumbrils and guillotines of revolution, evolution would have been absorbed. It is not necessary for scientific truth, once apprehended, to be completely understood. Niels Bohr said that anyone who says they have understood quantum mechanics has demonstrated that they do not understand quantum mechanics. But we know it is true because it works, within the fraction of a millimetre. Einstein likewise said he did not understand the nature of a photon. There is no need in science, and perhaps no need in religion, for a 'theory of everything'.

Darwin himself vacillated between a deep modesty about his central theories and a Faustian pride. On the one hand, John Stuart Mill was right:

> Mr Darwin's remarkable speculation on the origin of species is another unimpeachable example of a legitimate hypothesis . . . It is unreasonable to accuse Mr Darwin (as has been done) of violating the rules of induction. The rules of induction are concerned with the condition of proof. Mr Darwin has never pretended that his doctrine was proved. He was not bound by the rules of induction but by those of hypothesis.[25]

Equally, however, there can be no doubt that Darwin set out to destroy 'the old argument from design in Nature'.[26] He had thereby abandoned science *tout court* and set up his stall beside those of Herbert Spencer, Auguste Comte, Karl Marx and others – Dr Freud would not be slow to join the bazaar – of ersatz religions, theories of everything.

We began this chapter with Lord Salisbury offering to award Charles Darwin an honorary degree at the University of Oxford. I will conclude it with some words with which Salisbury addressed the British Association for the Advancement of Science in 1894. They are words which the childish pursuers of a theory to explain everything will never understand. They think, because they believe that the gradualist evolutionary theories explain all that there is to be known, that those who disbelieve this are obliged to come up with a rival theory. This, however, is not the case. In Salisbury's words, 'We live in a small bright oasis of knowledge surrounded on all sides by a vast unexplored region of impenetrable mystery.'[27]

16

Evolution Old and New

THE DARWIN–WEDGWOOD–ALLEN HABIT of marrying cousins, and of thereby keeping the money safely in the family, was not always a recipe for personal happiness. The twentieth-century Hensleigh Wedgwood, joint author of *The Wedgwood Circle* (1980) was of the view that 'apart from that of Emma and Charles, there were no completely happy marriages among the third generation, though they all maintained the pretence of happiness'.[1] Josiah III, known as Joe, was Emma's brother. He married Darwin's sister Caroline. They lived at a beautiful house called Leith Hill Place in Surrey, later the residence of their kinsman Ralph Vaughan Williams. Caroline had a nervous breakdown, and became an agoraphobic recluse, seldom leaving Leith Hill Place. Joe was a domestic tyrant, who made his three daughters, Margaret (Ralph Vaughan Williams's mother), Lucy and Sophy, line up each afternoon for a 'treat' – a spoonful of cream – but only if they had been 'good' in the morning.[2]

It was while staying at Leith Hill Place in June 1877 that Darwin was 'made very happy by finding two very old stones at the bottom of the field'. They were ancient slabs from a lime kiln. Caroline's gardeners were enlisted to lift them – they were five feet long and of considerable weight – to reveal the fine black mould of *worms*.[3] Darwin had had a passion for earthworms as a very young man, but his many other interests had forced him to place it to one side. Every now and again, the interest flickered into life. One of the things which endeared Frank's wife Amy Ruck (who died, as did the baby, giving birth to their first-born) to Darwin was her willingness to measure Welsh worm-holes.[4]

The strains of controversy, and in particular his inability to meet the objections of Mivart to his theory, had taken their toll on Darwin.

As so often in life when confronted with a challenge, he was rescued by a physical collapse. Back in August 1873, when he was only sixty-four, he had what Emma called a 'fit'. He suffered twelve hours of memory loss. When a doctor who specialized in nervous disorders came down from London, however, he pronounced that there were no obvious causes for the 'fit', opining that 'there was a great deal of work' in Darwin yet.[5]

This was not really true. Between answering his correspondence and meeting the many fans who had the temerity to go down to Down House unannounced to see the great man, he had no energy for further controversy – even though it would be forced upon him at the last. Being a naturalist was, first and foremost, what Darwin *was*. He was one of the greatest naturalists who has ever lived, with the most acute eye and an encyclopaedic knowledge of species. The worms were perfect objects of study in these twilight years. For although, when he was peering excitedly under Caroline and Joe's lime-kiln slabs, he was only sixty-eight, he appeared like the Ancient of Days in William Blake's drawings, with his thin stooping figure and his flowing beard.

It had been back in July 1874 that Darwin had begun the last truly fructiferous intellectual friendship of his life. By now he was used to receiving sacks full of sycophancy with his morning mail, as well as having to brace himself for the strains of controversy. In George John Romanes, however, he found a friendship during his last eight years on earth which was not only emotionally supportive but of intense intellectual interest.

When he first wrote to Darwin, Romanes was twenty-six, the child of a wealthy, landed Scottish family whose clergyman father was Professor of Greek at Kingston, Ontario. Romanes himself had prepared to take orders, but, entering Gonville and Caius College, Cambridge, he had been drawn first to mathematics, then to physiology. Frail health intervened, and ultimately prevented Romanes from following a medical career, and he began research on the nervous systems of invertebrates – chiefly jellyfish (medusae) and echinoderms (sea urchins). He set up a laboratory in his summer home in Dunskaith, at the entrance of the Cromarty Firth in Ross-shire, as well as in his London house near Regent's Park, 18 Cornwall Terrace.[6] He did not live long. He outlived Darwin by only twelve years, dying soon after his forty-sixth birthday.

There were three factors of fundamental importance about Romanes's approach to science which were helpful to Darwin, and, as Joel Schwartz has said, this was a time when Darwin most needed support.[7] One was his scientific research, particularly into the jellyfish and the sea urchins, specifically conducted with the hope of proving a Darwinian account of evolution. Darwin at this stage no longer had the energy to spend hours each day conducting experiments. He was trying to prove his theory of heredity, pangenesis; and Romanes conducted repeated experiments, grafting plant hybrids. He also helped Darwin with his continued work on animal intelligence, demonstrating a Darwinian breadth. He bought a pet monkey which was kept under constant observation; while also taking a keen interest in earthworms, suggesting to Darwin that worms could be called intelligent if they were 'taught by experience how best to manipulate some unknown exotic leaf'.[8]

These were the kind of inquiries which Darwin himself found most satisfying, and this is the tenor of nearly all their correspondence.

> Dear Mr Romanes. I hear from Mr Farrar that his gardener has raised some young plants of the cut-leaved vines & that they will be hardened off enough to travel in 3 weeks time – Unless I give further instructions, they will be dispatched by Railroad to your Scotch address, in (now) rather under 3 weeks . . . [7 April 1875]

> My dear Romanes [as intimacy grew, he dropped the 'Mr', though Romanes always called him 'Mr Darwin'] I am terribly sorry about the onions, as I expected great things from them . . . As tubers of potatoes graft so well, wd. it not be good to try other tubers as of dahlias & other plants? [24 September 1875]

> Dear Mr Darwin . . . you are afraid I am neglecting Pangenesis for Medusae . . . I should like to assure you that such is not the case. [18 December 1875][9]

The year 1877 found Romanes experimenting with rabbits and guinea pigs and telling Darwin that 'tame' rabbits, even if very hungry, will not eat nettles.[10]

This, the day-to-day pleasure of observing and discussing the results of experiments, was what Darwin liked best. Romanes, who left money for the foundation of the annual lectures at Oxford which bear his name, was always concerned with the bigger picture, and his friendship furnished Darwin with two other areas of support which even Huxley could not quite provide.

One was an out-and-out rejection of Christianity. Romanes, having abandoned his intention of following his father into a clergyman's career, had agonized about the relationship between faith and science. He had won the Burney Prize at Cambridge with his essay *On Christian Prayer and General Laws*. Like so many Victorian intellectuals, Romanes was a divided soul, who asked his Cambridge friend Francis Paget (later Dean of Christ Church) how he could help his children to be Christians while himself nursing doubts. To Darwin he presented a much more materialist face, and he looked to Darwin to bolster his own unbelief in religion. 'The entire structure of Pauline theology', Romanes was to write to his friend the Revd J. T. Gulick in 1891, 'has had its foundation undermined by Darwinian science: the "first man" having been politely removed, there is no longer any logical justification (according to this theology) for the "second man".' He went on, rather loftily, to ask, 'If the saviour of the world had so little knowledge of its reputed destroyer, as to believe that evil spirits were the cause of epilepsy, madness, deaf-mutism, etc., what is the value of his teaching on other matters?'

There was another reason why Romanes was a figure of such importance in Darwin's story. He was a faithful footsoldier in the laboratory. He was a religious doubter. And he was also better equipped than Darwin himself to express Darwinism's *metaphysical*, philosophical significance.

The very first letter which Romanes had written to Darwin in July 1874 had amounted to a paper on the philosophy of science; and in particular on the phenomenon of causation. 'I have so poor a metaphysical head that Mr Spencer's terms of equilibration &c always bother me & make everything less clear,' Darwin frankly conceded when he received Romanes's paper.

Romanes believed, as did Haeckel, that Darwinism had solved, not merely a scientific conundrum – how do species evolve? – but also an epistemological one – can we ever be certain of causation? Can we ever verify that one thing causes another thing? Can we legitimately infer, because matter behaved in a particular way in one set of circumstances, that it will do so again if identical circumstances were repeated? Berkeley and Hume, neither of them having any scientific knowledge worth speaking of, believed they had delivered the ultimate critique of Cartesian science. We think we know that water will boil at a certain temperature, that the sun will come up in the morning, but we can neither prove that these things will happen nor prove that even if we put the kettle on the gas just as we did yesterday, the water inside it will boil. Lyell, more than any other scientist perhaps, took on Hume by demonstrating that geology was a sequence of observable events: that one thing demonstrably led to another. Hume had denied the possibility of an *a priori* comprehensive system of knowledge. Darwin's theory of natural selection confirmed – or appeared to confirm – what Lyell's geology had all but proved. 'In succeeding where Hume had basically failed, Darwin was giving rise to the most radical revolution in modern thought: the classical tradition starting from Plato and Aristotle and its long-lasting consequences were knocked down and replaced by something new and equably workable.'[11] The modern scientific outlook, the belief that science was the most reliable road to the truth, had arrived. Darwin was not being falsely modest when he said that he had 'a poor metaphysical head'. Although passionate in his commitment to the scientific significance of his theories, it was his disciples, above all Haeckel and Romanes, who were more acutely aware of the wider intellectual ramifications.

By the time he was seventy years old, there had erupted a painful incident which awoke old spectres from his earliest days in Shrewsbury, and, beyond that time, to the era of his grandfather Erasmus Darwin.

Samuel Butler's name is perhaps obscure in the twenty-first century. In Edwardian England (he died in 1902) he was seen as the iconoclast who had, years before the Bloomsbury Set did so, blown the whistle on Victorian humbug, and in particular on the Victorian cult of the family and good breeding. His novel *The Way of All Flesh* was published posthumously, and it still has the power to shock with its claims, or demonstrations, that family life is based not on mutual love, but on hatred, and in particular on the hatred felt by children for their parents. It is also a book which mercilessly lays bare the need for the solid family structures of upper-middle-class Britain to maintain the many-petalled flower of the Establishment. 'That a man should have been bred well and breed others well; that his figure, head, hands, feet, voice, manner and clothes should carry conviction upon this point, so that no one can look at him without seeing that he has come of good stock and is likely to throw good stock himself, that is the *desiderandum*.'[12]

As Butler's most hostile biographer saw, '*The Way of All Flesh*, like *The Origin of Species*, said what a large number of people wanted said; and the fact that Shaw went out of his way to praise it so highly, and was himself a practising creative evolutionist, gave it just the necessary *cachet* of respectability.'[13]

Darwin and Butler probably saw many of the same truths in the artificial threads of convention which held together the Establishment. The differences between them were that, whereas Butler was a tormented homosexual who hated his family, Darwin was, as far as one can judge these things, seemingly happy with his sexuality (heterosexuality), and – whatever his difficulties with his own father – he was a devoted family man; indeed, his whole emotional life was bound up with his wife, his children, his siblings and his cousins. Dr Alison Pearn was absolutely right to emphasize, in an unpublished lecture in Philadelphia on Darwin's feuds, that 'Darwin's science was situated in a remarkably domestic environment, with loving family and friends as the context within which he thought and wrote: and that career in science was built on co-operation, and a finely honed ability to get on with those whose help and support he needed.'[14]

Butler called at Down House on Sunday 19 May 1872, according to Emma's diary. The occasion would seem to have been that Butler, who had befriended a young artist named Arthur Dampier May, hoped that May's dog drawings would be of interest to Darwin in his researches into the expression of emotion in animals' faces. It was to George Darwin, in his capacity as secretary, that Butler had written in the first instance.

15. Clifford's Inn

Fleet Street

Dear Darwin

My young friend May has brought me these this morning: he tells me to say that they are entirely at Mr Darwin's disposal, and that he shall be delighted in case he finds them in any way useful. I don't think the lower one satisfies him, but I should think that a suggestion would be attended to: he said he found it so far more difficult to get a dog into the fighting attitude than the fawning one, that he had less chance of studying. I send the drawings to you rather than to your father because it is no use troubling him at all unless you think them likely to please him. Would you like to meet the youth? he seems to me to shape uncommonly well.

Yours very truly

S. Butler[15]

The Butlers and the Darwins went back together through the generations. Dr Robert Darwin had been acquainted with Samuel Butler's grandfather, the Headmaster of Shrewsbury School – the model for the obnoxiously hypocritical Mr Pontifex in *The Way of All Flesh*. Darwin was at university with Butler's father. Samuel Butler was friendly with Darwin's sons and had stayed at Down House. And now came this offer of help with illustrations for *The Expression of Emotion*.[16]

An early sign that the milk of Butler's human kindness was not untainted with acid came with the novel he published the same year as that visit to Down with the drawings, 1872. *Erewhon* is a satirical attack on 'Victorian Values', and it had already been completed by the time he was offering Darwin May's dog drawings. *Erewhon* (anagram of 'Nowhere') is an all-too-recognizable picture of Victorian Britain, with its worship of the idol Ydgrun (Mrs Grundy – that is, fear of what-will-the-neighbours-say as the basis of morality). While mocking both Christianity and the Victorian humbugging allegiance to its doctrines, Butler also mocked Victorian materialism, and indeed the first part of his novel had been published nine years previously, in the Canterbury Province (as it then was) of New Zealand, and entitled 'Darwin among the Machines'. As well as having the temerity to attack Darwin, Butler also satirized the British universities, and Cambridge in particular. The Colleges of Unreason, in Erewhon, are too clearly the places which gave Darwin his education, and some of Darwin's sons their employment and their home.

Darwin himself was not entirely without humour, but he was made uneasy by Butler's squib. Emma wrote to Horace on 28 January 1873, 'We are just finishing Butler & find it v. striking & interesting in the latter part; but I am very angry w. him for giving such horrible stories about brutality to the dogs, which one cannot forget. I believe he is a humane man, but he tells it in rather a jocose way.'[17] It is clear from these references that, while they could see the cleverness of *Erewhon*, the Darwins did not like it. Above all, Darwin could sniff out, what he detested more than anything, disloyalty – personal disloyalty from a young friend with many family connections, and, further, disloyalty to the tribe. He could sniff out that Butler was a class traitor.

Much worse than *Erewhon*, from the Darwins' point of view, was to come. The occasion was the *Festschrift* compiled for Darwin's birthday by Ernst Krause in Germany for his journal *Kosmos*. The special issue of the periodical contained an article by Krause himself on Darwin's grandfather Dr Erasmus. It was entitled 'A Contribution to the History of the Descent Theory'. Krause here came to the nub of Darwin's mystery. He did so, not in a deliberately destructive spirit, but with a certain clumsiness. He acknowledged the genius of *both* the Darwins, Erasmus and Charles, but the very nature of the essay gave the lie to the now popular notion that, to use the words of Engels, '[Charles] Darwin discovered the law of evolution.'

Darwin, therefore, viewed the German tribute with mixed sensations, but he was in retrospective and reflective mode. He had lately completed his short *Autobiography*, to which so much notice has been given in this book, and he also wished to set the record straight, correcting the memoirs of two of Dr Erasmus's female contemporaries, Anna Seward of Lichfield and Mrs Schimmelpenninck, both of whom, he felt, did his grandfather an injustice.

Darwin wrote to his cousin Francis Galton – also a grandson of Dr Erasmus, and the custodian of the family papers. Murray was approached to see if he would publish an English translation of Krause's article, and a memoir of Dr Erasmus by Charles, based on archival research among these documents. George, among Darwin's sons, was the most inclined to ancestor-worship, and he was glad to be enlisted to help. Darwin himself revelled in the talk of pedigree. 'All your astronomical work is a mere insignificant joke compared with your Darwin discoveries,' said Father Charles to son George approvingly. 'O good Lord that we should be descended from a Steward of the Peverel; but what in the name of heaven does this mean?'[18] One thing it certainly meant – and this was the sort of 'discovery' which upper-middle-class students of pedigree were agog for in the nineteenth century – was that they were not mere arrivistes. There was the quiet reassurance that, in the background of their Victorian, frock-coated, carriage-driven lives, there were, as well as the graves of honest yeomen, the occasional frayed escutcheon aloft over a tomb in a country church, a family name found among justices of the peace or officers in the army, the comfort of knowing the predecessors in the family tree to be gentry.

George, in his enthusiasm, enlisted an American acquaintance, Colonel Chester, a keen genealogist, to draw up a pedigree stretching back 200 years. For a while even the earthworms were neglected as old Darwin drooped over the ancient wills and seventeenth-century titles. Meanwhile, Darwin set to work on his own short biographical sketch of his grandfather, the pioneer of evolutionary theory who, in the just completed *Autobiography*, Darwin claimed had made no impression on his young mind. He submitted what he wrote to the inspection of other family members. Etty Litchfield read it, and made him excise approving references to his grandfather's religious scepticism. Sometimes he strayed from his theme to include personal anecdotes – 'Henrietta, is this too egoistic to include?' he asked.[19]

One of the factors to spur Darwin on was Galton's essay on the heredity of genius. It placed Darwin in a peculiar position. On the one hand, his name was now synonymous with the notion that, rather than being individual creations, each being owed its characteristics to inheritance. Paradoxically, egotism required him to adapt this claim, by adding the aggressive old idea of the struggle for existence. If, like some still not perfected finch's beak, he inherited much of his grandfather's genius, he had also, with brighter plumage and sharper claws than the prototype, defeated the old man, pecked him off the branches. He dismissed Dr Erasmus's 'overpowering tendency to theorize and generalize': 'I fear that his speculations on this subject cannot be held to have much value.'

The reader was to be left in no doubt. The newer, younger evolved evolutionist left the older, eighteenth-century one a mere fossil, a curiosity who had nothing to say to the modern generation.

A strange cluster of coincidences and conjunctions now came together, like the strands of some elaborately plotted work of fiction. In May 1879, Samuel Butler had published an essay entitled *Evolution, Old and New, or the Theories of Buffon, Dr Erasmus Darwin, and Lamarck as Compared with That of Mr Charles Darwin*. This was scarcely an embarrassment to match that of 1858, when Wallace had popped up out of the blue with a theory of natural selection just as Darwin was about to reveal to the world an all but identical theory. Nevertheless, it was embarrassing that Butler's essay should have resurrected the figure of Dr Erasmus some five months *before* Darwin's own *Erasmus Darwin* which was published by Murray in November 1879.

Butler was a natural iconoclast. Some of his arch 'naughtiness' will seem forced to us. He was, for all that, an original mind who deserves his description in *The Dictionary of National Biography* of 'philosophical writer'.[20] Though the man who maintained – in 'The Authoress of the Odyssey' – that Homer was a woman, or that Shakespeare's sonnets were about the poet's love not for the Earl of Southampton but for a lad of the lower class – could be seen as an addict of heterodoxy, he had taken the trouble to read much evolutionary material, and his hits against Darwin were palpable.

In his expanded *Autobiography*, suppressed by the family until his granddaughter Nora Barlow published the full version in 1958, Darwin's account of what happened reads as follows: 'Owing to my having omitted to mention that Dr Krause had enlarged and corrected his article in German before it was translated, Mr Samuel Butler abused me with almost insane virulence. How I offended him so bitterly, I have never been able to understand.'[21]

Darwin omits to make two things clear. One is that he had published a much reduced version of Krause's article, in translation, without explaining to English readers that he had heavily doctored the German's original essay. Secondly, he has conveniently changed the chronology of Butler's 'attack'. Butler's essay on *Evolution, Old and New* appeared – as I have said – five months *before* Darwin's account of his grandfather. The previous year, 1878, Butler had published *Life and Habit*, in which he assessed Darwin Junior's contribution to evolutionary theory in a highly critical manner. He found Darwin's theory, of the mechanism of evolution, unconvincing.

> Anyone can make people see a thing if he puts it in the right way, but Mr Darwin made us see evolution, in spite of his having put it, in what seems to not a few, an exceedingly mistaken way. Yet his triumph is complete, for no matter how much any one now moves the foundation, he cannot shake the superstructure, which has become so currently accepted as to be above the need of any support from reason, and to be as difficult to destroy as it was originally difficult of construction. Less than twenty years ago, we never met with, or heard of, any one who accepted evolution . . . Yet, now, who seriously disputes the main principles of evolution?[22]

Darwin's reaction when under attack was now instinctive – unleash the Bulldog. 'I am astounded at Butler,' wrote Huxley – 'who I thought was a *gentleman*, though his last book appeared to me supremely foolish.' This had been the approach to Mivart's closely reasoned critiques of Darwin's work: the man's a cad, keep him out of the Athenaeum. 'Has Mivart bitten him and given him Darwinophobia? It is a horrid disease and I would kill any son of a [here there is a drawing of a, presumably female, dog] I found running loose with it without mercy. But don't you worry with these things. Recollect what old Goethe said about his Butlers and Mivarts: "Hat doch der Wallfisch seine Laus / Muss ich auch meine haben" ["Still, as the whale has his louse, I must also have mine"].'[23]

Butler repeated the story (about Darwin altering Krause's article) in the book *Unconscious Memory*, published in 1880, and seven years later, after Darwin was dead, he published his most detailed criticism of Darwinism in *Luck, or Cunning, as the Main Means of Organic Modification*, pointing out Darwin's many inconsistencies and contradictions. 'There is hardly an opinion on the subject of descent with modification which does not find support in some one passage or another of the "Origin of Species".' Here, of course, Butler put his finger on the reason why Darwin, so soon after the publication of

The Origin, was regarded widely as the man who had 'discovered' evolution. Having flirted with almost all explanations for evolution, ducked and dived when contradicted, and cheerfully reverted to Lamarckianism when it suited him, Darwin could be seen simply as Mr Evolution by the reading public. Butler noted, as we have done in this book, how Darwin shamelessly spoke of 'my theory', even when rehearsing the arguments of Lamarck's original theory, which antedated 'his' by fifty years. Butler noted that, in later editions of *The Origin*, this use of the word 'my' when applied to evolutionary theory was dropped, possibly under the influence of Haeckel's work.

Butler was not a serious scientist, and his defence of Lamarckianism seems now unconvincing. The paradox is that, in attempting to put forward a version of Lamarckianism with a theory that characteristics can be acquired by 'unconscious memory', Butler was proposing something very similar to Darwin's own theory of pangenesis. Another irony, pointed out by Løvtrup, is that Darwin's son Francis, the botanist, adopted a theory very similar to Butler's in his article 'The Analogies of Plant and Animal Life', in which he inquires 'whether among plants anything similar to memory or habit as it exists in animals, may be found'.[24]

By now, however, Darwin personally was beyond criticism in two specific senses. He had, firstly, adopted so many of the criticisms into the six editions of *The Origin of Species* that the theory itself was in tatters, and he was regarded, as has been said, simply as the Father of Evolution, the Man Who Had Discovered It. The specifics had become immaterial. The great majority of intelligent people now believed in evolution, as they have continued to do to this day in ever-growing numbers. So momentous a change in human belief and consciousness needed a prophetic figurehead, and that person was Darwin.

Darwin was inviolable, however, in a second sense. This was that he did not realize that his theory *per se* – the theory of evolution by natural selection – had been fatally criticized, not only by his supposed enemies such as Mivart, but by his supporters too. Huxley, for example, never committed himself to believing that all the innovations in living forms come about through the accumulation of many small steps. A major modification of the micro-mutationist position of Darwin's original idea has to be made. Suppose you accept the 'Darwinian' explanation for the slow 'evolution' of complex organisms such as the eye. Suppose you accept that each of the forty or so different forms of the eye, from the trilobite 450 million years ago which deployed an 'optimal design' for seeing under water, to the compound eyes of insects that comprise up to tens of thousands of lenses to give full 360-degree field vision to avoid being attacked by predators from behind[25] – suppose you believe that each of these seemingly purpose-built and complex optical phenomena to have evolved in the manner demonstrated by the Scandinavian Darwinists with their computer in 1979 – a process which would have taken 'only' half a million years or so! Even if you can imagine how all these different creatures, with their different, highly complicated optical capacities, were for hundreds of millions of years able to survive *without* seeing their predators clearly, there are other phenomena in nature which make 'Darwinian' explanations simply preposterous.

A very clear example occurred in 1974 with the discovery of 'Lucy', a three-and-a-half-million-year-old hominid, by Tom Gray and Donald Johanson. Lucy belonged to the species called *Australopithecus afarensis*. Compare Lucy's femur, or upper thighbone, with that of a chimpanzee. The sharp upward angle of her hip joint and her distinctively shaped pelvis relocate the centre of her body's gravity over her feet. She could stand upright. A Darwinian imagines apes spending millions of years standing half upright, or *gradually* a *bit* more upright every ten million years or so. No palaeontological evidence survives for such halfway creatures since, had they existed, they would undoubtedly have been eaten by predators. Lucy is one of the many indicators in the natural world that *leaps* occur in the story of evolution, and that Darwin's theory of gradualism was simply wrong and unscientific.

The other area in which Darwinism has been supplanted by scientific evidence is the rich and multifaceted science of genetics which has made such extraordinary strides since the identification of deoxyribonucleic acid (DNA) in 1944, and the subsequent discovery of its structure. Many branches of genetics have opened up since then, but among them is the measurement of genetic variation by techniques such as gel electrophoresis of enzymes. This way of measuring has shown that genetic variation is sometimes uncorrelated with reproductive success and adaptive evolution, leading biologists such as Motoo Kimura, for example, to develop a theory of molecular evolution. This theory asserts that much of the evolutionary change observed at the molecular level occurs via random genetic drift and is *unaffected by natural selection.*[26]

Remarkable as the discovery of Watson and Crick was, it remained, until the 1970s, a largely theoretical affair. 'DNA makes RNA (ribonucleic acid, a molecule with long chains of nucleotides – something vital for living beings) makes protein' was the essence of it. The details of what was encoded within DNA remained impenetrable. Christopher Wills, Professor of Biology at the University of California, made a by now famous analogy. He held *Webster's Third New International Dictionary* on his lap and calculated that it contained 27,000 letters per page; so, with 2,600 pages, this was more than seventy million letters. There are about three billion nucleotide molecules in the human genome. It would therefore take forty-three volumes the size of *Webster's* to carry this amount of information. Yet in each minuscule nucleotide molecule, this is what is carried. By 1970, molecular biologists had worked out that this was the level of information being carried by genes, but they saw no way of, as it were, opening the dictionary, unlocking the code.

The first task was to isolate the individual straws of nucleotide sequences within the haystack of the genome in such a way that they could be scrutinized in detail. Viruses, as the smallest of all organisms, do not have the space within their own cells to make the protein necessary for their own survival and replication, so they infect organisms larger than themselves such as bacteria (or human beings) and then borrow their protein-making machinery. The bacteria resent this invasion, so they create their own enzymes which chop up the genes into their own useless little fragments. The first of these enzymes – 'restriction enzymes' – was discovered in 1968. Over the following decade over 150 further such enzymes were discovered. Scientists were now, as it were, tearing pages out of *Webster's Dictionary* and reading their contents. The science which has come to be known as the New Genetics was born.

In 1970, two Americans, Howard Temin and David Baltimore, quite independently discovered another, very special enzyme, this time made by a certain type of virus. They found the insulin gene. Moreover, they discovered that the enzyme produced by a certain type of virus actually reverses the classic process of DNA makes RNA makes protein. So, if the RNA coding for the insulin protein could be isolated from the pancreas cell, the addition of 'reverse transcriptase', as they called it, would turn it back into the gene from which it originated. A strand of insulin gene had been picked out of the human haystack.

While Temin and Baltimore were making these momentous discoveries, two other scientists, Frederick Sanger in Cambridge (England) and Walter Gilbert of Harvard, were working out a way to calculate the precise sequence of nucleotides in any strand of DNA. In one decade, molecular biologists had moved from a situation where the details of DNA were completely unknown, locked away in the trillion-times-miniaturized forty-three volumes the size of *Webster*, to a position where they could actually study precise genes. It was now technically possible to pin down where, and whether, micro-mutations were taking place.

The New Genetics was not able to solve any of the metaphysical mysteries of who human beings are or where they come from. It was, however, able to demonstrate the inadequacy of Darwinism as an explanation for the evolutionary process; and this was because it demonstrated information and capacities not being 'evolved' at all, but being handed down, fixed, hard and as it were ready-made. The double helix gives rise to the infinite diversity of living beings, down to the last detail of every tree, fish, bird and human being. Yet it contains not the smallest trace of those details. What the New Genetics revealed to us is the paucity of genes. They 'multi-task'. The Pax6 gene brings all eyes, with all their complexity, into being. Pax6 in a mouse produces a mouse-eye, which is, like our eye, a camera-type; whereas the same gene in a fly produces sheets of lenses at different angles. Or again, the same gene, known as distal-less, orchestrates the formation of the diverse limbs of mouse, worm, butterfly and sea urchin.

These are phenomena which, since the 1980s, the human race has for the first time been able to observe. When Darwin wrote, he was of necessity simply speculating about how life-forms evolved or came into being. The discovery of 'master genes' multi-tasking and thereby determining the shape of a mouse's paws or the functioning eyes of the fruit fly is in complete contradiction to Darwin's theory that these came about as a series of infinitely gradual and random mutations. Whereas, until the 1980s, it was just about possible to imagine that evidence might one day come to light which would substantiate at least some of Darwin's theory of evolution, the science of New Genetics delivered its death blow. Adios, theory!

Different views can be taken on the question how far a thinker in one age can be held to account for the misapplication of those views in action. Rousseau and Voltaire would have been conceited enough to be pleased, in 1789, that some people attributed the outbreak of the Revolution to their influence. It would be a harsh critic who blamed Voltaire for the Terror, and the innumerable heads rolling into baskets from the guillotine in 1793. Likewise, much as Karl Marx, making his inebriated way home from the Reading Room of the British Museum to his house in Maitland Park, north London, dreamed of the uprising of the proletariat throughout all lands, it would be harsh to lay at his door the millions of murders perpetrated by Lenin, Trotsky, Stalin and Mao.

It remains fair, however, to say that Darwin was a direct and disastrous influence, not only on Hitler, but on the whole mid-twentieth-century political mindset. 'The overflow of our birthrate will give us our chance. Overpopulation compels a people to look out for itself . . . Necessity will force us to be always at the head of progress. All life is paid for with blood.'[27] These ideas, in Hitler's *Table Talk*, are Malthus filtered through Herbert Spencer, but they derive directly from Darwin. 'One may be repelled by this law of nature which demands that all living things should mutually devour one another. The fly is snapped up by a dragon-fly, which is itself swallowed by a bird, which itself falls victim to a larger bird.'[28]

What Darwin, his cousin Francis Galton and Spencer made into a disastrous commonplace was the notion that aggressive competition is the guiding principle behind the universe. Simone Weil, working for the Free French in wartime London, drew a direct connection, in her seminal work *The Need for Roots*, between the destructive work of these Victorian scientific thinkers and the catastrophes of the mid-twentieth century. 'The modern conception of science is responsible, as is that of history and that of art, for the monstrous conditions under which we live.'[29] Darwin the omnivore-naturalist, the great collector, the beetle-maniac, the pigeon-fancier, the slow patient lover of the earthworm, was not responsible for any catastrophe, still less for the 'modern' 'scientific' outlook which underpinned National Socialism. And it was this benign father-figure whom visitors saw when they came to Down House. Darwin the theorist was a different figure. Since his belief in the struggle for existence was simply that – a theory – and since it had been inspired not by a scientist but by an economist, it would prove to be a much more difficult idea to counteract. Factual error can be conquered by the simple presentation of facts. Darwinism, as is shown by the current state of debate, is resistant to argument because it is resistant to fact. Yet, as Karl Popper rightly wrote in *The Open Society and its Enemies*, 'if our civilization is to survive, we must break the habit of deference to great men'.[30] The worship of Darwin as a man, the attribution to him of insights and discoveries which were either part of the common scientific store of knowledge or were the discoveries of others, this is all necessary to bolster the religion of Darwinism.

In his conclusion to *Why Us?* (2009), a robust demolition of the Darwinian position, James Le Fanu quoted Douglas Futuyma of the University of New York State: 'Darwin made theological and spiritual explanations of life superfluous. Together with Marx's theory of history and Freud's attribution of human behaviour to influences over which we have little control, the theory of evolution was a crucial plank in the platform of materialist science.'

Le Fanu was not setting out to dispute the indisputable fact of evolution; rather, to question Darwin's childishly simplistic 'reason for everything'. Le Fanu accepts the idea that Marx, Freud and Darwin were all seeking to construct a 'platform of materialist science'. Freud and Marx have been toppled from their thrones in our own day. Le Fanu writes, 'Now it is the turn of Darwin, whose reputation can scarcely survive the devastating verdict of the findings of the recent past.'[31]

Ras, Darwin's brother, died in August 1881. They brought the body back to Downe to bury him in the churchyard. Darwin would not have George Ffinden, the local vicar, to conduct the service. Instead, he brought in John Wedgwood, the clergyman cousin who had married Charles and Emma at Maer all those years before.[32]

It was the last time in his life that Darwin attended church. Three months after Ras had died, Darwin published *The Formation of Vegetable Mould through the Action of Worms*. It was by far his most popular work, selling 3,500 copies within days of publication.[33]

With the book complete, Emma could persuade him to move out of his appallingly smelly study, with its makeshift privy behind the curtain. The room was gutted and overhauled, and a new study built. While his work was disrupted – and in reality it had come to an end – the old man read again the book which perhaps had a more profound influence upon him than any other – Alexander von Humboldt's *Personal Narrative*, volume three. He spent the winter reading it.[34] Here were the South American travels of the great German polymath, with the excited marginalia of the twenty-five-year-old Darwin. Humboldt had been, of all European scientists, the man whom Darwin wished to emulate, the most famous man in the world next to Napoleon. When the youthful Humboldt met the aged Frederick the Great, the King of Prussia had quipped, 'You are called Alexander, do you intend to conquer the world?' 'Yes, sir. With my brain,' was the confident reply. Darwin, in his generation, had achieved a comparable cosmic feat. Goethe, who was more than a little in awe of Humboldt, is said to have used his character and career, in part, to feed into his conception of Faust.

Faust, at the beginning of the drama, is hamstrung by the metaphysical problem of how we can know *anything*. David Hume's sceptical empiricism seemed to have made any knowledge impossible, until the radicalism of his thought awoke Kant from his 'dogmatic slumbers' and permitted the tiny sage of Königsberg to glimpse the 'Thing in Itself' – *Ding an sich*. What had come to birth, first in the brain of Kant, and then in the imagination of Goethe's Faust, was the concept of modern science, the idea that, whereas most truths communicate themselves to us imaginatively, or metaphorically, or through mythology, scientific language could describe an actual world, in a language which was authentically verifiable. A different way of looking at the world was born, with, in consequence, the application of the 'scientific' approach to truth being applied, not only to science, but to all other branches of human inquiry. Darwin's had been the first generation to be free to think in this way, and the consequence had been an unprecedented advance in science, and an unprecedented confusion in the areas of metaphysics, aesthetics, ethics and religion. Old Goethe had foreseen it all when he made Faust summon up mayhem, which could be resolved only by humanity soaking itself once again in the healing balm of the Imagination. (This is the subject of *Faust, Part Two*.) Science seen as an independent discovery of the *Ding an sich* becomes a phantasmagoric illusion.

As an old man, Darwin lamented, 'My mind seems to have become a kind of machine for grinding general laws out of large collections of facts.'[35] His old love of poetry, as we have seen, and music, had entirely deserted him. He had become an empty husk, a bore. Reading Humboldt, it must have been painfully apparent on every page that, whereas the German had been one of the great inspirations, not only to modern science, but to Romantic literature and liberal politics, Darwin had merely succeeded in making himself a Mr Gradgrind. The dead weight of materialist, upper-middle-class England, and Dr Robert Waring's bank, and the plutocratic class system, had eaten his soul. Humboldt died a poor man. Darwin wrote sadly on his copy of the great man's work, 'April 3, 1882, finished'. Presumably, he referred to the task of reading the book. But sixteen days later, on 19 April, Charles Darwin was dead.

The immediate cause of death, on the death certificate signed by C. H. Allfrey, the doctor of the nearby village of St Mary Cray, was 'Angina Pectoris Syncope'.[36] Darwin had been slipping downhill all winter, with the regular attendance of four doctors. In the event, the death was quiet and undramatic, with no famous last words.

Emma had expected, and wanted, Darwin to be buried in Downe churchyard, but their cousin Francis Galton had immediately alerted the President of the Royal Society, William Spottiswoode, and the unstoppable movement began to have Darwin interred in Westminster Abbey. Darwin's neighbour, Sir John Lubbock, moved in the House of Commons that this should happen, and the next morning a petition was offered to the Dean of Westminster, Dr George Bradley. It was sometimes supposed, in late twentieth-century biographies of Darwin, that there would be some difficulty about burying in the National Valhalla the man who had, apparently, proved God to be an unnecessary hypothesis. This view is to get things slightly the wrong way round. For a start, the fame and distinction in the eyes of his contemporaries was not in question. Tributes poured in from all over the world. Burying Darwin in the Abbey, like offering him an honorary degree at Oxford, was a demonstration that, far from being cowed by Darwin's agnosticism, the Establishment was determined to neuter its danger by bestowing upon it a laurel crown. In any case, Darwin's lack of belief in a Creator had never been publicly acknowledged, and he was not an avowed atheist, merely agnostic, like many of the clergy themselves – perhaps like Dr Bradley.

Darwin's coffin was therefore carried into the Abbey by pallbearers who reflected the Victorian reverence for him. Here was the man, after all, who had told them that their land-grabs in Africa, their hunger for stock-market wealth in the face of widespread urban poverty, their rigid class system and their everlasting wars were not things to be ashamed of, but actually part of the processes of nature. The very flax spores which went to the making of the dean's surplice and bands as he processed up the great Gothic aisle, the very stamen and petals of the flowers on the coffin, were there, according to Darwin, because flax and lilies, like Victorian stockbrokers, came into being only through fighting and struggling with weaker versions of themselves. The coffin was borne by Lord Derby, the Duke of Devonshire (Chancellor of the University of Cambridge) and the Duke of Argyll, whose scientific objections to Darwin were, for the occasion, held politely in abeyance. The American Ambassador, James Russell Lowell, and the scientists Spottiswoode, Lubbock, Huxley and Hooker made up the rest of the team of pallbearers. From the monastic stalls of the choir, Herbert Spencer, coiner of the phrase 'survival of the fittest', looked on, and must have felt they were fit indeed for their task. The seventy-nine-year-old Hensleigh Wedgwood was of the party following the coffin, but his sister Emma, Darwin's wife of so many years, remained behind at Down House.

17
Mutual Aid

On 21 September 2015, at Bonham's Auction House in New York, a short note, written by Charles Darwin to a complete stranger, was put up for auction. It was expected to fetch between $70,000 and $90,000. In the event it was sold for $197,000. That is over $4,800 per word, scribbled by an old man on a piece of paper on 24 November 1880.

The reason for the excitement in the saleroom, however, was obvious. Since the mighty Darwin Project had begun in Cambridge (England), nearly all Darwin's correspondence has been gathered up and put into print or online. A letter which had slipped through the net was a great rarity. This, moreover, short though it be, is no ordinary letter. Ever since he had published *The Origin of Species* in 1859, Charles Darwin had been secretive, as far as the public was concerned, about the true nature of his religious beliefs. More than that, he had been secretive about the extent to which 'his' theory was conceived as an alternative metaphysic, a substitute for the Christian view of the world. Perhaps Darwin had not set out, from the first keeping of notebooks on the species question, to undermine religious belief, but from a fairly early stage it had become clear to him that this was what he was doing, and certainly by the time that Annie had died in Malvern in 1851 he had turned implacably against the religion which his wife so persistently kept.

Instead of a Creator God, there was an inexorable process of blind nature. Rather than it being Love which ruled the sun and other stars, evolution could take place only as a result of a struggle, of warfare. 'It has been truly said that all nature is at war; the strongest ultimately prevail, the weakest fail . . . the severe and often recurrent struggle for existence will determine that those variations, however slight, which are favourable shall be preserved or selected, and those which are unfavourable shall be destroyed.' So Darwin, in 1868, in *Variation of Animals and Plants under Domestication*. While preparing the book for publication, he had mulled over the question of whether to come out as a non-believer. Asa Gray, his American admirer, had expressed the belief that there was a benignity in nature. Darwin, in *Variation*, rejects this belief, and so by implication rejects the loving God of Christianity. 'It is foolish to touch such subjects,' Darwin wrote to Hooker, 'but there have been so many allusions to what I think about the part which God has played in the formation of organic beings that I thought it shabby to evade the question.' Nevertheless, he did, in effect, evade the question. It would be hard, in his published writings, to find an outright admission that he considered 'his' theory incompatible with Christian belief.

Then, out of the blue, towards the close of 1880, when Darwin was past his seventieth birthday, he received a letter from a young lawyer named Francis McDermott. It was not a trick letter. It came from an obvious admirer, and it asked, 'If I am to have the pleasure of reading your books, I must feel that at the end I shall not have lost my faith in the New Testament. My reason in writing to you is to ask you in writing to give me a Yes or No to the question, Do you believe in the New Testament?'

Perhaps the very simplicity of the question, placed by one whose profession it was to pose questions in a court of law, prompted in Darwin an uncharacteristic candour. Perhaps, having reached old age, he knew he had nothing really to lose. So Darwin wrote back: 'November 24, 1880 – Dear Sir, I am sorry to have to inform you that I do not believe in the Bible as a divine revelation, & therefore not in Jesus Christ as the Son of God. Yours faithfully, Charles Darwin'.[1]

One hundred and thirty-five years after he wrote these words, Darwin's letter was sold for a sum which would buy one of the more modest village houses in Downe. The reason is clear. Darwin, both as a figurehead manipulated by twentieth- and twenty-first-century admirers, and in his own person, was the one who felt most concern to bring matters of religious belief into the very core of *what he was proposing as a scientist*. This is what makes Darwin such an unusual figure in the history of science. Galileo was attacked by the Church for his astronomical observations, but he had not intended his confirmation of Copernicus's calculations as an assault upon the faith. Stephen Hawking and Albert Einstein have both used the word 'God' in their scientific writings. Their theories do not stand or fall, however, by their theology. Many have believed Einstein's General Theory of Relativity without endorsing the letter he wrote in April 1929 to Rabbi Goldstein: 'I believe in Spinoza's God, who reveals himself in the lawful harmony of all that exists.'[2]

Darwin is different. His two-pronged theory, what he called 'his' theory, was believed, by himself, to be a refutation not so much of previous science as of Paley, an extremely minor theologian. Not surprisingly, the neo-Darwinians in our day still refer to Paley's analogy of the watch discovered by the walker in a field. Darwin's afterlife, especially in our own day, has been intimately connected with the central questions of religion – whether it makes sense to speak of a God, whether nature does reveal a heavenly harmony of the kind seen by Einstein, whether ethics, as understood since the times that Plato and Aristotle discussed them four centuries before Christ, make any sense in the face of a pitiless, purposeless 'river of life'; whether kindliness and co-operation and unselfishness have any role to play as we all elbow one another out of the way, driven by our selfish genes, in the scramble up Mount Improbable.

Hence the fact that, although it is not absolutely necessary to choose between Darwin and God (most modern Christians have subscribed to the Darwinian theory of evolution), it has come in the early years of the twenty-first century to look like that. Darwinism or Intelligent Design? Evolution or creation? Darwin or God? Make your choice. Hence the excitement, the box-office appeal, the sale-room value of Darwin's candid expression of disbelief in the inspired word of the Scriptures, and in the divine Sonship of Jesus.

Let us go back six years before Darwin wrote that letter to a young lawyer. In 1874, while Horace Darwin was graduating from Trinity College, Cambridge, while Francis, at Down House, helped Charles Darwin with his voluminous correspondence and with a revised edition of *The Descent of Man*, Prince Peter Kropotkin (Pyotr Alexeyevich Kropotkin), a thirty-four-year-old Russian intellectual, was being incarcerated in the Peter and Paul Fortress in St Petersburg for subversive activity against the state. Although he was treated comparatively gently on account of his high birth, Kropotkin was to spend two years in prison before he escaped. It gave him time to catch up on his reading. Kropotkin was a gentle communist-anarchist who believed that human societies progressed by mutual co-operation, rather than by the autocratic repression practised by the tsars, or by the ruthless market capitalism practised in Victorian England. His primary academic interest, apart from political philosophy, was geography. He had explored the glacial regions of Finland and Sweden for the Russian Geographical Society, and in his prison cell he was allowed to work up his papers on the Ice Age. He also gave his mind to the Darwinian theories.

In particular, Kropotkin was influenced by an article which he read in prison by the Dean of St Petersburg University, a zoologist called Professor Karl Kessler. What Kessler proposed was that co-operation played a larger role in evolution than did competition.[3] When Kropotkin eventually came to live in England, he wrote a series of essays in the periodical the *Nineteenth Century* in which he expounded his evolutionary ideas, and in particular his responses to Huxley's robust articles, in the same periodical, on the struggle for existence. Kropotkin's study of animal life in the Arctic extremes of Sweden and Finland had persuaded him that animals are not in a state of perpetual competition with one another. Species might be so, but individual members of species co-operate in innumerable ways, and this is one of the reasons that they survive. In 1902, Kropotkin worked up his essays into a book entitled *Mutual Aid: A Factor in Evolution*.

> In the animal world we have seen that the vast majority of species live in societies, and that they find in association the best arms for the struggle for life: understood, of course, in its widest Darwinian sense – not as a struggle for the sheer means of existence, but as a struggle against all natural conditions unfavourable to the species. The animal species . . . in which the individual struggle has been reduced to its narrowest limits . . . and the practice of mutual aid has attained the greatest development . . . are invariably the most numerous, the most prosperous, and the most open to further progress. The mutual protection which is obtained in this case, the possibility

> of attaining old age and of accumulating experience, the higher intellectual development, and the further growth of sociable habits, secure the maintenance of the species, its extension, and its further progressive evolution. The unsociable species, on the contrary, are doomed to decay.[4]

Kropotkin was the recipient, in his lifetime, of adulation and derision in almost equal measure. After the Revolution of 1917, despite his detestation of Bolshevism, he returned to Russia in the hope of better days. The crowds who came out to greet his return to Mother Russia numbered tens of thousands. When he died of pneumonia on 8 February 1921, Lenin grudgingly allowed a public funeral and, again, crowds of thousands massed. It was the last great anarchist demonstration in Russia, with demonstrators carrying anti-Bolshevist banners.

What of his opinions from a purely scientific point of view? His views have seemed ever more plausible with the years. Stephen Jay Gould, while having some reservations about individual points in *Mutual Aid* – in particular Kropotkin's reliance on the naturalistic fallacy in contradicting Social Darwinism – was by no means wholly dismissive.[5] Perhaps even more striking has been the conversion of Edward O. Wilson to the notion of mutual aid as a factor in evolution.

Wilson's retraction of belief in the 'selfish gene', however, did not come without cost. Indeed, his apostasy was condemned by Richard Dawkins in *Prospect* magazine as 'an act of wanton arrogance'.[6] Dawkins claims that Wilson has muddled two concepts. One is the fact that individuals benefit from living in groups. 'Of course they do,' Dawkins replied. 'Penguins huddle for warmth. That's not group selection: every individual benefits. Lionesses hunting in groups catch more and larger prey than a lone hunter could; enough to make it worthwhile for everyone.'

This, according to the orthodox neo-Darwinism of Dawkins, deriving from the work especially of W. D. Hamilton in the 1960s, is to miss the truly 'Darwinian' thing about genes. Genes are, if viewed in this light, the 'units of natural selection' and it is within the individual that these genes are fighting for mastery.

Darwin and Huxley, largely guided by Herbert Spencer rather than by scientific observation, depicted a world at war, with pine trees, sheep, garden songbirds all furiously in conflict, with weaker versions of themselves like the ruthless Victorians' social striving for bigger carriages, richer wives and larger share portfolios. Observation – sheer common sense – eventually saw that this might have been the way that a very strange, indeed anomalous, social group in British history behaved, but it is *not* how nature works. Ah, reply the neo-Darwinists, you are confusing co-operation – which is in any case enlightened selfishness – with the way in which living organisms themselves actually operate. It is inside us that the struggle is going on. The Kingdom of Darwin is within you. The genes themselves which determine whether you have red or brown hair, whether you are a nice person or a nasty person, are at war, and it is the selfish genes which win. The discoveries of genetics, in the Hamilton–Dawkins thesis, show us in microcosm the old Victorian struggle, with the strong driving out the weak.

Wilson's reply to Dawkins that his original thesis, made in *Nature* in collaboration with Martin Nowak and Corina Tarnita, is that the old kin-selection theory – in which fit genes are competing against unfit – needs drastic modification. 'The central issue of the book, which he urges others not to read, is the replacement of inclusive fitness theory (kin selection theory) by multilevel selection theory (ie, individual and group selection combined) . . .' Wilson claimed that they had demonstrated 'that while inclusive fitness theory sometimes works, its mathematical basis is unsound, and inclusive fitness itself is an unattainable phantom measure. Multilevel selection is mathematically sound, analytically clear, and works well for real cases – including human social behaviour.' He adds, 'The science in our argument has, after 18 months, never been refuted or even seriously challenged – and certainly not by the archaic version of inclusive fitness from the 1970s recited in *Prospect* by Professor Dawkins.'[7] John Hands, in his magisterial *Cosmosapiens: Human Evolution from the Origin of the Universe* (2015), put it well: 'When it was discovered in the 1970s that some 98 per cent of the human genome did not consist of genes, defined then as protein-coding sequences of DNA, this was referred to as "junk DNA", which seemed a bold, if hubristic, view to take of the vast majority of our DNA that didn't fit the model.'[8]

We start to see that the debate is concerned less with specific facts about genes and more with the concept we entertain of science itself. Followers of Hamilton, and to a smaller extent of Dawkins, are likely to seem simplistic in their accounts of human behaviour, with their hope of identifying specific genes which will somehow 'explain' all the complexity and sheer capriciousness of human behaviour, including human baseness and the heights of imaginative genius or altruistic unselfishness. While genetics might explain the human propensity to pass on, let us say, spina bifida, it could never explain more nebulous but palpable and recognizable qualities such as courage or enterprise, or the sheer intuitive quick-wittedness which enables human beings to make genuine scientific discoveries. These things are mysterious. If someone tried to explain, in purely genetic terms, why Arthur Rubinstein or Alfred Cortot were incomparably better pianists than, say, their siblings or their parents, you would know that they understood neither science nor music. To believe that there is a 'Rubinstein' gene which enabled one person to play Beethoven sonatas better than another person is not borne out by any scientific inquiry.

Between Dawkins and Wilson there lurks a deep divide; and it is expressed by their attitudes to Darwin. Partly in his own person, and partly as conveyed to the world by Thomas Huxley, his representative on earth, Darwin is the man who laid the groundwork for a purely materialist concept of human science, a way of explaining humanity to itself without resort to any words like mystery or genius or character, let alone resort to supernatural pictures of the world.

There is some paradox here, of course, since the ardent neo-Darwinians are not troubled by the fact that the real Darwin said very little about any of these things. To this extent, the neo-Darwinians bear a close, and almost pathetic, resemblance to those Christian fundamentalists who believe that the historical Christ had strong views about, let us say, homosexuality, when no record survives of his ever having spoken upon the subject. Clearly, Darwin lived before the science of genetics had properly speaking got going and he did not know the pioneering work of Mendel. But, beyond that, he said very little – even in *The Descent of Man* – about what science could plausibly tell us about human beings. Still less did he tell us very much about what science can tell us about the origins of the human race. Indeed, the more one reads Darwin, the more paradoxical it seems that he, rather than the author of *Vestiges* or Spencer or Huxley, should have been chosen, by some process of selection, natural or otherwise, to be the torch-bearer of the modern branches of human life-sciences. If he were merely the hero of pigeon-fanciers or of earthworm specialists, you would understand it. As the man who helped human beings understand who and what they are, he wrote almost nothing and, to give him his dues, it is not even clear that he would have considered important such debates as were so acrimoniously played out between Dawkins and Wilson. Coexistent in his costive and paradoxical character was a variety of positions and viewpoints. The towering ambition which wanted to be a universal genius lived side by side with the simple naturalist-clergyman he never quite became; beside the overpowering ambition to promote 'his' theory come what may, there existed humility and honesty, and a despondent readiness, if contrary evidence proved overwhelming, to say, 'Adios theory!'

'Much of what Darwin said is, in detail, wrong. Darwin if he read [*The Selfish Gene*] would scarcely recognize his own original theory in it.'[9] So Richard Dawkins in 1976.

Scientific theories, however, stand or fall by science. As we return to the Natural History Museum and confront Boehm's monumental statue of Darwin seated on the grand staircase, are we looking at one of the great scientists such as Newton, Einstein, Mendel, Watson and Crick whose discoveries altered the state of human knowledge?

The truthful answer is no.

Was he as comprehensive in his vision of science and of science in its place in the whole human story as Goethe, or Alexander von Humboldt, or even William Whewell or Pierre Teilhard de Chardin? Definitely not. Was he even as broad-ranging, or as humane, in his understanding of the implications of science as Alfred Russel Wallace? Not at all. Was he as incisive as Thomas Huxley in defending his own ideas? No.

Was Charles Darwin one of the greatest naturalists who ever lived? Undoubtedly! Did his collection of specimens, made as a young man while voyaging with HMS *Beagle*, hugely expand the possibilities of taxonomic and geological research in a vast range of fields? Beyond question. Did he enrich for ever the study of barnacles, pigeons and earthworms? Indubitably. Is he the man who 'discovered' evolution? Absolutely not.

The ever-generous Wallace wrote, in 1898,

> We may best attain to some estimate of the greatness and completeness of Darwin's work by considering the vast change in educated public opinion which it rapidly and permanently effected. What that opinion was before it appeared is shown by the fact that neither Lamarck, nor Herbert Spencer, nor the author of the *Vestiges*, had been able to make any impression on it. The very idea of progressive development of the species from other species, was held to be a 'heresy' by such great and liberal-minded men as Sir John Herschel and Sir Charles Lyell; the latter writer declaring, in the earlier editions of his great work, that the facts of geology were 'fatal to the theory of progressive development'. The whole literary and scientific worlds were violently opposed to all such theories, and altogether disbelieved in the possibility of establishing them. It had been so long the custom to treat species as special creations, and the mode of their creation as 'the mystery of mysteries', that it had come to be considered not only presumptuous, but almost impious, for any individual to profess to lift the veil from what was held to be the greatest and most mysterious of Nature's secrets.[10]

This is undoubtedly the key to why Darwin achieved his mythic status in the history of science. More than any other exposition of the evolutionary idea, it was *The Origin of Species*, in 1859, which changed the minds of the educated world, and changed them, as Wallace said, for ever. As someone who was so extremely generous to Darwin, and so modest, Wallace would probably have wanted to say that he alone would not have been able to persuade the educated public, any more than would Blyth. It is questionable whether he is entirely right in his dismissal of *Vestiges*. This surely was the book which at the very least laid the groundwork for the acceptance of progressivist or evolutionary theory. One has only to consider Tennyson's anguish in his best-selling poem *In Memoriam*, the work which more than any other was a mirror of the mid-Victorian soul, to see that the fossil evidence, casting whole species 'as rubbish to the void', had established in people's minds the fact that species become extinct. Lamarck's theories of how new forms arise in nature had equally been popularized by Chambers, and they would – as we have seen in this book – eventually be largely accepted by Darwin, somewhat to Wallace's surprise. So Darwin, in the late 1850s, was pushing at an open door. Nevertheless, it was his hand, more than any others, which gave that door a push. And his readable, modest and readily comprehensible prose – more than his grandfather's rather windy heroic couplets, or Huxley's later bombast – which persuaded the Victorians that 'special creation' was wrong and 'evolution' was right. To this extent, Wallace was absolutely right to single Darwin out as the greatest spokesman of evolutionary theory, even though, as the neo-Darwinians are honest enough to acknowledge, so many of the details of the first edition of *The Origin* are plumb wrong.

There is, moreover, another reason why Darwin's position as the guardian of the evolutionary truth has remained inviolate since 1859. This is the extreme simplicity of the idea of descent by slow modification. All serious scientists, and most intelligent opinion in the world, now accept the fact of evolution. The question which remains is – how? (Leave aside the far more problematic – why?) The why? question might be answerable by metaphysicians but it is probably beyond science to provide an answer. Almost as soon as Darwin had gone into print with answers to the how? question, other life-scientists sought to modify, or to contradict, either his entire thesis or parts of it. None of them, however, came up with a complete, coherent rival system. None of them provided *an answer to everything*. Darwinism or neo-Darwinism in its purest form offers the kind of metaphysical consolation which is offered by religious ideas of creationism. For it is a catch-all explanation. Suppose, as many scientists have thought since Darwin began to publish, suppose that natural selection accounts for the elimination of species but cannot actually account for the emergence of new ones. Or suppose that natural selection accounts for the emergence of *some* species but not of others, and not in all cases . . . Suppose that some adaptations are plausibly explained by the inheritance of acquired characteristics, as Lamarkianism proposes. Suppose in other cases, which palaeontologists for the most part would aver, that nature does sometimes leap, that new species or radically modified versions of an old species do suddenly appear, without seemingly the gradualism demanded by orthodox Darwinism. What then? Suppose that the true state of things, is that some of these explanations fit some of the evidence some of the time. It is not very snappy, is it? You could not write a Richard Dawkins-style best-seller by telling the public that the current state of scientific knowledge is actually quite ambiguous, quite confused. As Jean Gayon (Professor of Philosophy and History of Life-Sciences at the University of Paris) put it in his paper 'From Darwin to Today in Evolutionary Biology' (2009), 'in spite of its tremendously increased theoretical and experimental basis, evolutionary biology remains today a largely descriptive and historical science'.[11] Thomas Nagel, foremost philosopher in the United States, made a similar point, as was indicated in the Prelude to his book, *Mind and Cosmos* (2012), in which he questioned the reductionist simplicities of neo-Darwinism, reminding his readers 'how little we really understand about the world'. He added, 'An understanding of the universe as basically prone to generate life and mind will probably require a much more radical departure from the familiar forms of naturalistic explanation than I am at present able to conceive.'[12] In other words, there are no catch-all explanations.

Religion, it must in fairness be said, never offered any explanations either. In purely scientific terms, saying that God was responsible for this or that phenomenon is both tautologous (for the Judaeo-Christian Creator God must, by definition, be held responsible for all that is) and unhelpful. It is not a saying which explains *how* nature operates. Life-science is about *how*. The lure of neo-Darwinism is that by linking itself to a very simple-minded reading of genetics it can give explanations for the development of all life-forms, and not only life-forms but of that much more mysterious phenomenon: mind, or consciousness. The trouble is that these explanations are often far-fetched, and many of them are unsubstantiated. While hitching themselves to the memory of Darwin, their hero, they actually do him a disservice. For if Darwin had one great virtue as a scientific theorist, it was – however unwillingly – a preparedness to change his mind in the face of evidence. Had he been given Methuselah-glands (or genes) and lived to see the development of modern genetics and the discovery of the double-helix structure of DNA, it is by no means certain that Darwin would have been a Darwinian. In one of the more comprehensive surveys of Victorian science, Philip G. Fothergill reminded us as long ago as 1952 that 'at the time of Darwin's death, Wallace was the only one who believed that natural selection was the sole causal agent in the evolution of species'.[13] The synthesis propounded by the neo-Darwinians, of genetic facts with the reductionist fantasy that genes or memes can somehow explain everything, and that these explanations are compatible with Darwin's 1859 micro-mutational theories, is only attractive to a certain type of temperament – the same sort of temperament which, in religion, became a neo-Thomist or a Marxist in the 1930s – someone whose mind craves a grand narrative. It may be that those of us who have learnt to live without a grand narrative have no need for Darwinism, any more than we need neo-Thomism or Freudianism or Marxism. An 'uncertainty principle' works well in areas other than physics.

John Keats wrote to his brothers in December 1817 (when Darwin was a boy of eight), 'At once it struck me what quality went to form a Man of Achievement, especially in Literature, and which Shakespeare possessed so enormously – I mean Negative Capability, that is, when a man is capable of being in uncertainties, mysteries, doubts, without any irritable reaching after fact and reason.'[14]

This 'irritable' reaching after fact is, of course, the foundation of the modern obsession with science. Those who live in twenty-first-century Western countries deride the misuse of science by the totalitarian regimes of an earlier age, particularly those of the Soviet Communists and the Nazis. Our sense of moral superiority to Hitler has the danger of blinding us to the extent to which we, too, can place undue belief in science, not when it is supplying us with factual data so much as when it is actually merely feeding us myths. Of these myths, one of the most potent is the Darwinian belief that 'all of nature is a constant struggle between power and weakness, a constant struggle of the strong over the weak'.[15]

One of the most cheering aspects of the current stage of the evolutionary conversation is the extent to which hardline Darwinism is now set to one side. (The sentence I quoted at the end of the last paragraph was, of course, spoken not by Darwin or Huxley but by Adolf Hitler in a speech entitled 'World Jewry and World Markets, the Guilty Men of the World War'.) One of the most charming recent books to explore this theme both in evolutionary science and in anthropology is Penny Spikins's *How Compassion Made Us Human: The Evolutionary Origins of Tenderness, Trust and Morality* (2015). Spikins points out our tendency to impose our own preconceptions upon the imagined past. As a university teacher she has conducted a simple experiment, asking her first-year students to imagine a Neanderthal. In over 300 cases tested, although half the students were women, no one imagined a female Neanderthal, and no one imagined a Neanderthal child. (She, perhaps fancifully, herself imagines Neanderthal and human children playing together.) 'Very often we assume that our ancestors would have been powerful, invulnerable, striding forwards alone, much like the image we see of human evolution – a man walking forwards, getting bigger, stronger and more upright as he goes.'[16]

This, of course, is precisely the 'Darwinian' picture of human evolution which was propagated by Huxley and is to be seen in natural history museums round the world. Yet, as Spikins reminds us, 'evolution, both biologically and socially, has been about groups of people, not all of them strong and powerful'.[17] Anthropology, a science which was in its infancy when Darwin wrote, and one in which he took surprisingly little interest, is full of examples of kindness and mutual dependency, these being essential ingredients in the survival of human groupings. Common sense would help us see this if we simply considered the difficulty of survival in extreme weather conditions and the often inhospitable parts of the globe where hunter-gatherers have, time out of mind, sustained life. Modern anthropologists studying peoples as far flung as the Hadza of East Africa or the Inuit of Canada or the Jo'huansi do not find 'violent apes', all grabbing for themselves along the pattern of Victorian carriage-folk. On the contrary, after an animal is killed, in all these communities, the food is stored and held in common. Polly Weissner, working among the Jo'huansi, chronicled 297 meals eaten by eight families at Xamsa village: 197 of these meals were shared, or provided by other families.[18] The Malthus–Darwin picture of killer apes fighting their way to the top is simply untrue. Jane Goodall's life and experience among the chimpanzees presents a mixed picture. The apes are aggressive and fight, but likewise they have their own form of morality, and constantly perform acts of unselfishness for one another.[19]

At the same time, the essential mysteriousness of how we, as human beings, differ from the apes is something which scientific and academic inquiry continues to unfold. Genetic evidence makes plain what, formerly, could only be felt by hunch. (One thinks of Sedgwick's marginalia to Chambers!) The analysis of recent skeletal remains in the Denisova Cave in the Altai mountains of Russia provides the best evidence yet of the genetics of any Pleistocene-era individual. These skeletons are of individuals decisively different from Neanderthals, while possessing similar genomes – a fact which suggests a common ancestry. A minute fraction of human beings trace a portion of their ancestry to these Denisovans, suggesting that they interbred with humans at some stage (4 per cent of the population of Papua New Guinea and some of the native peoples of Australia possess some of this DNA). Yet again, genetic research reveals the enormous difference between these individuals and those who would evolve into humans. We all (including those Papua New Guineans and Australians who have the tiniest smattering of Denisovan DNA) have chromosome 2, which is an amalgam of two different chromosomes in the other great apes. Some time in the evolutionary past, two separate chromosomes fused into one, giving humans a karyotype of forty-six chromosomes, where chimpanzees, bonobos and gorillas have forty-eight.[20]

To everyone who considers the evidence, a slightly different perspective will occur. That goes without saying. What has been evident to thoughtful human beings ever since they considered the matter is how very different the human race is from the other higher primates. However we wish to stretch the meaning of words and to say that other creatures possess an aesthetic sense, or are able to use language, or have music, or laughter, the simple fact is that these things are among the palpable features which make us 'human'.

In the generation after Darwin's death, we can trace the ways in which his ideas began to feed into the common consciousness. Two of the most brilliant of English author-illustrators went into print with a masterpiece exactly twenty years after Darwin's death: Rudyard Kipling published the *Just So Stories* and Beatrix Potter *The Tale of*

Peter Rabbit. Kipling's stories, of which perhaps the best is the story of how the elephant got its trunk – a comparatively squat nose, seized by a crocodile in the Limpopo river, was subsequently found useful in all sorts of ways – gently lampoons the irritable reaching after fact and reason and the essentially illusory nature of evolutionary 'explanations' for the state of nature. Potter, in her quiet way a deeper writer even than Kipling, subverts the differences between the species. Peter Rabbit, like the succession of creatures whose tales she told in subsequent volumes, both is and is not an animal. He looks like a rabbit, but he wears a little blue coat, and he speaks English and has all the instincts of a naughty English schoolboy. Potter, raised a Unitarian and subsequently fairly obviously an agnostic, comically posits the Darwinian/later-Victorian viewpoint of the kinship and fluidity of the species. Whereas Potter, with exquisite irony, reveals a world where mice can be expert tailors, and where foxes eat their duck roasted with sage and onion stuffing (or try to do so), the reductionist neo-Darwinian would want to make fellow humanity into a mere 'violent ape', and in the very act of so doing in fact resurrects the essential human mystery, which has been present ever since human self-consciousness dawned. Self-consciousness is perhaps the essence of the mystery, since it is impossible to envisage it in the fellow creatures with which we share so many other characteristics. We can search for nits in one another's hair, just as the chimpanzees do, but the thought of a chimpanzee playing – or being – Hamlet would be as much an inversion of reality as a rabbit wearing a blue coat and contemplating the death of his father at the hand, or rake, of Mr McGregor.

The Bible gets its creation myths out of the way in the first couple of chapters. The rest of its thousand-odd pages are concerned with the chaos of human experience. The Darwinians are otherwise. In the attempt to give a more plausible Just So story to explain our origins, they have made origins the be-all and end-all of their discourse. Darwin himself saw eye to eye with Wallace about how natural selection worked, but the pair parted company after *The Descent of Man*. As far as evolution *via* natural selection went, Wallace was more Darwinian than Darwin. When it came to reflecting upon the mysterious phenomenon of what makes us human, Wallace was much closer to Darwin's enemy Mivart. Darwin, naturally enough, was horrified. We can now see that both men were scientists who in the natural selection idea were trying to come up with the solution to a specifically scientific problem, and in the other matter – the definition of human nature – came unstuck. So might any of us. The world, however, moves on, and huge generalized descriptions of 'who we are' are not really necessary. Darwin the hero of the present-day Darwinians seems an almost risible figure, because the picture of humanity which *The Descent* provides is so inadequate. He strayed from territory in which he was a master – the field of observant natural history – into generalizations about the human race, and immediately found himself reversing the very tenets of his supposedly scientific hypothesis. That is to say, having as a scientist proposed a theory in which natural selection proceeded by the elimination of the weak by the strong, he gazed, as an appalled Victorian gentleman, upon the opposite process at work. If they were not controlled, and if necessary sterilized, there was a distinct danger of the meek inheriting the earth, by the very processes – selection by sex – which Darwin's own class were either too repressed or too bored to rival. It would be unfair to saddle Darwin with all the blame for the sorry history of eugenics, and for the habit of mind which produced not only the eugenic movement but the tyrannies of the twentieth century. It would not be fanciful, however, to see in the deification of Darwin, both by his few disciples in the nineteenth century and by the many in the twentieth and twenty-first, something emblematic. Science made such enormous strides from the late eighteenth century onwards because it established its intellectual independence, its right to be science. Among the many prodigious advances, perhaps none matched the truly extraordinary accuracy with which the science of genetics was able to map the structure of DNA and the progress of characteristics from one generation of species to the next.

Beside these advances, the claims of the Darwinians, and the equally strident dismissal of these claims by the anti-Darwinian Bible-bashers, are really just 'noises off'. The very many accurate observations made by Charles Darwin the naturalist, observations which are scattered through his multifarious oeuvre, not least in his vast correspondence, are the lasting monument to his titanic stature in nineteenth-century science. The other stuff has already been shown to be ephemeral. Because we, the human race, have entered so many new worlds since the 1780s, it is hard to know whether the advances in science happened because of the changes in societal outlook or the other way around, or indeed whether they happened hand in hand. Huxley saw Darwin as the man who finally made it impossible for intelligent people to disbelieve in evolution. We now realize that, sooner or later, those who had read Lamarck, Lyell, Mendel, Huxley himself would have eventually come to see the truth. In that story, Darwin played a part. The idea, however, that he was alone responsible for the scales falling from the eyes of the human race is a piece of mythology, as implausible as many of the more ancient mythologies which his disciples believed themselves to have demolished. Two ideas, in the discussion of evolution, have been shown in this book to be distinctive to Darwin. One is that evolution always proceeds by infinitesimally gradual and small processes, that nature never leaps. This idea is, at the very least, proved time and again to be questionable. The second idea, that nature is everlastingly at war, and that evolutionary progress happens by conflict, is almost the reverse of the truth. Huxley, when he first heard of Darwin's thesis, exclaimed, 'How extremely stupid not to have thought of that!'[21] After nearly 150 years of thinking about these two ideas, many human beings have come to recognize that such sudden realizations of the truth are often indications of collective psychology. Huxley and the other Victorians who leapt at the idea of the survival of the fittest as the explanation of everything did so for other than scientific reasons. We can see that now. The passage of time makes it seem 'extremely stupid' not to modify such ideas. The great fact of evolution is now indisputable. Darwin and his associates played a major role in this story. Theirs, however, was not the monopoly of truth, and in the perspective of time their errors seem as towering as their achievement.

Acknowledgements

MY FIRST DEBT is to those who have worked, over so many years, on the magnificent Darwin Correspondence Project, based in the Cambridge University Library. The fruits of their labours, in multitudinous printed volumes, and online, make researching the life of Charles Darwin possible for all of us. Especial thanks, in my own case, are due to Professor James Secord; to Dr Alison Pearn, who has helped with many inquiries, with Sam Evans who also provided material. Many thanks to Dr Marina Franca-Spada for useful suggestions, and to Professor Jonathan Haslam and Dr Karina Urbach for being kind enablers. Dr James Le Fanu read the typescript in more than one version, with great patience. John Hands also read it, and made innumerable helpful suggestions and corrections. Charles Colville read the draft typescript too, and offered encouragement. Any remaining howlers are my fault alone. As usual, it would not have been possible to write without the London Library and the British Library; many thanks to the staff in both places. Amy Boyle typed and retyped, and I thank her for her patience. In differing ways, I owe so much to my agents, Clare Alexander and Leah Middleton, and to my publishers, especially to Roland Philipps, Nick Davies and Caroline Westmore. Peter James has been a searchingly punctilious and patient copy-editor; and I am grateful to Douglas Matthews for compiling the index. It was while I was writing my last book about the Victorians for John Murray, over twenty years ago, that Ruth Guilding first took me to Down House on an early summer day, when the garden was in bloom. Many thanks to her for that, as for much else. I suppose it was during that afternoon that this book began its long gestation.

Illustration Credits

Alamy: 1 above right/portrait by Joseph Wright of Derby/Active Museum, 4 below/Natural History Museum, 8 below/Natural History Museum, 9 above right/Granger Historical Picture Archive, 13 below right/Natural History Museum. Bridgeman Images: 12 below right/Bibliothèque Nationale Paris. Reproduced by kind permission of the Syndics of Cambridge University Library: 3 above right, 9 centre left and below right. Reproduced by kind permission of the Master and Fellows of Christ's College, Cambridge: 3 below. Edinburgh University Library, Scotland: 2 above left/ Bridgeman Images. Mary Evans Picture Library: 12 above left. Getty Images: 3 centre left/photo DeAgostini, 5 below right, 7 above/ Universal History Archive, 8 above left, 11 above left and right, 12 above right, 13 centre left and 16 below/photo Michael Nicholson/ Corbis, 15 above right/Universal Images Group. Natural History Museum/photos Mary Evans Picture Library: 2 centre left, 6 below, 8 above right, 15 above left. Karl Pearson, *The Life, Letters and Labours of Francis Galton*, vol. 1, 1914: 13 above right. Gwen Raverat, *Period Piece: A Cambridge Childhood*, Faber & Faber, 1952, illustration of 'Aunt Etty ordering dinner in her patent anti-cold mask', p. 123. © Estate of Gwen Raverat. All rights reserved, DACS. 2017: 11 below. Courtesy of Shropshire Council, Shropshire Museums: 1 centre left/portrait by W. W. Ouless, *c.*1830. Photo © Wedgwood Museum/WWRD: 1 below/family portrait by George Stubbs, 1780. Wellcome Library, London: text p.v, 2 below right, 4 above left and right, 5 above right and centre left, 6 above, 10 centre left and above right, 12 below left, 14 above/*Punch*, 28 December 1861, 14 below/Charles Henry Bennett 1863, 15 below, 16 above/Alexander von Humboldt, *Vues des Cordilleres*, vol. 2, plate 25, 1810–1813.

Reproduced with permission from John van Wyhe, ed. 2002– The Complete Work of Charles Darwin Online (http://darwin-online.org.uk/): 7 below left and right, 10 below.

Notes

Abbreviations

Autobiography	Charles Darwin, *The Autobiography of Charles Darwin, 1809–1882 with Original Omissions Restored*, ed. by his granddaughter Nora Barlow, New York: W. W. Norton, 1969
Browne, *Power*	Janet Browne, *Charles Darwin: The Power of Place*, London: Jonathan Cape, 2002
Browne, *Voyaging*	Janet Browne, *Charles Darwin: Voyaging*, London: Jonathan Cape, 1995
C	*The Correspondence of Charles Darwin*, Cambridge: Cambridge University Press, 1985 onwards. The first number following *C.* indicates the volume number, the second the page number: so *C.* 1. 16 is an abbreviation for *The Correspondence of Charles Darwin*, volume 1, page 16
CUL	Cambridge University Library
DAR	Darwin Archive, Cambridge University Library
DNB	*Dictionary of National Biography*, 1885 onwards, with Supplements
Descent	Charles Darwin, *The Descent of Man and Selection in Relation to Sex*, 2 vols, London: John Murray, 1871. The citation style follows that for *C* (*Correspondence*) above
Diary	*Charles Darwin's Beagle Diary*, ed. Richard Darwin Keynes, Cambridge: Cambridge University Press, 1988, paperback 2001
Expression	Charles Darwin, *The Expression of the Emotions in Man and Animals*, London: John Murray, 1872
LLD	Francis Darwin (ed.), *The Life and Letters of Charles Darwin, Including an Autobiographical Chapter*, 3 vols, London: John

Murray, 1887. The citation style follows that for *C* (*Correspondence*) above

Notebooks — *Charles Darwin's Notebooks, 1836–1844*, ed. Paul H. Barrett et al., Cambridge: Cambridge University Press for the British Museum (Natural History), 1987

ODNB — *Oxford Dictionary of National Biography*, ed. H. C. G. Matthew et al., Oxford: Oxford University Press, 2004 onwards

Origin (1859) — Charles Darwin, *On the Origin of Species by Means of Natural Selection or The Preservation of Favoured Races in the Struggle for Life*, London: John Murray, 1859

Origin (1872) — Charles Darwin, *On the Origin of Species by Means of Natural Selection or The Preservation of Favoured Races in the Struggle for Life*, 6th edn, London: John Murray, 1872

Voyage — Charles Darwin, *The Voyage of the Beagle*, ed. Janet Browne and Michael Neve, London: Penguin, 1989

Prelude

1. Dawkins, 'The Descent of Edward Wilson'.
2. *Guardian*, 7 November 2014.
3. Quoted Le Fanu, p. 125.
4. Gould, 'Is a New and General Theory of Evolution Emerging?', *Palaeobiology*, 6 (1), 1980, p. 119.
5. Eldredge, *Reinventing Darwin*, p. 95.
6. Nagel, p. 7.

Chapter 1: A Symbol

1. Thomas Huxley, 'The Darwin Memorial' (1885), *Collected Essays*, vol. 2, p. 249.
2. *Autobiography*, pp. 104–5.
3. I owe these insights to Paul White's *Thomas Huxley: Making the 'Man of Science'* (2003), pp. 58–62.
4. Balzac, *The Wild Ass's Skin*, p. 41.
5. By three independent researchers, Correns, Tschermak and de Vries. Correns accused de Vries of trying to pass off Mendel's discoveries as his own, and it was partly the violence of their spat which alerted the scientific world to the truly revolutionary nature of Mendel's discovery.

6. Dawkins, *The Blind Watchmaker*, p. 6.
7. Thomas Huxley, 'The Darwin Memorial' (1885), *Collected Essays*, vol. 2, p. 252.
8. *New Statesman*, 20 April 2011.
9. Owen, *On the Nature of Limbs*, p. 86.
10. Stephen Jay Gould, 'Evolution's Erratic Pace', *Natural History*, 86(5): 14.
11. A 'Bulldog' is the nickname given to the University policemen at Oxford and Cambridge. Huxley did not mean it as a canine image, but a metaphor in which he saw himself as dragooning the young, and disciplining the heretics. He often so referred to himself in the 1870s. Leonard Huxley, vol. 1, p. 363.

Chapter 2: The Old Hat

1. Watson, pp. 32, 466, 477 and *passim*.
2. Hilton, p. 211.
3. Hall and Clutton-Brock, p. 77.
4. Ibid., p. 220.
5. Annan in Plumb, p. 244. A revised version of the essay appears in Annan, *The Dons*, pp. 304–41.
6. Lancaster, pp. 1–8.
7. Ibid., p. 8.
8. Many, probably most, British people are scarcely aware that these institutions, once so central to intellectual and political life, exist. A 'club' as generally understood, means a music venue, a place to dance and take recreational drugs. We are talking here about 'clubland'.
9. Francis, *Herbert Spencer*, p. 22.
10. Erasmus Darwin, *The Temple of Nature*, vol. 2, pp. 41–6.
11. Francis, *Herbert Spencer*, p. 318.
12. Erasmus Darwin, *The Botanic Garden*, vol. 2, pp. 303–4, 307–8.
13. King-Hele, p. 316.
14. Austen, *Pride and Prejudice*, p. 202.
15. William Hazlitt, 'My First Acquaintance with Poets', in *The Complete Works of William Hazlitt*, ed. P. P. Howe, after the edition of A. R. Waller and Arnold Glover, 21 vols, London: J. M. Dent, 1931–4 (first published in *The Liberal*, No. 3, 1823).
16. O'Brien, p. 34 and *passim*.
17. Thomson, p. 30.
18. Ibid., p. 27.

19. Ibid., p. 20.
20. *LLD*. 1. 28.
21. *LLD*. 1. 27.
22. Bowlby, p. 62.
23. Butler, *Notebooks*, p. 112.
24. Ibid., p. 127.
25. Ibid., p. 106.
26. *LLD*. 1. 29.

Chapter 3: What He Owed to Edinburgh

1. *Autobiography*, p. 24.
2. Mack, p. 229.
3. This was Isaac Williams – see Tyerman, p. 187.
4. *Autobiography*, p. 27.
5. Desmond et al., p. 16.
6. *LLD*. 1. 33.
7. *LLD*. 1. 35.
8. Wedgwood and Wedgwood, p. 183.
9. *C*. 1. 2.
10. *C*. 1. 4.
11. *LLD*. 1. 32.
12. Cosh, p. 513.
13. Ibid., p. 517.
14. *LLD*. 1. 27.
15. *C*. 1. 16.
16. *C*. 1. 18.
17. *C*. 1. 23.
18. *C*. 1. 26.
19. Thomson, p. 37.
20. *LLD*. 1. 37.
21. Thomson, p. 39.
22. *LLD*. 1. 42.
23. *LLD*. 1. 44.
24. Desmond et al., p. 19.
25. Austen, *Jane Austen's Letters*, p. 280.
26. Ibid.
27. *C*. 1. 36.
28. *C*. 1. 39.
29. *C*. 1. 134, but the phrase is highlighted by Thomson, p. 43.

30. Eiseley, *Darwin's Century*, p. 66.
31. Cross and Livingstone, p. 1419.
32. Hutton, quoted Eiseley, *Darwin's Century*, p. 73.
33. *Encyclopaedia Britannica*, 11th edn, 1911, vol. 28, p. 523.
34. As Eiseley saw so eloquently, *Darwin's Century*, *passim*.
35. *Autobiography*, p. 26.
36. Thomson, p. 53.
37. Grant, p. 433, quoted in Thomson, p. 52.
38. Eldredge, *Eternal Ephemera*, p. 22.
39. Ibid.
40. Ibid., p. 24.
41. *Autobiography*, p. 50.
42. Scott, p. 275.
43. Thomson, p. 83.
44. *Autobiography*, p. 49.
45. Thomson, p. 83.
46. Notebook preserved (DAR18) in Cambridge University Library, transcribed Thomson, p. 66.
47. See Denton, p. 103.
48. Eiseley, *Darwin's Century*, p. 30.
49. Ibid., p. 34.
50. Boswell, vol. 2, p. 75.
51. Ibid., p. 260.
52. Boswell, vol. 5, p. 111.
53. Eiseley, *Darwin's Century*, p. 44.
54. Boyle, p. 639.
55. Lewes, p. 295.
56. Erasmus Darwin, *Zoonomia*, vol. 1, p. 506, quoted King-Hele, p. 300.
57. Coleridge, vol. 1, p. 177.
58. Quoted Eiseley, *Darwin's Century*, p. 48.
59. Ibid., p. 49.
60. See Hands, p. 263.
61. Ibid., p. 372.
62. See p. 15 *supra*.
63. Coleridge, vol. 4, pp. 574–5.
64. *LLD*. 1. 45.
65. *C*. 1. 48.
66. *C*. 1. 72.
67. *C*. 1. 325.
68. *C*. 1. 48.

Chapter 4: Cambridge: Charles Darwin, Gent

1. *C.* 1. 302.
2. Paul White, vol. 1, p. 5.
3. *LLD.* 1. 46.
4. *C.* 1. 58.
5. *C.* 1. 56.
6. Browne, *Voyaging*, p. 15.
7. *LLD.* 1. 45.
8. Barlow, *Darwin and Henslow*, p. 4 and *passim*.
9. *DNB*, vol. 9, p. 586.
10. *LLD.* 1. 52.
11. *LLD.* 1. 49.
12. *C.* 1. 70.
13. *C.* 1. 62.
14. *C.* 1. 66.
15. *C.* 1. 62.
16. *C.* 1. 68.
17. *C.* 1. 70.
18. *C.* 1. 73.
19. *C.* 1. 71.
20. Searby, p. 595.
21. *C.* 1. 101.
22. Agnes Mary Clerke, *Encyclopaedia Britannica*, 13th edn, 1926, vol. 13, p. 874.
23. Norman Kemp Smith in Hume, *Hume's Dialogues*, p. v.
24. Quoted Kemp Smith in ibid., p. 22.
25. Leslie Stephen, *English Thought*, vol. 1, p. 311, quoted Kemp Smith in ibid., p. 30.
26. Ibid., p. 216.
27. Ibid., p. 227.
28. Ibid., p. 211.
29. Paley, p. 132.
30. Ibid., p. 25.
31. *LLD.* 1. 307.
32. Quoted Hands, p. 264.
33. *LLD.* 1. 309.
34. Paley, p. 193.
35. Blake, p. 147.
36. Auden, p. 173.
37. *LLD.* 1. 309.

38. *LLD*. 1. 56.
39. Thomson, p. 128.
40. A. P. Martin, vol. 1, p. 19.

Chapter 5: The Voyage of the *Beagle*

1. Anderson, p. 385.
2. *C*. 1. 127.
3. *C*. 1. 128.
4. *C*. 1. 131.
5. *C*. 1. 134
6. James Taylor, pp. 36–7.
7. FitzRoy, pp. 13–17.
8. *Voyage*, p. 9.
9. Boswell, vol. 1, p. 348.
10. *Voyage*, pp. 426–32; Moorehead, p. 26.
11. *C*. 1. 136.
12. I owe the list of comparisons to Thomson, p. 139.
13. *C*. 1. 142
14. *C*. 1. 139
15. *C*. 1. 141.
16. *C*. 1. 144.
17. Greville, p. 130.
18. Ibid., p. 142.
19. *C*. 1. 151.
20. *C*. 1. 154.
21. *C*. 1. 155.
22. *C*. 1. 161.
23. *C*. 1. 163.
24. *C*. 1. 201. I have been unable to find of what Dr Darwin's anti-nausea receipt of raisins consisted.
25. *C*. 1. 205 (all FitzRoy quotations above).
26. Thomson, p. 143.
27. *C*. 1. 226.
28. Barlow, *Charles Darwin and the Voyage of the Beagle*, p. 155.
29. Ibid., p. 25, footnote quoting his *Autobiography*.
30. *Autobiography*, p. 77, quoted *C*. 1. 239.
31. Ibid., p. 101.
32. Mrs Lyell, *Life*, vol. 1, p. 26.
33. Ibid., p. 271.

34. Ibid., p. 316
35. Browne, *Voyaging*, p. 190
36. *C.* 1. 205.
37. Browne, *Voyaging*, p. 203.
38. *Diary*, p. 36.
39. Ibid., p. 37.
40. Ibid.
41. *C.* 1. 238.
42. *Diary*, p. 43.
43. Ibid., p. 49.
44. *C.* 1. 218.
45. *C.* 1. 227.
46. *C.* 1. 235.
47. *Diary*, p. 52.
48. *C.* 1. 226.
49. *Diary*, p. 51.
50. Zweig, p. 166.
51. Hemming, *Amazon Frontier*, p. 572.
52. *Diary*, p. 62.
53. Ibid., p. 51.
54. See Kirsten Schultz, *Tropical Versailles* (2001).
55. Ibid., p. 83.
56. Freyre, p. 75.
57. Ibid., p. 71.
58. *Diary*, p. 66.
59. Ibid.
60. Ibid., p. 67.
61. Ibid., p. 70.
62. *C.* 1. 232.
63. *C.* 1. 237.
64. *C.* 1. 237.
65. *C.* 1. 232–9.
66. *C.* 1. 237.
67. *C.* 1. 236.
68. *C.* 1. 233.
69. *C.* 1. 247.
70. *Diary*, p. 79.
71. *Voyage*, p. 50.
72. *Diary*, p. 79.
73. Ibid., p. 80.
74. Hugh Thomas, p. 743.

75. *Descent*, p. 54.
76. Ibid., p. 128.
77. Ibid., p. 173.
78. Seibt, p. 248.
79. *Diary*, p. 81.
80. Ibid.
81. Ibid., p. 85.
82. Ibid., p. 89.
83. Boswell, vol. 3, p. 200.
84. Ibid., p. 203.
85. *Diary*, p. 91.
86. Ibid., p. 97.
87. Ibid., p. 100.
88. Ibid.
89. Ibid., p. 103.
90. Ibid., p. 104.
91. Ibid., p.109.
92. Ibid.
93. Quoted ibid., p. 106.
94. Eldredge, *Eternal Ephemera*, pp. 92–3.

Chapter 6: 'Blackbirds . . . gross-beaks . . . wren'

1. *Diary*, p. 133.
2. Cook, p. 26.
3. *Diary*, p. 122.
4. Ibid., p. 133.
5. *Voyage*, p. 190.
6. *Diary*, p. 122.
7. *C.* 1. 304.
8. *Diary*, p. 223.
9. Ibid., p. 145.
10. Ibid.
11. Ibid., p. 146.
12. *C.* 1. 327.
13. *C.* 1. 363.
14. Moorehead, p. 110.
15. *Diary*, p. 166.
16. Ibid., p. 171.
17. Ibid., p. 190.

18. Ibid., p. 191.
19. Ibid., p. 198.
20. Cook, p. 49.
21. Milton, p. 228.
22. Ibid.
23. Ibid., p. 229.
24. Trevelyan, p. 37.
25. *Diary*, p. 226.
26. Ibid.
27. Ibid., p. 227.
28. Moorehead, p. 83.
29. Quoted ibid., p. 85.
30. *Diary*, p. 235.
31. Ibid., p. 237.
32. *C.* 1. 286.
33. *C.* 1. 166.
34. *C.* 1. 266.
35. *Origin* (1859), p. 490.
36. *C.* 1. 381.
37. *Diary*, p. 245.
38. Ibid., p. 249.
39. Ibid., p. 253.
40. Ibid., p. 250.
41. *C.* 1. 308.
42. *C.* 1. 357.
43. *C.* 1. 371.
44. *C.* 1. 533.
45. *C.* 1. 460.
46. *Diary*, p. 250.
47. Tennyson, vol. 2, p. 370, *In Memoriam*, l. iv.
48. See Livio, p. 83. 'We know today that the age of the Earth is about 4.54 billion years. *This is about fifty times longer than Kelvin's estimate.*'
49. *Voyage*, p. 403.
50. Ibid., p. 401.
51. *Diary*, p. 261.
52. *C.* 1. 411.
53. *Diary*, p. 253.
54. Ibid., p. 255.
55. Ibid., p. 261.
56. Ibid., p. 257.
57. Ibid., p. 315.

58. Ibid.
59. Woodruff, pp. 745, 747.
60. *Diary*, p. 263.
61. *C.* I. 418.
62. *C.* I. 418.
63. *Autobiography*, p. 75.
64. Ibid.
65. *Diary*, p. 280.
66. *C.* I. 363.
67. *Diary*, p. 295.
68. Ibid., p. 296.
69. Ibid., p. 297.
70. *C.* I. 461.
71. *Diary*, p. 353.
72. Ibid., p. 359.
73. Ibid., p. 353.
74. Ibid., p. 359.
75. Quoted ibid., p. 360.
76. Barlow, *Charles Darwin and the Voyage of the Beagle*, p. 246.
77. Hands, p. 259.
78. Desmond and Moore, p. 220.
79. Hands, p. 259.
80. Ibid., p. 314.
81. Desmond and Moore, p. 172.
82. *Diary*, p. 347.
83. *Charles Darwin's Notebooks*, ed. Chancellor and Van Whye, pp. 415–16.
84. Ibid., p. 431.
85. Quoted Eiseley, *Darwin and the Mysterious Mr X.*, p. 214.
86. Barlow, *Charles Darwin and the Voyage of the Beagle*, p. 247.
87. Ibid.
88. *Origin* (1859), p. 434.
89. Quoted Denton, p. 53.
90. Stove, p. 94.
91. *C.* I. 472.
92. *Diary*, p. 367.
93. Ibid.
94. A friend who read this sentence said, 'But Darwin went on to have ten children.' I did not write 'no libido'. Having ten children over a longish period could be achieved with infrequent coition.
95. *Diary*, p. 369.
96. Ibid., p. 379.

97. Ibid., p. 384.
98. Ibid., p. 393.
99. Ibid., p. 384.
100. Ibid.
101. Ibid.
102. Ibid., p. 396.
103. Browne, *Voyaging*, p. 9.
104. John Ritchie, p. 1.
105. Marshall, p. 12.
106. *Diary*, p. 396.
107. That is, William Kirby and William Spence, *An Introduction to Entomology* (1815–26).
108. *Diary*, p. 403.
109. Ibid., p. 413.
110. *C.* 1. 503.
111. *Diary*, p. 447.

Chapter 7: The Ladder by Which You Mounted

1. *Autobiography*, p. 41.
2. Desmond and Moore, p. 189.
3. *C.* 2. 14.
4. *C.* 2. 16.
5. *C.* 2. 58.
6. *C.* 2. 58.
7. *C.* 2. 58.
8. Quoted Eiseley, *Darwin and the Mysterious Mr. X*, p. 83.
9. Ibid., p. 64.
10. Quoted ibid.
11. Eiseley, *Darwin and the Mysterious Mr X*, pp. 40–1.
12. *LLD.* 1. 229.
13. Mrs Lyell, vol. 1, p. 461.
14. Quoted Bailey, p. 119.
15. *Sic* in Mrs Lyell, vol. 1, p. 459. But at this date FitzRoy was still at sea and the court martial was not until May 1836, so the date must be wrong. See *Sydney Herald*, 19 May 1836, p. 4, 'Naval Courts Martial' (Trove Digitised Newspapers).
16. Bailey, p. 121.
17. Ibid., p. 122.
18. Charles Lyell, vol. 2, p. 36.

19. Ibid., p. 11.
20. Ibid., p. 12.
21. Ibid., p. 37.
22. Ibid., p. 44.
23. *C.* 2. 128.
24. *C.* 2. 155.
25. *C.* 2. 430.
26. *C.* 2. 431.
27. *C.* 2. 38
28. *Notebooks*, p. 170.
29. Ibid., p. 171.
30. Ibid., p. 188.
31. Ibid., p. 189.
32. Notebook B dates from July 1837–January 1838, C from February–July 1838, D from July–October 1838, E from September 1838–July 1839.
33. *Notebooks*, pp. 196–7.
34. Ibid., p. 227.
35. Ibid.
36. Ibid., p. 167.
37. Litchfield, vol. 1, p. 244.
38. Martineau, *Selected Letters*, p. 41.
39. *Autobiography*, p. 120.
40. Cobbett, p. 46.
41. See Malthus, p. 24.
42. See Gregory Clark, *A Farewell to Alms: A Brief Economic History of the World* (2007), pp. 29ff.
43. Quoting Malthus, p. 529.
44. Notebook E.3, *Notebooks*, p. 397.
45. *Quarterly Review*, October 1825, 32: 420–2.
46. Letter to Milnes, Martineau, *Letters*, p. 95.
47. Martineau, *Life in the Sickroom*, pp. 8–9.
48. Martineau, *Letters*, p. 90.

Chapter 8: Lost in the Vicinity of Bloomsbury

1. *Notebooks*, p. 574.
2. Ibid., p. 565.
3. Ibid., p. 567.
4. Ibid.
5. Ibid., p. 574.

6. Ibid., p. 585.
7. Notebook N.97, Ibid., p. 591.
8. *Notebooks*, p. 577.
9. *C.* 2. 443.
10. *C.* 2. 445.
11. *C.* 2. 444.
12. *C.* 2. 115.
13. *C.* 2. 115.
14. *C.* 2. 114.
15. *C.* 2. 132.
16. *C.* 2. 136.
17. *C.* 2. 145.
18. *C.* 2. 163.
19. *C.* 2. 172.
20. *C.* 2. xix.
21. *C.* 2. 123.
22. *C.* 2. 172.
23. *C.* 2. 135.
24. *C.* 2. 165.
25. Colp, p. 17.
26. *Notebooks*, p. 175; Colp, p. 17.
27. *C.* 2. 443.
28. *C.* 2. 129.
29. Ashton, p. 42.
30. Ibid., pp. 42–3.
31. *C.* 2. 147.
32. *C.* 2. 147.
33. *C.* 2. 153.
34. *C.* 2. 149.
35. *C.* 2. 195.
36. *C.* 2. 131. The form 'Cuvington' appears at C. 2. 135.
37. *C.* 2. 171.
38. *C.* 2. 169.
39. *C.* 2. 159, 236.
40. *C.* 2. 236.
41. *C.* 2. 197.
42. *C.* 2. 256.
43. *C.* 2. 254–5.
44. *C.* 2. 237.
45. *Quarterly Review*, October 1825, 32: 194.
46. Ibid., 201.

47. *Athenaeum*, June 1839, 607: 446.
48. *C.* 2. 199.
49. Agnes Mary Clerke, *Encyclopaedia Britannica*, 13th edn, 1926, vol. 13, p. 875. See also Andrea Wulf's superb *The Invention of Nature: The Adventures of Alexander von Humboldt, the Lost Hero of Science* (2015).
50. *C.* 2. 221.
51. *C.* 2. 218.
52. Browne, *Voyaging*, p. 417.
53. *C.* 2. 249.
54. *C.* 2. 253.
55. *C.* 2. 255.
56. *C.* 2. 399.
57. *C.* 2. 262
58. Browne, *Voyaging*, p. 429, quoting Darwin mss., CUL, August 1840.
59. Litchfield, vol. 2, p. 249.
60. *Encyclopaedia Britannica*, 15th edn, 1974, vol. 1, p. 142.
61. Browne, *Voyaging*, p. 433.
62. Bowler and Pickstone, vol. 6, p. 178.
63. *C.* 2. 284.
64. *C.* 2. 387.
65. Eiseley, *Darwin and the Mysterious Mr. X*, p. 59.
66. Ibid., p. 171.
67. *Notebooks*, p. 341.
68. Ibid., p. 342.
69. Notebook D.35, *Notebooks*, p. 342.
70. Ibid.
71. *C.* 4. 139.
72. *Autobiography*, p. 61.
73. Rupke, *Richard Owen*, p. 1.
74. *C.* 2. 22.
75. *C.* 2. 76.
76. Rupke, *Richard Owen*, p. 51.
77. Hare, vol. v, p. 358.
78. *Notebooks*, p. 210.
79. *LLD.* 1. 74.
80. *Notebooks*, p. 554.
81. And all other references to Down(e) C. 2. 324–5.
82. *C.* 2. 326.
83. *C.* 2. 334.
84. Browne, *Voyaging*, p. 443.
85. Ibid.

Chapter 9: Half-Embedded in the Flesh of their Wives

1. Keynes, p. 73.
2. Ibid., p. 74.
3. Boulter, p. 45.
4. C. 3. 214.
5. Darwin and Wallace, p. 114.
6. Ibid., p. 115.
7. *Origin* (1859), p. 83.
8. Keynes, p. 113.
9. Ibid., pp. 84–99.
10. Boulter, p. 34.
11. *C.* 3. 43.
12. *C.* 3. 51.
13. Secord, p. 169.
14. Ibid., p. 235.
15. Ibid., p. 234.
16. Chambers, *Vestiges*, p. 226.
17. Ibid., p. 148.
18. Ibid., p. 182.
19. Ibid., p. 195.
20. Ibid., p. 203.
21. Quoted in the sequel to Chambers's *Vestiges*, *Explanations*, p. 92.
22. Chambers, *Vestiges*, p. 366.
23. Ibid., p. 368.
24. Secord, p. 246.
25. Desmond and Moore, p. 323.
26. Hilton, pp. 474–5.
27. *C.* 3. 103.
28. *C.* 3. 108.
29. Quoted Browne, *Voyaging*, p. 467.
30. *C.* 4. 384.
31. *C.* 4. 238.
32. Stott, p. xxi.
33. *C.* 4. 230.
34. *C.* 4. 369.
35. *C.* 4. 369.
36. *C.* 4. 399.
37. *C.* 4. 128.
38. *C.* 4. 147.
39. *C.* 4. 147.

40. *C.* 4. 147.
41. *C.* 4. 181.
42. Wedgwood and Wedgwood, p. 250.
43. Brian Smith, p. 188.
44. Ibid., p. 196.
45. Gully, p. 60.
46. Ibid., p. 177.
47. Ibid., p. 181.
48. *C.* 4. 375.
49. Keynes, p. 239.
50. *C.* 4. 226.
51. *C.* 4. 226.
52. Keynes, p. 140.
53. *C.* 4. 224.
54. Brian Smith, p. 210.
55. *C.* 4. 244.
56. *C.* 4. 239.
57. *C.* 4. 385.
58. Keynes, p. 147.
59. Colp, p. 45.
60. Ibid.
61. Ibid., p. 47.
62. *Autobiography*, pp. 86–7.
63. Keynes, p. 245.
64. Ibid.
65. Wittgenstein, p. 223.
66. Since Keynes (2001).
67. Select Committee on the State of Children Employed in the Manufactories of the United Kingdom, 1816, quoted ibid., p. 135.
68. Ibid., p. 154.
69. Ibid., p. 148.
70. Ibid., p. 20.
71. Ibid., p. 148.
72. Ibid., p. 152.
73. Ibid., p. 154.
74. Ibid., p. 156.
75. The letters are lost: ibid., p. 156.
76. Gully, p. 564.
77. Haight, p. 99.
78. Ibid., p. 107.
79. Keynes, p. 159.

80. Newman, p. 217.
81. Ibid., p. 223.
82. St Aubyn, p. 293.
83. *C.* 5. 13.
84. *C.* 5. 18.
85. *C.* 5. 24.
86. *C.* 5. 28.
87. Keynes, p. 182.
88. Ibid., p. 183.
89. *C.* 5. 29.
90. Litchfield, vol. 2, p. 86.

Chapter 10: An Essay by Mr Wallace

1. *C.* 5. 219.
2. *C.* 5. 64.
3. *LLD* 2. 188–9.
4. Blyth, p. ii.
5. *C.* 5. 400.
6. *C.* 5. 372.
7. *C.* 5. 453.
8. *C.* 5. 438.
9. *C.* 5. 440.
10. *C.* 6. 217.
11. *C.* 6. 195.
12. *C.* 6. 178.
13. *C.* 6. 199.
14. *C.* 6. 260.
15. *C.* 6. 260.
16. *C.* 6. 260.
17. *C.* 6. 217.
18. Wedgwood and Wedgwood, p. 2.
19. *C.* 6. 85.
20. Browne, *Voyaging*, p. 522.
21. Ibid.
22. *C.* 6. 45.
23. Browne, *Voyaging*, p. 525.
24. *C.* 6. 41.
25. *C.* 6. 276.
26. *C.* 6. 268.

27. *C.* 6. 268.
28. *C.* 6. 269.
29. *C.* 6. 45.
30. Hope Simpson, pp. 22, 29.
31. Browne, *Voyaging*, p. 535.
32. Ibid.
33. Ibid., p. 533.
34. Ibid., p. 532 and *passim*.
35. *C.* 6. 225.
36. *C.* 6. 92.
37. *C.* 6. 432, 433.
38. *C.* 6. 432, 433.
39. *C.* 6. 448.
40. *C.* 6. 492.
41. *C.* 6. 446.
42. *C.* 6. 514.
43. Dickens, *Nicholas Nickleby*, Chapter 7.
44. Ibid., Chapter 3.
45. Hemming, *Naturalists in Paradise*, p. 16.
46. *C.* 7. 119.
47. See Sir Gavin de Beer's Foreword to Darwin and Wallace, *Evolution by Natural Selection*.
48. *C.* 7. 123.

Chapter 11: A Poker and a Rabbit

1. *C.* 7. 123.
2. *C.* 7. 113.
3. *C.* 7. 113.
4. *C.* 7. 102.
5. *C.* 7. 116.
6. *C.* 7. 121.
7. *C.* 7. 121.
8. *C.* 7. 127.
9. *C.* 7. 129.
10. *C.* 7. 125.
11. Browne, *Power*, p. 41.
12. Ibid.
13. Ibid., p. 42.
14. C. 7. 138.

15. Browne, *Power*, p. 47.
16. *C.* 7. 138.
17. *C.* 7. 138.
18. *C.* 7. 141.
19. *C.* 7. 149.
20. Tennyson, vol. 2, p. 315, *In Memoriam*, l. 4. Tennyson wrote chiefly in reaction to Lyell, and had finished *In Memoriam before* he read *Vestiges*. See Tennyson, vol. 2, p. 371.
21. 'In Memoriam' Prologue, l. 4. Tennyson, vol. 2, p. 315.
22. DAR 245: 300.
23. Litchfield, p. 175.
24. *C.* 7. 270.
25. Quoted Desmond and Moore, p. 473.
26. Browne, *Power*, p. 73.
27. Paston, p. 170.
28. *C.* 7. 290.
29. *C.* 7. 290.
30. *C.* 7. 295.
31. Browne, *Power*, p. 75.
32. Ibid., p. 72.
33. *C.* 7. 264.
34. *C.* 6. 310.
35. Desmond and Moore, p. 476.
36. *C.* 7. 350.
37. *C.* 7. 365.
38. *C.* 6. 376.

Chapter 12: Is It True?

1. *Origin* (1859), p. 2.
2. Notebook of 1837, quoted Eiseley, *Darwin's Century*, p. 352.
3. *Origin* (1859), p. 79.
4. Ibid., p. 84.
5. Ibid., p. 109.
6. Wallace, 'On the Law Which Has Regulated the Introduction of New Species', *Annals and Magazine of Natural History*, September 1855, 16(2), quoted Denton, p. 87.
7. Slotten, p. 117.
8. Denton, p. 78.
9. *Origin* (1859), p. 173.

10. Denton, p. 21.
11. *Origin* (1859), p. 186.
12. Nilsson and Pelger, pp. 53–8.
13. Summarized in Dawkins, *Climbing Mount Improbable*, pp. 149 ff.
14. Ibid., p. 154.
15. Hands, p. 307.
16. *Origin* (1859), p. 490.
17. Vorzimmer, p. 20.

Chapter 13: The Oxford Debate and its Aftermath

1. *C.* 8. 244.
2. *C.* 8. 272.
3. *C.* 8. 266.
4. *C.* 8. 299.
5. Rupke, *Richard Owen*, p. 239.
6. Yanni, p. 65.
7. Hesketh, p. 82.
8. Ibid., p. 90.
9. *C.* 8. 227.
10. Hesketh, p. 7.
11. Ibid., p. 85.
12. Secord, p. 50.
13. Hesketh, p. 105.
14. Irvine, p. 94.
15. Ibid., p. 93.
16. Litchfield, vol. 2, p. 181.
17. Browne, *Power*, p. 149.
18. Litchfield, vol. 2, p. 180.
19. *C.* 8. 451
20. *C.* 8. 451.
21. Irvine, p. 99.
22. *C.* 8. 366.
23. *C.* 10. 330.
24. Doyle, *The Memoirs of Sherlock Holmes*, p. 227: '"There is nothing in which deduction is so necessary as in religion," said he, leaning with his back against the shutters. "It can be built up as an exact science by the reasoner. Our highest assurance of the goodness of Providence seems to me to rest in the flowers. All other things, our powers, our desires, our food, are really necessary for our existence in the first instance. But this rose

is an extra. Its smell and its colour are an embellishment of life, not a condition of it. It is only goodness which gives extras.'"

25. Ronse De Craene, p. viii.
26. Sanderson, pp. 257–9.
27. Barlow, *Darwin and Henslow*, p. 225.
28. Browne, *Power*, p. 153.
29. *C.* 9. 98.
30. *C.* 9. 155.
31. Litchfield, vol. 2, p. 247.
32. Campbell and Matthews, 'Darwin's Illness Revealed', *Postgrad Med J*, 2005, 81: 248–51.
33. *C.* 9. 133.
34. Browne, *Power*, p. 229, quoting DAR 112(A): 79–82.
35. *C.* 11. 620.
36. Colp, p. 83.
37. Browne, *Power*, p. 249.
38. Ibid., p. 235.
39. *C.* 9. 99.

Chapter 14: Adios, Theory

1. Christopher Wills, *Exons, Introns and Talking Genes: The Science behind the Human Genome Project* (1992), quoted Le Fanu, p. 129.
2. Rose, p. 33.
3. Francis Darwin and Seward, ii, 340. Thanks for this to Dr Alison Pearn.
4. Olby and Gautrey, pp. 7–20.
5. How Darwin arrived at his conclusions about inheritance, who his precursors were in forming his theory and how his theory of 'blending' was dissected by other scientists is fully dealt with in Vorzimmer, Chapters 2 and 6.
6. Ibid., p. 30.
7. Ibid., p. 65.
8. Ibid., p. 130.
9. *C.* 17. 37.
10. *C.* 10. 129–30.
11. *C.* 17. 381.
12. *C.* 17. 519.
13. *C.* 15. 66.
14. Leonard Huxley, *The Life and Letters of T. H. Huxley*, vol. 1, pp. 267–8.

15. Spencer, *Principles of Biology*, vol. 1, p. 453.
16. Løvtrup, p. 33.
17. *LLD*, 3, 193.
18. *C.* 18. 70.
19. An especially good book on him is Mark Francis's *Herbert Spencer and the Invention of Modern Life* (2007).
20. See my *God's Funeral* (1999).
21. Leonard Huxley, *The Life and Letters of T. H. Huxley*, vol. 1, p. 212.
22. Mivart, *Genesis*, pp. 14–15.
23. Ibid., p. 21.
24. *Origin* (1872): pp. 186–7.
25. Ibid., pp. 167–8.
26. Mivart, *Genesis*, p. 37.
27. Ibid., p. 38.
28. *Origin* (1872), p. 174.
29. Gruber, p. 85.
30. Joe D. Burchfield, in *The Oxford Companion to the History of Modern Science*, ed. J. L. Heilbron, p. 225.
31. Quoted Browne, *Power*, p. 315.
32. See Hull, *Darwin and his Critics* (1973), and Vorzimmer.
33. *Origin* (1859), p. 62.
34. Ibid., p. 63.
35. Stove, p. 58.
36. *Descent*. 1. 132–3.
37. Stove, p. 181.
38. Ibid., p. 69.
39. Browne, *Power*, p. 350.
40. *LLD*. 3. 359.
41. *Origin* (1859), p. 79.
42. Sitwell, p. 7.
43. First published as *Le Phénomène humain* (1955).
44. *Descent*. 2. 405.
45. Ibid.
46. Ibid., 1. 34.
47. Ibid., 1. 67.
48. *Voyage of the* Beagle, ed. Browne and Neve, p. 173.
49. *Descent*. 1. 67.
50. Spencer, *First Principles*, p. 287.
51. *Descent*. 1. 60.
52. Spencer, *First Principles*, p. 287.
53. *Descent*. 1. 156.

54. Chorley, Dunn and Beckinsale, vol. 1, p. 447.
55. *Descent.* 1. 180.
56. Ibid., 178–9.
57. Charles Lyell (*Principles of Geology*, vol. 2, p. 489) suggested that the Spanish Inquisition lowered the general levels of intelligence in Europe.
58. *Descent.* 2. 394.
59. Ibid.
60. *Origin* (1859), p. 62.
61. Ibid., p. 63.
62. Paradis and Williams, pp. 204–5, quoted in Stove, p. 56.
63. Stove, p. 3
64. Ibid., p. 56.
65. Ibid., p. 19.
66. Bertrand Russell, p. 697.
67. Secord, p. 236.
68. *C.* 17. xix.
69. Gruber, p. 172.
70. Mivart, *Genesis*, pp. 2–3.
71. Browne, *Power*, p. 354.
72. Gruber, p. 92.
73. Baden Powell, a clergyman-mathematician who died two weeks before the Oxford debate, aged thirty-three. He was an enthusiastic supporter of Darwin's book, but asked Darwin why the *Origin*, having borrowed ideas from himself, Baden Powell, made no acknowledgement of the fact. Darwin's excuse, 'My health was so poor', lacks conviction.
74. Leonard Huxley, *The Life and Letters of T. H. Huxley*, vol. 2, p. 128.
75. *C.* 19. 591.
76. *Descent.* 2. 403.
77. Stove, p. 8.
78. Quoted ibid., p. 16.
79. Browne, *Power*, p. 357.
80. Hudson, p. 298.
81. Raverat, p. 119.
82. Ibid., pp. 125–6.
83. Ruth Schwartz Cowan, 'Sir Francis Galton', *ODNB*, vol. 21, p. 348.
84. *C.* 22. 389.
85. *C.* 22. 389.
86. *C.* 22. xxi.
87. Desmond, Moore and Browne, p. 117.
88. *Descent.* 2. 439.

89. Bland and Hall, pp. 214–15.
90. Ibid., p. 214.
91. Ibid.
92. Foucault, p. 149.
93. *C.* 22. 467.
94. *C.* 22. 517.
95. Weindling, pp. 316–17.
96. Kragh, '*From Here to Eternity*', pp. 246–7.
97. Ruse, 'A Reappraisal of Ernst Haeckel', pp. 711–12.
98. Richards, *The Romantic Conception of Life*, p. 172.
99. Ibid., p. 174.
100. Ibid.
101. Haeckel, p. 5.
102. Richards, *The Romantic Conception of Life*, p. 97.
103. Ibid., p. 254.
104. Goldschmidt, p. 35.
105. Richards, *The Romantic Conception of Life*, p. 351.
106. Ibid.

Chapter 15: Immense Generalizations

1. Roberts, p. 119.
2. Browne, *Power*, p. 338.
3. *C.* 20. 41.
4. *Expression*, p. 45.
5. Ibid., p. 51.
6. Ibid., p. 73.
7. Ibid., p. 155.
8. *LLD.* 1. 33.
9. *Expression*, p. 364.
10. Wallace, quoted Browne, *Power*, p. 318.
11. Ibid.
12. Blum, p. 59.
13. Ibid., p. 72.
14. Ibid., pp. 68–9.
15. *C.* 22. 25–6.
16. Raverat, p. 219.
17. Ibid., p. 209.
18. Browne, *Power*, p. 344.
19. Wheen, p. 363.

20. Browne, *Power*, p. 403.
21. Wheen, p. 364.
22. Hunt, p. 328.
23. Ibid., p. 331.
24. Ruskin, p. 310.
25. Mill, *A System of Logic*, vol. 2, p. 18.
26. *LLD*. 1. 309.
27. Roberts, p. 595.

Chapter 16: Evolution Old and New

1. Wedgwood and Wedgwood, p. 263.
2. Ibid.
3. Browne, *Power*, p. 447.
4. Ibid., p. 446.
5. Ibid., p. 400.
6. See Joel Schwartz, pp. 1–9, and C. 22. xxviii ff.
7. Joel Schwartz, p. 9.
8. Ibid., p. 260.
9. Ibid., p. 131.
10. Ibid., p. 3.
11. Di Gregorio, *From Here to Eternity*, p. 555.
12. Butler, The Way of All Flesh, p. 405.
13. Muggeridge, p. 242.
14. Thanks to Dr Pearn, of the Darwin Correspondence Project, for showing me this lecture.
15. *C*. 20. 231.
16. Pearn's lecture, Philadelphia, 2012.
17. DAR 258: 572.
18. Litchfield, vol. 2, p. 237.
19. Proof sheets are in CUL, quoted Browne, *Power*, p. 471.
20. *DNB* Supplement, 1901–11, p. 285.
21. *Autobiography*, pp. 134–5.
22. Butler, *Life and Habit*, pp. 275–6.
23. Quoted Løvtrup, p. 273.
24. Ibid., p. 274.
25. See Le Fanu, pp. 94–5.
26. Heilbron, p. 283.

27. Hitler, p. 262.
28. Ibid., p. 141.
29. Weil, p. 227.
30. Quoted Løvtrup, p. 423.
31. Le Fanu, p. 262.
32. Browne, *Power*, p. 489.
33. Ibid., p. 490.
34. Wulf, p. 282.
35. *LLD*. 1. 101.
36. Browne, *Power*, p. 495.

Chapter 17 Mutual Aid

1. Bonham's sale catalogue, 21 September 2015, New York.
2. Isaacson, p. 388.
3. Hands, p. 267.
4. Kropotkin, *Mutual Aid*, Conclusion.
5. Gould, 'Kropotkin was No Crackpot', pp. 12–21.
6. Dawkins, 'The Descent of Edward Wilson'.
7. In *Prospect*, July 2012, Issue 196.
8. Hands, p. 278.
9. Dawkins, *The Selfish Gene*, p. 201.
10. Wallace, *The Wonderful Century*, pp. 140–1.
11. Hodge and Radick, p. 297.
12. Nagel, p. 127.
13. Fothergill, p. 111.
14. Keats, p. 72.
15. Weikart, p. 210.
16. Spikins, pp. 32–3.
17. Ibid., p. 33.
18. Ibid., p. 52.
19. See Jane Goodall, *Through a Window: My Thirty Years with the Chimpanzees of Gombe* (1990).
20. Hawks, pp. 77–9.
21. Leonard Huxley, vol. 1, p. 189.

Bibliography

Ackerman, S., *Discovering the Brain*, Washington, DC: National Academy Press, 1992

Allman, W. F., *The Stone Age Present: How Evolution Has Shaped Modern Life: From Sex, Violence and Language to Emotions, Morals, and Communities*, New York: Simon & Schuster, 1994

Amundson, Ron, 'Two Concepts of Constraint: Adaptationism and the Challenge from Development Biology', *Philosophy of Science*, 1994, 61: 556–78

——, 'Historical Development of the Concept of Adaptation', in Michael Rose and George V. Lauder (eds), *Adaptation*, New York: Academic Press, 1996, 11–53

——, *The Changing Role of the Embryo in Evolutionary Thought: Roots of Evo-Devo*, Cambridge: Cambridge University Press, 2005

Anderson, Katharine (ed.), *The Narrative of the Beagle Voyage, 1831–1836*, London: Pickering & Chatto, 2012

Andreski, Stanislav (ed.), *Herbert Spencer: Structures, Function and Evolution*, London: Michael Joseph, 1971

Annan, Noel, 'The Intellectual Aristocracy', in J. E. Plumb (ed.), *Studies in Social History: A Tribute to G. M. Trevelyan*, London/New York/Toronto: Longmans Green, 1955, 241–87

——, *The Dons*, London: HarperCollins, 1999

Anon, Review of *Evolution and Ethics* by T. H. Huxley, 1893, *Athenaeum*, 3430: 119–20

Appel, Toby A., *The Cuvier–Geoffroy Debate: French Biology in the Decades before Darwin*, New York: Oxford University Press, 1987

Armstrong, Patrick H., *All Things Darwin: An Encyclopedia of Darwin's World*, Westport, CT/London: Greenwood Press, 2007

Ashton, Rosemary, *Victorian Bloomsbury*, New Haven/London: Yale University Press, 2012

Atkins, Sir Hedley, *Down: The Home of the Darwins*, London: Phillimore for the Royal College of Surgeons, 1976

Auden, W. H., *A Certain World: A Commonplace Book*, London: Faber & Faber, 1971

Austen, Jane, *Pride and Prejudice* (first published 1813), Oxford: Clarendon Press, 1932

——, *Jane Austen's Letters*, Oxford/New York: Oxford University Press, 1995

Bagehot, Walter, *Physics and Politics* (first published 1869), Boston: Beacon Press, reprint 1956

Bahn, P., and J. Vertut, *Images of the Ice Age*, New York: Facts on File, 1988

Bailey, Edward, *Charles Lyell*, London/Edinburgh: Thomas Nelson, 1962

Balzac, Honoré de, *The Wild Ass's Skin* (trans. Herbert J. Hunt), Harmondsworth: Penguin, 1977

Barlow, Nora, *Charles Darwin and the Voyage of the Beagle*, London: The Pilot Press, 1945

—— (ed.), *Darwin and Henslow: The Growth of an Idea – Letters 1831–1860*, London: John Murray, Bentham-Moxon Trust, 1967

Barrow, John D., and F. J. Tipler, *The Anthropic Cosmological Principle*, New York: Oxford University Press, 1986

Bartholomew, Michael, 'Huxley's Defence of Darwin', *Annals of Science*, 1975, 32: 525–35

Barton, Ruth, 'Evolution: The Whitworth Gun in Huxley's War for the Liberation of Science from Theology', in D. Oldroyd and I. Langham (eds), *The Wider Domain of Evolutionary Thought*, Dordrecht: Reidel, 1983, 261–87

Bashford, Alison, and Philippa Levine (eds), *The Oxford Handbook of the History of Eugenics*, Oxford: Oxford University Press, 2010

Beer, Gillian, *Darwin's Plots: Evolutionary Narrative in Darwin, George Eliot and Nineteenth-Century Fiction*, London: Routledge & Kegan Paul, 1983

Behe, Michael, *Darwin's Black Box: The Biochemical Challenge*, New York: Free Press, 1996

Bell, Sir Charles, *The Hand: Its Mechanism and Vital Endowments as Evincing Design*, London: William Pickering, 1833

Blake, William, *Poetry and Prose of William Blake*, ed. Geoffrey Keynes, London: Nonesuch Press, 1939

Bland, Lucy, and Lesley A. Hall, 'Eugenics in Britain: The View from the Metropole', in Alison Bashford and Philippa Levine (eds), *The Oxford Handbook of the History of Eugenics*, Oxford: Oxford University Press, 2010, 214–15

Blauberg, E. E., *Petr Alexeevich Kropotkin*, Moscow: Institut Philosophii RAN, 2012

Blum, Deborah, *Ghost Hunters*, London: Century, 2007

Blyth, Edward, *Catalogue of the Birds in the Museum Asiatic Society*, Calcutta: J. Thomas, Baptist Mission Press, 1849

Bonney, T. G., *Charles Lyell and Modern Geology*, London: Cassell, 1895
Boswell, James, *Boswell's Life of Johnson* (first published 1799), ed. George Birkbeck Hill, 6 vols, Oxford: Clarendon Press, 1934
Boulter, Michael, *Darwin's Garden: Down House and the Origin of Species*, London: Constable, 2008
Bowlby, John, *Charles Darwin: A Biography*, London: Hutchinson, 1990
Bowler, Peter J., 'Darwinism and the Argument from Design: Suggestions for a Reevaluation', *Journal of the History of Biology*, 1977, 10: 29–43
——, *Evolution: The History of an Idea*, Berkeley: University of California Press, 1984
——, *Theories of Human Evolution: A Century of Debate, 1844–1944*, Baltimore: Johns Hopkins University Press, 1986
——, *Life's Splendid Drama*, Chicago: University of Chicago Press, 1996
——, and John V. Pickstone (eds), *The Cambridge History of Science*, vol. 6: *The Modern Biological and Earth Sciences*, Cambridge: Cambridge University Press, 2009
Boyle, Nicholas, *Goethe: The Poet and the Age*, vol. 1: *The Poetry of Desire (1749–1790)*, Oxford: Clarendon Press, 1991
Brasier, M. D., *Darwin's Lost World: The Hidden History of Animal Life*, Oxford: Oxford University Press, 2009
Brooke, J. H., *Science and Religion: Some Historical Perspectives*, Cambridge: Cambridge University Press, 1991
Brown, L. R., *The Emerson Museum: Practical Romanticism and the Pursuit of the Whole*, Cambridge: Cambridge University Press, 1997
Browne, Janet, *Charles Darwin: Voyaging*, London: Jonathan Cape, 1995 [abbreviated as Browne, *Voyaging*]
——, *Charles Darwin: The Power of Place*, London: Jonathan Cape, 2002 [abbreviated as Browne, *Power*]
Burrow, J. W., *Evolution and Society: A Study in Victorian Social Theory*, Cambridge: Cambridge University Press, 1966
Butler, Samuel, *Life and Habit*, London: Trübner, 1878
——, *Luck or Cunning, as the Means of Organic Modification*, London: A. C. Fifield, 1886
——, *Unconscious Memory*, London: Jonathan Cape, 1920
——, *The Way of All Flesh* (first published 1903), London: Chapman & Hall, 1920
——, *Notebooks*, London: Jonathan Cape, 1952
Byrne, R., *The Thinking Ape: Evolutionary Origins of Intelligence*, Oxford: Oxford University Press, 1995
Cadbury, D., *The Dinosaur Hunters*, London: Fourth Estate, 2000
Camardi, Giovanni, 'Richard Owen: Morphology and Evolution', *Journal of the History of Biology*, 2001, 34: 481–515

Campbell, Anthony K., and Stephanie B. Matthews, 'Darwin's Illness Revealed', *Postgraduate Medical Journal*, 2005, 81: 248–51

Carpenter, W., *Nature and Man: Essays Scientific and Philosophical*, New York: Appleton, 1889

Chadwick, Owen, *The Victorian Church*, 2 vols, London: Adam & Charles Black, 1966

[Chambers, Robert], *Vestiges of the Natural History of Creation*, London: John Churchill, 1844

——, *Explanations: A Sequel to 'Vestiges'*, London, John Churchill, 1845

Chorley, Richard J., Antony J. Dunn and Robert P. Beckinsale, *The History of the Study of Landforms*, 8 vols, London: Methuen, 1964

Clark, Gregory, *A Farewell to Alms: A Brief Economic History of the World*, Princeton: Princeton University Press, 2007

Clifford, W. K., 'Cosmic Emotion', *Nineteenth Century*, 1877, 1: 411–29

Cobbett, William, *Rural Rides*, ed. G. D. H. and Margaret Cole, London: Peter Davies, 1930

Cochran, Gregory, and Henry Harpending, *The 10,000 Year Explosion: How Civilization Accelerated Human Evolution*, New York: Basic Books, 2009

Coleridge, Samuel Taylor, *Collected Letters*, ed. Earl Leslie Griggs, 4 vols, Oxford: Clarendon Press, 1966–71

Colp, Ralph, Jr, *To be an Invalid: The Illness of Charles Darwin*, Chicago/London: University of Chicago Press, 1977

Cook, James, *The Journals of Captain James Cook on his Voyages of Discovery*, ed. J. C. Beaglehole, Woodbridge: The Boydell Press, 1999

Corballis, M. C., *The Lopsided Ape: Evolution of the Generative Mind*, New York: Oxford University Press, 1991

Correspondence of Charles Darwin, The, Cambridge: Cambridge University Press, 1985– [abbreviated as *C*]

Corsi, Pietro, *Science and Religion: Baden Powell and the Anglican Debate, 1800–1860*, Cambridge: Cambridge University Press, 1988

Cosh, Mary, *Edinburgh: The Golden Age*, Edinburgh: John Donald, 2003

Courtney, W. L., 'Professor Huxley as a Philosopher', *Fortnightly Review*, 1895, 64: 17–22

Cross, F. L., and E. A. Livingstone, *The Oxford Dictionary of the Christian Church*, 2nd edn, Oxford/New York: Oxford University Press, 1974

Crowe, M. J., *The Extraterrestrial Life Debate 1750–1900: The Idea of Plurality of Worlds from Kant to Lowell*, Cambridge: Cambridge University Press, 1986

Cunningham, Conor, *Darwin's Pious Idea: Why the Ultra-Darwinists and Creationists Both Get It Wrong*, Grand Rapids, MI: William B. Erdmans, 2010

Darwin, Charles, *On the Origin of Species by Means of Natural Selection or The Preservation of Favoured Races in the Struggle for Life*, London: John Murray, 1859 [abbreviated as *Origin* (1859)]

——, *The Descent of Man and Selection in Relation to Sex*, 2 vols, London: John Murray, 1871 [abbreviated as *Descent*]

——, *The Expression of the Emotions in Man and Animals* London: John Murray, 1872 [abbreviated as *Expression*]

——, *On the Origin of Species by Means of Natural Selection or The Preservation of Favoured Races in the Struggle for Life*, 6th edn, London: John Murray, 1872 [abbreviated as *Origin* (1872)]

——, *The Origin of Species . . . A Variorum Text*, ed. Morse Peckham, Philadelphia, PA: University of Pennsylvania Press, 1959

——, *The Autobiography of Charles Darwin, 1809–1882 with Original Omissions Restored*, ed. by his granddaughter Nora Barlow, New York: W. W. Norton, 1969 [abbreviated as *Autobiography*]

——, *The Red Notebook of Charles Darwin*, ed. with an Introduction and Notes by Sandra Herbert, British Museum (Natural History), Ithaca, NY/ London: Cornell University Press, 1980

——, *The Descent of Man and Selection in Relation to Sex* (first published 1871), ed. John Tyler Bonner and Robert M. May, Princeton: Princeton University Press, 1981

——, *Charles Darwin's Notebooks* (first published 1836, 1844), ed. Paul H. Barrett et al., Cambridge: Cambridge University Press for the British Museum (Natural History), 1987 [abbreviated as *Notebooks*]

——, *The Works of Charles Darwin*, ed. Paul Barrett and R. B. Freeman, 29 vols, New York: New York University Press, 1987–2016

——, *Charles Darwin's Beagle Diary*, ed. Richard Darwin Keynes, Cambridge: Cambridge University Press, 1988, paperback 2001 [abbreviated as *Diary*]

——, *The Voyage of the Beagle*, ed. Janet Browne and Michael Neve, London: Penguin, 1989 [abbreviated as Brown, *Voyage*]

——, *Charles Darwin's Notebooks from the Voyage of the Beagle*, ed. Gordon Chancellor and John van Wyhe, Cambridge: Cambridge University Press, 2009

——, and Thomas Henry Huxley, *Autobiographies*, ed. Gavin de Beer, London/ New York/Toronto: Oxford University Press, 1974

——, and Alfred Russel Wallace, *Evolution by Natural Selection*, with a Foreword by Sir Gavin de Beer, Cambridge: Cambridge University Press, 1958

Darwin, Erasmus, *The Botanic Garden*, London: J. Johnson, 1789

——, *The Temple of Nature*, London: J. Johnson, 1803

Darwin, Francis (ed.), *The Life and Letters of Charles Darwin*, 3 vols, London: John Murray, 1887 [abbreviated as *LLD*]

——, and A. C. Seward (eds), *More Letters of Charles Darwin*, 2 vols, London: John Murray, 1903

Dawkins, Richard, *The Blind Watchmaker*, Harlow: Longman, 1976

——, *The Selfish Gene*, Oxford: Oxford University Press, 1976

——, *The Extended Phenotype: The Long Reach of the Gene*, New York: W. H. Freeman, 1982

——, *Climbing Mount Improbable*, London: Viking, 1996

——, *The Greatest Show on Earth*, London: Black Swan, 2010

——, 'The Descent of Edward Wilson', *Prospect*, June 2012, Issue 195, http://www.prospectmagazine.co.uk/science-and-technology/edward-wilson-social-conquest-earth-evolutionary-errors-origin-species

——, *River Out of Eden*, London: Weidenfeld & Nicolson, 2014

Dean, D., *Gideon Mantell and the Discovery of Dinosaurs*, Cambridge: Cambridge University Press, 1999

Denton, Michael, *Evolution: A Theory in Crisis*, London: Burnett Books, 1985

Desmond, Adrian, 'Designing the Dinosaur: Richard Owen's Response to Robert Edmond Grant', *Isis*, 1979, 70: 224–34

——, *Archetypes and Ancestors: Paleontology in Victorian London, 1850–1875*, Chicago: University of Chicago Press, 1982

——, *The Politics of Evolution: Morphology, Medicine and Reform in Radical London*, Chicago: University of Chicago Press, 1989

——, *Huxley: The Devil's Disciple*, London: Michael Joseph, 1994

——, and James Moore, *Darwin*, London: Michael Joseph, 1991

——, James Moore and Janet Browne, *Charles Darwin*, Oxford: Oxford University Press, 2007

Dewey, J., 'Evolution and Ethics', *Monist*, 1898, 8: 321–41

——, *Reconstruction in Philosophy* (first published 1920), enlarged edn, Boston: Beacon Press, 1948

——, *The Quest for Certainty: A Study in the Relation of Knowledge and Action*, vol. 4 of *The Later Works of John Dewey* (first published 1929), ed. Jo Ann Boydston, Introduction by Stephen Toulmin, Carbondale, IL: Southern Illinois University Press, 1984

Di Gregorio, Mario A., *T. H. Huxley's Place in Natural Science*, New Haven: Yale University Press, 1984

——, 'A Wolf in Sheep's Clothing: Carl Gegenbaur, Ernst Haeckel, the Vertebral Theory of the Skull, and the Survival of Richard Owen', *Journal of the History of Biology*, 1995, 28: 247–80

——, *From Here to Eternity: Ernst Haeckel and Scientific Faith*, Göttingen: Vandenhoeck & Ruprecht, 2005

Dickens, Charles, *Nicholas Nickleby*, vol. 1, London: Chapman & Hall, 1920

Doyle, Arthur Conan, *The Memoirs of Sherlock Holmes*, Oxford: Oxford University Press, 1993

Drummond, H., *The Ascent of Man* (first published 1894), 7th edn, New York: James Pott, 1898

Dubois, R., *The Wooing of Earth: New Perspectives of Man's Use of Nature*, New York: Scribner's, 1980

Eiseley, Loren, *Darwin's Century*, London: Victor Gollancz, 1959

——, *Darwin and the Mysterious Mr. X*, London: Dent, 1979

Eldredge, N., *Time Frames: The Rethinking of Darwinian Evolution and the Theory of Punctuated Equilibria*, New York: Simon & Schuster, 1985

——, *Unfinished Synthesis: Biological Hierarchies and Modern Evolutionary Thought*, New York: Oxford University Press, 1985

——, *Dominion: Can Nature and Culture Co-Exist?* New York: Henry Holt, 1995

——, *Reinventing Darwin: The Great Debate at the High Table of Evolutionary Theory*, New York: John Wiley, 1995

——, *Eternal Ephemera: Adaptation and the Origin of Species from the Nineteenth Century Through Punctuated Equilibria and Beyond*, New York: Columbia University Press, 2015

Ellegård, A., *Darwin and the General Reader*, Gothenburg: Göteborgs Univ. Årsskrift, 1958

Engels, F., *Dialectics of Nature* (first published 1927), ed. and trans. Chelems Dutt, New York: International Publishers, 1940

FitzRoy, Robert, *Narrative of the Surveying Voyages of His Majesty's Ships Adventure and Beagle Between the Years 1826 and 1836*, London: Henry Colburn, 1839

Flew, Antony, *Evolutionary Ethics*, New York: St Martin's Press, 1967

Foster, Sir Michael, Obituary notice for Thomas Henry Huxley, *Proceedings of the Royal Society of London*, 1896, 59: xlvi–lxvii

Fothergill, Philip G., *Historic Aspects of Organic Evolution*, London: Hollis and Carter, 1952

Foucault, Michel, *The History of Sexuality: An Introduction*, Harmondsworth: Penguin, 1987

Francis, Mark, 'Naturalism and William Paley', *History of European Ideas*, 1989, 10: 203–20

——, *Herbert Spencer and the Invention of Modern Life*, Stocksfield: Acumen, 2007

Freeman, Michael, *Victorians and the Prehistoric Tracks to a Lost World*, New Haven/London: Yale University Press, 2004

Freud, Sigmund, *Civilization and its Discontents* (first published 1931), ed. and trans. James Strachey, New York: W. W. Norton, 1962

Freyre, Gilberto, *The English in Brazil*, Recife, Brazil: Boulevard Books, 2011

Galton, Francis, *Essays in Eugenics*, London: The Eugenics Education Society, 1909

——, *Hereditary Genius: An Enquiry into its Laws and Consequences* (first published 1892), ed. C. D. Darlington, 2nd edn, New York: World Publishing, 1962

Gerhardt, John, and Marc Kirschner, *Cells, Embryos, and Evolution*, Malden: Blackwell, 1997

Ghiselin, M. T., *The Triumph of the Darwinian Method* (first published 1969), Chicago: University of Chicago Press, reprint 1984

Gibbon, E., *The History of the Decline and Fall of the Roman Empire*, ed. John Bury, 7 vols, 2nd edn, London: Methuen, 1901–2

Gilbert, Scott F., 'Owen's Vertebral Archetype and Evolutionary Genetics: A Platonic Appreciation', *Perspectives in Biology and Medicine*, 1980, 23: 475–88

Gillespie, Charles C., *Genesis and Geology: The Impact of Scientific Discoveries upon Religious Beliefs in the Decades before Darwin*, New York: Harper & Row, 1951

Gillespie, N. C., 'Divine Design and the Industrial Revolution: William Paley's Abortive Reform of Natural Theology', *Isis*, 1990, 81: 214–29

Gladstone, William E., *An Academic Sketch*, First Romanes Lecture, 24 October 1891, Oxford: Clarendon Press, 1982

Glick, T. F. (ed.), *The Comparative Reception of Darwinism*, Chicago: University of Chicago Press, 1988

Goldschmidt, Richard, *Portraits from Memory: Recollections of a Zoologist*, Seattle: University of Washington Press, 1956

Goodall, Jane, *The Chimpanzees of Gombe: Pictures of Behavior*, Cambridge, MA: Belknap Press, 1986

——, *Through a Window: My Thirty Years with the Chimpanzees of Gombe*, Boston: Houghton Mifflin, 1990

Gould, Stephen Jay, *Ever Since Darwin*, New York: W. W. Norton, 1977

——, 'Is a New and General Theory of Evolution Emerging?', *Palaeobiology*, 1980, 6(1): 19–130

——, *Panda's Thumb*, New York: W. W. Norton, 1980

——, *Dinosaur in a Haystack: Reflections in Natural History*, London: Jonathan Cape, 1996

——, 'Kropotkin was No Crackpot', *Natural History*, June 1997, 106: 12–21

——, *The Structure of Evolutionary Theory*, Cambridge, MA: Harvard University Press, 2002

Grant, Alexander, *The Story of Edinburgh University during its First Three Hundred Years*, London: Longmans, Green, 1884

Greene, John C., 'Biology and Social Theory in the Nineteenth Century:

Auguste Comte and Herbert Spencer', in Marshall Clagett (ed.), *Critical Problems in the History of Science*, Madison, WI: University of Wisconsin Press, 1959, 419–47

——, *The Death of Adam: Evolution and its Impact on Western Thought*, Ames, IA: Iowa State University Press, 1959

——, 'Darwin as Social Evolutionist', *Journal of the History of Biology*, 1977, 10: 1–27

Greville, Charles, *Leaves from the Greville Diary: A New and Abridged Edition*, London: Eveleigh Nash & Grayson, 1929

Grey, William R., *Enigmas of Life*, London: William Rathbone, 1874

Gruber, Jacob W., *A Conscience in Conflict: The Life of St George Jackson Mivart*, London: Greenwood Press, 1980

——, and John Thackray (eds), *Richard Owen Commemoration*, London: Natural History Museum, 1992

Guizot, François Pierre Guillaume, *History of Civilization from the Fall of the Roman Empire to the French Revolution*, trans. William Hazlitt, 2 vols, London: Macmillan, 1885

Gully, James, *The Water Cure in Chronic Disease*, London: John Churchill, 1846

Haight, Gordon S., *George Eliot: A Biography*, Oxford: Clarendon Press, 1968

Haeckel, Ernst, *The Evolution of Man*, London: C. K. Paul, 1879

Hall, Brian K., *Homology: The Hierarchical Basis of Comparative Biology*, San Diego: Academic Press, 1994

——, *Evolutionary Developmental Biology*, 2nd edn, Dordrecht: Kluwer, 1999

——, 'Balfour, Garstang and de Beer: The First Century of Evolutionary Embryology', *American Zoologist*, 2000, 718–28

Hall, Stephen, and Juliet Clutton-Brock, *Two Hundred Years of British Livestock*, London: British Museum, 1989

Hands, John, *Cosmosapiens: Human Evolution from the Origin of the Universe*, London: Duckworth Overlook, 2015

Hanham, Harold, 'Introduction', in *On Scotland and the Scotch Intellect*, Chicago: University of Chicago Press, 1970, xiii–xxxvii

Hankins, Thomas L., *Science and the Enlightenment*, Cambridge: Cambridge University Press, 1985

Hare, Augustus, *The Story of my Life*, 6 vols, London: George Allen, 1900

Hartley, David, *Observations on Man, his Frame, his Duty, and his Expectations* (first published 1749), Gainesville, FL: University of Florida Press, reprint 1966

Hawks, John, 'The Denisova Genome: An Unexpected Window into the Past', in A. R. Sankhyan (ed.), *Recent Discoveries and Perspectives in Human Evolution: Introduction*, BAR International Series, vol. 2719, Oxford: Archaeopress, 2015, 77–9

Hazlitt, William, *The Complete Works of William Hazlitt*, ed. P. P. Howe (after the edition of A. R. Waller and Arnold Glover), 21 vols, London: J. M. Dent, 1931–4

Healey, Edna, *Emma Darwin: The Inspirational Wife of a Genius*, London: Headline, 2001

Hearnshaw, F. J. C. (ed.), *The Social and Political Ideas of Some Representative Thinkers of the Victorian Age* (first published 1933), New York: Barnes & Noble, reprint 1967

Heilbron, J. L. (ed.), *The Oxford Companion to the History of Modern Science*, Oxford: Oxford University Press, 2003

Helfand, Michael S., 'T. H. Huxley's "Evolution and Ethics": The Politics of Evolution and the Evolution of Politics', *Victorian Studies*, 1977, 2: 159–77

Hemming, John, *Amazon Frontier*, London, Macmillan, 1987

——, *Naturalists in Paradise: Wallace, Bates and Spruce in the Amazon*, London: Thames & Hudson, 2015

Herbert, Sandra, *Charles Darwin: Geologist*, Ithaca: Cornell University Press, 2005

Hesketh, Ian, *Of Apes and Ancestors: Evolution, Christianity and the Oxford Debate*, Toronto/Buffalo/London: University of Toronto Press, 2009

Hilton, Boyd, *A Mad, Bad and Dangerous People? England 1783–1846*, Oxford: Clarendon Press, 2006

Himmelfarb, Gertrude, *Victorian Minds*, New York: Knopf, 1968

——, *Marriage and Morals among the Victorians*, New York: Knopf, 1986

Hitler, Adolf, *Hitler's Table Talk, 1941–1944*, trans. Norman Cameron and R. H. Stevens, New York: Enigma Books, 2000

Hobbes, Thomas, *Leviathan, or The Matter, Forme, & Power of a Commonwealth, Ecclesiasticall and Civill* (first published 1651), ed. C. B. Macpherson, New York: Penguin, 1968

Hodge, Jonathan, and Gregory Radick, *The Cambridge Companion to Darwin*, Cambridge: Cambridge University Press, 2003

Hofstadter, Richard, *Darwinism in American Thought*, rev. edn, Boston: Beacon Press, 1955

Holmes, Richard, *The Age of Wonder*, London: Harper Press, 2008

Hope Simpson, J. B., *Rugby since Arnold*, London/New York: Macmillan/St Martin's Press, 1967

Hudson, Derek, *Munby: Man of Two Worlds*, London: John Murray, 1972

Hull, David L., *Darwin and his Critics*, Cambridge, MA: Harvard University Press, 1973

——, 'Darwin and the Nature of Science', in D. S. Bendall (ed.), *Evolution from Molecules to Men*, Cambridge: Cambridge University Press, 1983, 63–80

Hume, David, *Hume's Dialogues Concerning Natural Religion*, ed. and introduced by Norman Kemp Smith, 2nd edn, London/Edinburgh: Thomas Nelson, 1947

——, *Enquiries Concerning Human Understanding and Concerning the Principles of Morals* (first published 1777), ed. L. A. Selby-Bigge, 3rd edn (rev. P. H. Nidditch), Oxford: Oxford University Press, 1975

——, *A Treatise of Human Nature* (first published 1739–40), ed. L. A. Selby-Bigge, 2nd edn, Oxford: Oxford University Press, 1978

Humphrey, N., *Consciousness Regained: Chapters in the Development of Mind*, New York: Oxford University Press, 1983

——, 'The Uses of Consciousness', *57th James Arthur Lecture on the Evolution of the Human Brain*, New York: American Museum of Natural History, 1987

Hunt, Tristram, *The Frock-Coated Communist: The Revolutionary Life of Friedrich Engels*, London: Allen Lane/Penguin, 2009

Hutton, James, *The Theory of the Earth*, Edinburgh: Transactions of the Royal Society of Edinburgh, 1795

Huxley, Julian, 'Eugenics and Society', in *Man in the Modern World: Selected Essays*, New York: New American Library, 1948

——, *Evolution: The Modern Synthesis* (first published 1942), 2nd impression, London: George Allen & Unwin, 1963

——, *Evolution in Action* (first published 1953), Harmondsworth: Penguin, 1963

——, and Thomas H. Huxley, *Touchstone for Ethics: 1893–1943*, New York: Harper & Brothers, 1947

Huxley, Leonard, *Life and Letters of Thomas Henry Huxley*, 2 vols, London: Macmillan, 1900

Huxley, Thomas H., *Lessons in Elementary Physiology* (first published 1866), new edn, London: Macmillan, 1881

——, Introduction to Théodore Rocquain, *The Revolutionary Spirit*, trans. J. D. Hunting, London: Swan Sonnenschein, 1891

——, 'An Apologetic Irenicon', *Fortnightly Review*, 1892, 52: 557–71

——, *Evolution and Ethics and Other Essays*, London: Macmillan, 1894

——, *Collected Essays*, 9 vols, London: Macmillan, 1893–4

——, *T. H. Huxley's Diary of the Voyage of HMS Rattlesnake*, ed. Julian Huxley, Garden City, NY: Doubleday, 1936

——, *The Crayfish: An Introduction to the Study of Zoology* (first published 1879), Cambridge, MA: MIT Press, reprint 1974

——, et al., 'A Modern Symposium: The Influence upon Morality of a Decline in Religious Belief', *Nineteenth Century*, 1877, 1: 531–46

Irvine, William, *Apes, Angels and Victorians*, London: Weidenfeld & Nicolson, 1956

Isaacson, Walter, *Einstein: His Life and Universe*, New York: Simon & Schuster, 2007

Johanson, D. C., L. Johanson and B. Edgar, *Ancestors: In Search of Human Origins*, New York: Villard Books, 1994

Jollie, M., 'Segment Theory and the Homologizing of Cranial Bones', *American Naturalist*, 1981, 118: 785–802

Jones, Gareth Steadman, *Outcast London: A Study in the Relationship between Classes in Victorian Society* (first published 1971), New York: Penguin, reprint 1984

Keats, John, *The Letters of John Keats*, ed. Maurice Buxton Forman, London: Oxford University Press, 1947

Kevles, Daniel J., *In the Name of Eugenics: Genetics and the Uses of Human Heredity* (first published 1984), Berkeley: University of California Press, reprint 1986

Keynes, Randal, *Annie's Box: Charles Darwin, his Daughter, and Human Evolution*, London: Fourth Estate, 2001

King-Hele, Desmond, *Erasmus Darwin: A Life of Unequalled Achievement*, London: Giles de la Mare, 1999

Kirby, William, and William Spence, *An Introduction to Entomology; or Elements of the Natural History of Insects*, 4 vols, London: Longman, Rees, Orme, Brown & Green, 1815–26

Kohn, David, 'Theories to Work By: Rejected Theories, Reproduction, and Darwin's Path to Natural Selection', *Studies in the History of Biology*, 1980, 4: 67–170

Kragh, Helge, *An Introduction to the Historiography of Science*, Cambridge: Cambridge University Press, 1987

——, '*From Here to Eternity: Ernst Haeckel and Scientific Faith* – by Mario A. Di Gregorio', *Centaurus*, 2007, 49(3): 246–7

Kropotkin, P. A., *Memoirs of a Revolutionist*, London: Smith, Elder, 1899

——, *Mutual Aid: A Factor of Evolution* (first published 1914), Boston: Extending Horizons Books, reprint 1955

——, *Mutual Aid* (first published 1902), London: Freedom Press, 1987

Lancaster, Osbert, *All Done from Memory* (first published 1953 in an autographed edn limited to 45 copies), London: John Murray, 1963

Landau, Misia, 'Human Evolution as Narrative', *American Scientist*, 1984, 72: 262–8

Lankester, E. R., *Degeneration: A Chapter in Darwinism*, London: Macmillan, 1880

——, *The Kingdom of Man*, London: Archibald Constable, 1907

Larson, Edward J., *Evolution's Workshop: God and Science on the Galápagos Islands*, London/New York: Penguin, 2001

Le Fanu, James, *Why Us? How Science Rediscovered the Mystery of Ourselves*, London: Harper Press, 2009

Le Guyader, H., *Geoffroy Saint-Hilaire: A Visionary Naturalist*, trans. Marjorie Grene, Chicago: University of Chicago Press, 2004

Leakey, R., and R. Lewin, *Origins Reconsidered: In Search of What Makes Us Human*, New York: Doubleday, 1992

Leiss, William, *The Domination of Nature*, New York: Braziller, 1972

Lenoir, Timothy, *The Strategy of Life*, Chicago: University of Chicago Press, 1982

Lewes, George Henry, *The Life and Works of Goethe* (first published 1855), London: J. M. Dent (Everyman Library), 1959

Lewin, R., *The Origin of Modern Humans*, New York: Scientific American Library, 1993

Lightman, Bernard, *The Origins of Agnosticism: Victorian Unbelief and the Limits of Knowledge*, Baltimore: Johns Hopkins University Press, 1987

—— (ed.), *Victorian Science in Context*, Chicago/London: University of Chicago Press, 1997

Litchfield, H. E. (ed.), *Emma Darwin: A Century of Family Letters*, 2 vols, London: John Murray, 1915

Livio, Mario, *Brilliant Blunders: From Darwin to Einstein – Colossal Mistakes by Great Scientists That Changed our Understanding of Life and the Universe*, New York: Simon & Schuster, 2013

Lovejoy, A. O., *The Great Chain of Being*, Cambridge, MA: Harvard University Press, 1936

——, *Essays in the History of Ideas*, New York: G. P. Putnam's Sons, 1948

Løvtrup, Søren, *Darwinism: The Refutation of a Myth*, London/New York/Sydney: Croom Helm, 1987

Lubbock, John, *The Origins of Civilization and the Primitive Condition of Man*, London: Longmans, Green, 1870

Lucretius, *On the Nature of Things*, trans. H. A. J. Munro, London: G. Bell, 1929

Lyell, Charles, *Principles of Geology*, 2 vols, reprinted from the 6th edn, London: John Murray, 1875

Lyell, Mrs (ed.), *Life, Letters & Journals of Sir Charles Lyell*, 2 vols, London: John Murray, 1881

McCalman, Iain, *Darwin's Armada*, London/New York: Simon & Schuster, 2009

McGrew, W. B., *Chimpanzee Material Culture: Implications for Human Evolution*, Cambridge: Cambridge University Press, 1992

Mack, Edward C., *Public Schools and British Opinion since 1860*, Westport, CT: Greenwood Press, 1941

Malthus, Thomas R., *An Essay on the Principle of Population as It Affects the Future Improvement of Society . . . and A Summary View of the Principle of Population* (first published 1798), ed. Antony Flew, New York: Penguin, 1970

Manier, Edward, *The Young Darwin and his Cultural Cycle*, Dordrecht: Reidel, 1978

Marsh, George Perkins, *Man and Nature; or, Physical Geography as Modified by Human Action* (first published 1864), ed. D. Lowenthal, Cambridge, MA: Harvard University Press, reprint 1965

Marshack, A., *The Roots of Civilization* (first published 1971), rev. edn, Mount Kisco, NY: Moyer Bell, 1991

Marshall, A. J., *Darwin and Huxley in Australia*, Sydney/London/Auckland: Hodder & Stoughton, 1970

Martin, A. P., *Life and Letters of the Rt. Honourable Robert Lowe, Viscount Sherbrooke*, 2 vols, London: Longmans, Green, 1893

Martineau, Harriet, *Life in the Sickroom: Essays by an Invalid*, London: Houlston, 1844

——, *Harriet Martineau: Selected Letters*, ed. Valerie Sanders, Oxford: Clarendon Press, 1990

Mayr, Ernst, 'Cause and Effect in Biology', *Science*, 1961, 134: 1501–6

——, *The Growth of Biological Thought*, Cambridge, MA: Harvard University Press, 1982

Midgeley, Mary, *Evolution as a Religion* (first published 1985), rev. edn, London/New York: Routledge Classics, 2002

Mill, John Stuart, *A System of Logic*, 2 vols, London: Parker, 1862

——, *Essays on Ethics, Religion, and Society*, vol. 10 of *Collected Works of John Stuart Mill*, ed. J. M. Robson, Toronto: University of Toronto Press, 1969

Miller, Martin, *Kropotkin*, Chicago/London: University of Chicago Press, 1976

Milton, John, *Complete Poetry & Selected Prose*, ed. E. H. Visiak, London: Nonesuch Press, 1948

Mivart, St George, *On the Genesis of Species*, London: Macmillan, 1871

——, 'Evolution in Professor Huxley', *Popular Science Monthly*, 1893, 44: 319–33

——, 'Darwin's *Descent of Man*', in *Darwin and his Critics* (first published 1871), ed. David Hull, Chicago: University of Chicago Press, 1973

Moore, George E., *Principia Ethica* (first published 1903), Cambridge: Cambridge University Press, reprint 1984

Moore, James, *The Post-Darwinian Controversies: A Study of the Protestant Struggle to Come to Terms with Darwin in Great Britain and America, 1870–1900*, Cambridge, Cambridge University Press, 1979

Moorehead, Alan, *Darwin and the Beagle*, London: Hamish Hamilton, 1969

Morgan, C. Lloyd, *Animal Life and Intelligence*, London: Edward Arnold, 1891

——, *Habit and Instinct*, London: Edward Arnold, 1896
——, *Emergent Evolution*, New York: Henry Holt, 1923
——, *The Emergence of Novelty*, London: Williams & Norgate, 1933
Muggeridge, Malcolm, *The Earnest Atheist: A Study of Samuel Butler*, London: Eyre & Spottiswoode, 1936
Müller-Wille, Staffan, 'Linnaeus' Concept of a "Symmetry of All Parts"', *Jahrbuch für Geschichte und Theorie der Biologie*, 1995, 2: 41–7
Myers, Greg, 'Nineteenth-Century Popularizations of Thermodynamics and the Rhetoric of Social Prophecy', *Victorian Studies*, 1985, 29: 35–66
Nagel, Thomas, *Mind and Cosmos: Why the Materialist Neo-Darwinian Conception of Nature is Almost Certainly False*, Oxford: Oxford University Press, 2012
Newman, Francis William, *Phases of Faith; or Passages from the History of my Creed*, London: John Chapman, 1850
Nilsson, Dan-E., and Susanne Pelger, 'A Pessimistic Estimate of the Time Required for an Eye to Evolve', *Proceedings of the Royal Society of London*, B, 1994, 256: 53–8
Noland, Richard, 'T. H. Huxley on Culture', *Personalist*, 1964, 45: 94–111
Nyhart, Lynn, *Biology Takes Form*, Chicago: University of Chicago Press, 1995
O'Brien, P., *Warrington Academy, 1757–86: Its Predecessors and Successors*, Wigan: Owl Books, 1957
Oken, Lorenz, *Über die Bedeutung der Schädelknochen*, Jena: Göpferdt, 1807
Olby, Robert, and Peter Gautrey, 'The Eleven References to Mendel before 1900', *Annals of Science*, 1968, 24: 7–20
Ospovat, Dov, 'Perfect Adaptation and Teleological Explanation: Approaches to the Problem of the History of Life in the Mid-Nineteenth Century', *Studies in History of Biology*, 1978, 2: 33–56
——, *The Development of Darwin's Theory: Natural History, Natural Theology, and Natural Selection, 1838–1859*, Cambridge, Cambridge University Press, 1981
Owen, Richard, *Lectures on the Comparative Anatomy and Physiology of the Invertebrate Animals*, London: Longman, Brown, Green & Longmans, 1843
——, 'Report on the Archetype and Homologies of the Vertebrate Skeleton', *Reports of the British Association for the Advancement of Science*, 1847, 169–340
——, *The Archetype and Homologies of the Vertebrate Skeleton*, London: Richard and John E. Taylor, 1848
——, *On the Nature of Limbs*, London: John Van Voorst, 1849
Padian, Kevin, 'Form Versus Function: The Evolution of a Dialectic', in Jeff Thomason (ed.), *Functional Morphology in Vertebrate Paleontology*, Cambridge: Cambridge University Press, 1995

——, 'A Missing Hunterian Lecture on Vertebrae by Richard Owen, 1837', *Journal of the History of Biology*, 1995, 28: 333–68

——, 'The Rehabilitation of Sir Richard Owen', *Bio-Science*, 1997, 47: 446–52

Paley, William, *The Works of William Paley DD Archdeacon of Carlisle*, London: William Smith, 1842

Paradis, James, *T. H. Huxley: Man's Place in Nature*, Lincoln, NE: University of Nebraska Press, 1978

——, and George C. Williams, *T. H. Huxley's Evolution and Ethics with New Essays on its Victorian and Sociobiological Context*, Princeton: Princeton University Press, 1989

Paston, George, *At John Murray's: Records of a Literary Circle, 1843–1892*, London: John Murray, 1932

Pearn, Alison, 'Darwin Hates You: Owen, Mivart and Butler, Darwin's Failed Friendships in Theory and Practice', lecture at the Three Societies Meeting, Philadelphia, 2012

Pearson, Karl, *The Scope and Importance of the State of the Science of National Eugenics* (first published 1907), 2nd edn, London: Dulau, 1909

Pfeiffer, J. E., *The Creative Explosion: An Inquiry into the Origins of Art and Religion*, New York: Harper & Row, 1982

Pinker, S., *The Language Instinct*, New York: William Morrow, 1994

Plumb, J. E. (ed.), *Studies in Social History: A Tribute to G. M. Trevelyan*, London/New York/Toronto: Longmans, Green, 1955

Prothero, David, *Evolution: What the Fossils Say and Why It Matters*, New York: Columbia University Press, 2007

Quetelet, Lambert Adolphe, *A Treatise on Man and the Development of his Faculties* (first published 1835), trans. R. Knox, New York: Burt Franklin, reprint 1968

Raff, Rudolf A., *The Shape of Life*, Chicago: University of Chicago Press, 1996

Raverat, Gwen, *Period Piece: A Cambridge Childhood*, London: Faber & Faber, 1952

Reid, Sir Wemyss (ed.), *The Life of William Ewart Gladstone*, 2 vols, New York: G. P. Putnam's Sons, 1899

Richards, Eveleen, 'A Question of Property Rights: Richard Owen's Evolutionism Reassessed', *British Journal for the History of Science*, 1987, 20: 129–71

——, 'A Political Anatomy of Monsters, Hopeful and Otherwise', *Isis*, 1994, 85: 377–411

Richards, Robert, 'Instinct and Intelligence in British Natural Theology: Some Contributions to Darwin's Theory of the Evolution of Behavior', *Journal of the History of Biology*, 1981, 14: 193–230

——, 'Darwin and the Biologizing of Moral Behavior', in William R. Woodward and Mitchell G. Ash (eds), *The Problematic Science: Psychology in Nineteenth-Century Thought*, New York: Praeger, 1982, 43–64

——, *Darwin and the Emergence of Evolutionary Theories of Mind and Behavior*, Chicago: University of Chicago Press, 1987

——, *The Romanic Conception of Life: Science and Philosophy in the Age of Goethe*, Chicago: University of Chicago Press, 2002

Ridley, M., *Evolution*, Boston: Blackwell Scientific Publications, 1993

Ritchie, D. G., *Darwinism and Politics*, 2nd edn, London: Swan Sonnenschein, 1891

Ritchie, John, 'Lachlan Macquarie', *ODNB*, vol. 36

Ritvo, Harriet, *The Animal Estate: The English and Other Creatures*, Cambridge, MA: Harvard University Press, 1987

Roberts, Andrew, *Salisbury: Victorian Titan*, London: Weidenfeld & Nicolson, 1999

Roget, Peter Mark, *Animal and Vegetable Physiology Considered with Respect to Natural Theology*, 2 vols, London: William Pickering, 1834

Romanes, G. J., *Darwin and After Darwin: An Exposition of the Darwinian Theory and Discussion of Post-Darwinian Questions*, 3 vols, Chicago: Open Court, 1892–7

——, *The Life and Letters of George John Romanes, Written and Edited by his Wife*, London: Longmans, Green, 1896

Ronse De Craene, Louis P., *Floral Diagrams: An Aid to Understanding Flower Morphology and Evolution*, Cambridge/New York: Cambridge University Press, 2010

Rose, Michael R., *Darwin's Spectre*, Princeton: Princeton University Press, 1998

Rosenberg, Alex, and Robert Arp, *Philosophy of Biology: An Anthology*, Chichester: Wiley-Blackwell, 2010

Rowe, T., 'Definition and Diagnosis in the Phylogenetic System', *Systematic Zoology*, 1987, 36: 208-11

Rupke, Nicolaas, 'Richard Owen's Hunterian Lectures on Comparative Anatomy and Physiology, 1837–1855', *Medical History*, 1985, 29: 237–58

——, 'Richard Owen's Vertebrate Archetype', *Isis*, 1993, 84: 231–51

——, *Richard Owen: Victorian Naturalist*, New Haven: Yale University Press, 1994

Ruse, Michael, *The Darwinian Revolution: Science Red in Tooth and Claw*, Chicago: University of Chicago Press, 1979

——, *Taking Darwin Seriously: A Naturalistic Approach to Philosophy*, New York: Blackwell, 1986

——, 'A Reappraisal of Ernst Haeckel', *Lancet*, 2009, 373(9665): 711–12

——, *The Philosophy of Human Evolution*, Cambridge: Cambridge University Press, 2012

Ruskin, John, *Praeterita*, London/New York/Toronto: Everyman's Library, 2005

Russell, Bertrand, *A History of Western Philosophy*, London: Allen & Unwin, 1979

Russell, Edwin S., *Form and Function: A Contribution to the History of Animal Morphology* (first published 1916), Chicago: University of Chicago Press, reprint 1982

St Aubyn, Giles, *Souls in Torment: The Conflict between Science and Religion in Victorian England*, London: New European Publications, 2010

Sanderson, Michael J., 'Back to the Past: A New Take on the Timing of Flowering Plant Diversification', *New Phytologist*, 2015, 207(2): 257–9

Sankhyan, A. R. (ed.), *Recent Discoveries and Perspectives in Human Evolution: Introduction*, BAR International Series, vol. 2719, Oxford: Archaeopress, 2015

Savage-Rumbaugh, S., and R. Lewin, *Kanzi: Ape at the Brink of the Human Mind*, New York: John Wiley, 1994

Sawyer, Paul, 'Ruskin and Tyndall: The Poetry of Matter and the Poetry of Spirit', in James Paradis and Thomas Postlewait (eds), *Victorian Science and Victorian Values: Literary Perspectives*, New Brunswick, NJ: Rutgers University Press, 1985, 217–46

Schick, K. D., and N. Toth, *Making Silent Stones Speak: Human Evolution and the Dawn of Technology*, New York: Simon & Schuster, 1993

Schultz, Kirsten, *Tropical Versailles*, New York/London: Routledge, 2001

Schwartz, Benjamin, *In Search of Wealth and Power: Yen Fu and the West*, Cambridge, MA: Harvard University Press, 1964

Schwartz, Joel S., *Darwin's Disciple: George John Romanes*, Philadelphia: American Philosophical Society, 2010

Schweber, S. S., 'The Origin of the *Origin* Revisited', *Journal of the History of Biology*, 1977, 10: 229–316

Scott, Walter, *The Journal of Sir Walter Scott*, ed. W. E. K. Anderson, Oxford: Clarendon Press, 1972

Searby, Peter, *A History of the University of Cambridge*, vol. 3, Cambridge: Cambridge University Press, 1997

Secord, James A., *Victorian Sensation: The Extraordinary Publication, Reception and Secret Authorship of Vestiges of the Natural History of Creation*, Chicago/London: University of Chicago Press, 2000

Seibt, Gustav, *Mit einer Art von Wut: Goethe in der Revolution*, Munich: Beck, 2014

Shermer, Michael, *In Darwin's Shadow: The Life and Science of Alfred Russel Wallace*, Oxford: Oxford University Press, 2002

Sieveking, A., *The Cave Artists*, London: Thames & Hudson, 1979

Sinker, A. P., *Introduction to Lucretius*, Cambridge: Cambridge University Press, 1937

Sitwell, Osbert, *Left Hand, Right Hand*, London: Macmillan, 1957

Sloan, Phillip R., 'Introduction: On the Edge of Evolution', in Richard Owen (ed.), *The Hunterian Lectures in Comparative Anatomy, May and June 1837*, Chicago: University of Chicago Press, 1992, 3–72

——, 'Whewell's Philosophy of Discovery and the Archetype of the Vertebrate Skeleton', *Annals of Science*, 2003, 60: 39–61

Slotten, Ross A., *The Heretic in Darwin's Court: The Life of Alfred Russel Wallace*, New York: Columbia University Press, 2004

Smith, Adam, *The Theory of Moral Sentiments* (first published 1759), Oxford: Oxford University Press, reprint 1976

——, *An Inquiry into the Nature and Causes of the Wealth of Nations* (first published 1776), 2 vols, Oxford: Oxford University Press, reprint 1976

Smith, Brian S., *A History of Malvern*, Leicester: Leicester University Press, 1964

Smith, Charles H., and George Beccaloni, *Natural Selection and Beyond: The Intellectual Legacy of Alfred Russel Wallace*, Oxford: Oxford University Press, 2008

Smith, John Maynard, *The Theory of Evolution* (first published 1958), rev. edn, Cambridge: Cambridge University Press, 1993

Smith, Norman Kemp, *The Philosophy of David Hume*, London: Macmillan, 1941

Spencer, Herbert, *Social Statistics* (first published 1851), rev. edn, New York: Appleton, 1892

——, *The Principles of Biology*, rev. edn, 2 vols, London: Williams and Norgate, 1898

——, 'Evolutionary Ethics', *Athenaeum*, 1893, 3432: 193–4

——, *An Autobiography*, 2 vols, London: Williams & Norgate, 1904

——, *Essays: Scientific, Political and Speculative*, 3 vols, New York: Appleton, 1904

——, *First Principles* (first published 1862), 3rd edn, London: Williams & Norgate, 1910

——, *The Principles of Ethics* (American edn 1897), 2 vols, Indianapolis: Liberty Classics Press, reprint 1978

Spikins, Penny, *How Compassion Made Us Human: The Evolutionary Origins of Tenderness, Trust and Morality*, Barnsley: Pen and Sword Books, 2015

Stanley, Oma, 'T. H. Huxley's Treatment of Nature', *Journal of the History of Ideas*, 1957, 18: 120–7

Stearn, W. T., *The Natural History Museum in South Kensington*, London: Heinemann, 1981

Stent, Gunther S. (ed.), 'Introduction', in *Morality as a Biological Phenomenon: The Presuppositions of Sociobiological Research*, Berkeley: University of California Press, 1980

Stephen, Leslie, 'Ethics and the Struggle for Existence', *Contemporary Review*, 1893, 64: 157–70

——, 'The Huxley Memorial', *Nature Magazine*, 1895, 53: 183–6

——, *English Thought in the Eighteenth Century*, 2 vols, London: John Murray, 1927

Stocking, George W., Jr, 'What's in a Name? The Origins of the Royal Anthropological Institute (1837–73)', *Man*, 1971, 6: 369–91

——, *Victorian Anthropology*, New York: Free Press, 1987

Stott, Rebecca, *Darwin and the Barnacle*, London: Faber & Faber, 2003

Stove, David, *Darwinian Fairytales*, Aldershot: Avebury, 1995

Tattersall, Ian, *The Human Odyssey: Four Million Years of Human Evolution*, New York: Prentice-Hall, 1993

——, *The Fossil Trail: How We Know What We Think We Know about Human Evolution*, New York: Oxford University Press, 1995

——, *The Last Neanderthal: The Origin, Success, and Mysterious Extinction of our Closest Human Relative*, New York: Macmillan, 1995

——, *Becoming Human: Evolution and Human Uniqueness*, Oxford: Oxford University Press, 1998

——, E. Delson and J. A. Van Couvering (eds), *Encyclopedia of Human Evolution and Prehistory*, New York: Garland Publishing, 1988

Taylor, James, *Voyage of the Beagle: Darwin's Extraordinary Adventure in FitzRoy's Famous Survey Ship*, London: Conway, 2008

Taylor, Jonathan, *Science and Omniscience in Nineteenth-Century Literature*, Brighton: Sussex Academic Press, 2007

Teilhard de Chardin, Pierre, *The Phenomenon of Man*, London: Collins, 1959

——, *Le Phénomène humain*, Paris: Editions du Seuil, 1955

Tennyson, Alfred, *The Poems of Tennyson*, ed. Christopher Ricks, 3 vols, London: Longman, 1987

Thomas, Hugh, *The Slave Trade: The History of the Atlantic Slave Trade, 1440–1870*, London: Picador, 1997

Thomas, Keith, *Man and the Natural World*, New York: Pantheon Books, 1983

Thomson, Keith, *The Young Charles Darwin*, New Haven/London: Yale University Press, 2009

Tillyard, E. M. W., *The Elizabethan World Picture*, New York: Random House, 1942

Toulmin, Stephen, *The Return to Cosmology: Postmodern Science and the Theology of Nature*, Berkeley: University of California Press, 1982

Toynbee, Arnold, *The Industrial Revolution* (first published 1884), Boston: Beacon Press, reprint 1956

Trevelyan, George Otto, *The Life and Letters of Lord Macaulay*, London: Longmans, Green, 1881

Turner, Frank, *Between Science and Religion: The Reaction to Scientific Naturalism in Late Victorian England*, New Haven: Yale University Press, 1974

Tyerman, Christopher, *A History of Harrow School, 1324–1991*, Oxford/New York: Oxford University Press, 2000

Tylor, Edward, *Primitive Culture*, 2 vols, London: John Murray, 1871

Vorzimmer, Peter J., *Charles Darwin: The Years of Controversy – The Origin of Species and its Critics, 1859–82*, London: University of London Press, 1972

Wallace, Alfred Russel, 'The Origin of Human Races and the Antiquity of Man Deduced from the Theory of Natural Selection', *Journal of the Anthropological Society*, 1864, 2: 58–88

——, Untitled review of *The Descent of Man in Relation to Sex* by Charles Darwin, *The Academy*, 1871, 2: 177–83

——, 'Romanes versus Darwin: An Episode in the History of the Evolution Theory', *Fortnightly Review*, 1886, 40: 300–16

——, *Darwinism: An Exposition of the Theory of Natural Selection with Some of its Applications*, London: Macmillan, 1889

——, *The Wonderful Century*, London: Swan Sonnenschein, 1898

——, *Studies, Scientific & Social* (2 vols), London: Macmillan, 1900

——, *Man's Place in the Universe: A Study of the Results of Scientific Research in Relation to the Unity or Plurality of Worlds*, London: Chapman & Hall, 1903

Wallman, J., *Aping Language*, Cambridge: Cambridge University Press, 1992

Watson, J. Steven, *The Reign of George III, 1760–1815*, Oxford: Clarendon Press, 1960

Wedgwood, Barbara and Hensleigh, *The Wedgwood Circle, 1730–1897*, London: Studio Vista, 1980

Weikart, Richard, *From Darwin to Hitler: Evolutionary Ethics, Eugenics, and Racism in Germany*, New York: Palgrave Macmillan, 2004

Weil, Simone, *The Need for Roots* (first French edn, *L'Enracinement*, 1949), London: Routledge & Kegan Paul, 1952

Weindling, Paul, 'German Eugenics and the Wider World', in Alison Bashford and Philippa Levine (eds), *The Oxford Handbook of the History of Eugenics*, Oxford: Oxford University Press, 2010, 316–17

Wells, J., *Icons of Evolution*, Washington, DC: Regnery, 2000

Wheen, Francis, *Karl Marx*, London: Fourth Estate, 1999

White, Paul, *Thomas Huxley: Making the 'Man of Science'*, Cambridge: Cambridge University Press, 2003

White, R., *Dark Caves, Bright Visions: Life in Ice Age Europe*, New York: American Museum of Natural History and W. W. Norton, 1986

Whitehead, Alfred North, *Science and the Modern World*, New York: New American Library, 1925

Wichler, Gerhard, *Charles Darwin: The Founder of the Theory of Evolution and Natural Selection*, Oxford: Pergamon Press, 1961

Willey, Basil, *The Eighteenth-Century Background: Studies on the Idea of Nature in the Thought of the Period*, Boston: Beacon Press, 1940

Williams, Raymond, *Problems in Materialism and Culture: Selected Essays*, London: Verso, 1980

Wills, Christopher, *Exons, Introns and Talking Genes: The Science behind the Human Genome Project*, New York: Basic Books, 1991

Wilson, A. N., *God's Funeral*, London: John Murray, 1999

——, *The Victorians*, London: Hutchinson, 2002

Wilson, L. G., *Charles Lyell: The Years to 1841: The Revolution in Geology*, New Haven/London: Yale University Press, 1977

Winsor, Mary P., 'Non-Essentialist Methods in Pre-Darwinian Taxonomy', *Biology and Philosophy*, 2003, 18: 387–400

Wittgenstein, Ludwig, *Philosophical Investigations*, trans. Elizabeth Anscombe, New York: Macmillan, 1953

Woodruff, A. W., 'Darwin's Health in Relation to his Voyage to South America', *British Medical Journal*, 1965, 1: 745–50

Worster, Donald, *Nature's Economy: A History of Ecological Ideas*, Cambridge, Cambridge University Press, 1994

Wright, R., *The Moral Animal: Why We are the Way We are: The New Science of Evolutionary Psychology*, New York: Pantheon Books, 1994

Wulf, Andrea, *The Invention of Nature: The Adventures of Alexander von Humboldt, the Lost Hero of Science*, London: John Murray, 2015

Yanni, Carla, *Nature's Museums: Victorian Science and the Architecture of Display*, London: Athlone, 1999

Young, Robert, *Darwin's Metaphor: Nature's Place in Victorian Culture*, Cambridge: Cambridge University Press, 1985

Zweig, Stefan, *Brazil: Land of the Future*, London: Cassell, 1942

Index

NOTE: Works by Charles Darwin (CD) appear directly under title; works by others under author's name